IEE Control Engineering Series 29
Series Editors: Professors H. Nicholson, B. H. Swanick and D. P. Atherton

INDUSTRIAL DIGITAL CONTROL SYSTEMS

Other volumes in this series:

INDUSTRIAL DIGITAL CONTROL SYSTEMS

Edited by
K. Warwick and D. Rees

Peter Peregrinus Ltd on behalf of the Institution of Electrical Engineers

Published by: Peter Peregrinus Ltd., London, United Kingdom

© 1986 Peter Peregrinus Ltd.

While the author and the publishers believe that the information and guidance
given in this work is correct, all parties must rely upon their own skill and
judgment when making use of it. Neither the author nor the publishers assume
any liability to anyone for any loss or damage caused by any error or omission
in the work, whether such error or omission is the result of negligence or any
other cause. Any and all such liability is disclaimed.

British Library Cataloguing in Publication Data

Industrial digital control systems.—(IEE
    control engineering series; 29)
    1. Digital control systems
    I. Warwick, K.   II. Rees, D.   III. Series
    629.8'3   TJ216

ISBN 0-86341-081-2

Printed in England by Short Run Press Ltd., Exeter

Contents

List of Contributors

P.A. Witting
Department of Electrical and Electronic
  Engineering
Polytechnic of Wales

C.G. Proudfoot
Unilever Research
Port Sunlight Laboratories

G.K. Steel
Department of Electrical and Electronic
  Engineering
University of Aston

D. Rees
Department of Electrical and Electronic
  Engineering
Polytechnic of Wales

J.E. Marshall
School of Mathematics
University of Bath

K. Warwick
Department of Engineering Science
University of Oxford

R.A. Wilson
School of Engineering
Hatfield Polytechnic

I. Postlethwaite
Department of Engineering Science
University of Oxford

D.W. Clarke
Department of Engineering Science
University of Oxford

D.J. Sandoz
Vuman Ltd
University of Manchester

P.A.L. Ham
NEI Parsons Ltd
Newcastle-upon-Tyne

M.F.G. Morrish
CEGB South West Region
Bristol

J.N. Wallace
CEGB South West Region
Bristol

E.H. Higham
Cranfield Institute of Technology

A.L. Dexter
Department of Engineering Science
University of Oxford

Preface

The purpose of this book is to provide an introduction to the techniques involved in the design and application of digital control systems. The book is intended for Engineers, Managers and Engineering students who have a basic knowledge of mathematics, computers and control systems and who wish to become widely informed about using the latest advances in digital control technology. It must be emphasised that what follows is concerned neither with the proferring of highly abstract theoretical ideas nor with detailed descriptions of specialised industrial plant. Indeed the book is based on a continuingly successful series of annual IEE vacation schools which are extremely well supported by industrialists, academics and students alike. The book provides an overall balanced perspective of the theme of digital control and may be used as reference articles, either forming the main text or providing introductory supplementary material for courses on real-time control, digital control and industrial control systems.

Digital hardware has become a widely accepted basis for controller implementation during recent years. The revolution has been due to the availability of low cost computing power in the form of the microprocessor, which has allowed the gap between control theory and practice to be dramatically reduced, permitting the application of many control algorithms to real processes. This technological development has resulted in the theory of digital control systems taking on a more realistic and usable format in order to meet practically imposed discrete time requirements, resulting in the real-time implementation of modern control concepts such as system identification, optimal and multivariable control. The object of this book is to present these advances in digital control systems theory in a readily digestible way by linking the theoretical developments with practical implementations of adaptive and self tuning control, computer aided control system design and expert systems. An unique feature of the book is the presentation of case studies given in the later chapters. These concern themselves solely with the implementation of the techniques described, an approach which is entirely in character with the rest of the book. The tutorial nature of the case studies serves as a general introduction to the application of digital control schemes for industrial purposes.

The several contributions which constitute this book give a representative account of the different aspects involved in the design of Industrial Digital Control Systems and allow the reader to either consider the book as a whole for the purpose of a taught course or to use it as an introductory reference text to particular topics by consulting the relevant chapters.

The book commences with a brief overview and historical account of the background behind the use, in terms of both theory and practice, of digital control schemes for systems. Modelling of systems in a discrete time framework is discussed and various design methods are summarised, in particular classical, adaptive, optimal and state-space approaches are considered. The theoretical basis of the book and therefore the knowledge required can be gauged from Chapter 2 in which the foundations of digital signals and systems are put forward. The relationships between continuous time s-plane representations and discrete time z-plane representations are looked at from the point of view of system input/output transfer functions with special attention given to the position of both open and closed loop transfer function poles and zeros.

Chapters 3 and 4 then concern themselves with laying the foundation for the design of digital controllers. In Chapter 3 the main emphasis is placed on the three term (P+I+D) controller and its application in the process control environment. This popular and widely encountered control structure retains a relatively simple and hence readily understandable relationship with the system signals, each of the three terms having a straightforward affect on the signals under control. The PID controller is developed here stage by stage and various pointers are given as to the methods available for the selection of the controller settings. A highlight of this chapter is the consideration given to implementation problems encountered with the controller, such as choice of sampling rate and numerical accuracy. Chapter 4 however concentrates on various digital control schemes seen either as alternatives to the PID method, or as a necessary approach for some more commonly encountered variations on the basic feedback control requirement. Hence Dahlin, Kalman and predictive control algorithms are discussed whilst the use of feedforward compensation is also considered.

A frequently encountered problem in the implementation of controllers in general is the occurence of time delays in the system characteristics, indeed requirements for the remote control of manipulators in space or in hazardous environments ensures that problems with time delays will have to be overcome for many years hence. In Chapter 5 several examples are considered in which inherent time delays are apparent, and the control of these systems is looked at by means of the Smith predictor method. Alternatives to the Smith predictor are also considered and the effects of mismatch and disturbances on the closed loop system response are discussed.

The state-space approach to system modelling and controller design is seen as an important complementary procedure which can often reveal characteristics of both the system and controller which are not obvious from a straightforward polynomial construction. The fundamental aspects involved in state space design are given in Chapter

6 within which the ideas of system controllability and observability are subsequently introduced. By the end of Chapter 6 the main building blocks of digital control theory have been considered, and in the next three chapters 7, 8, and 9, the major problems in digital system modelling and control are considered from a more theoretical perspective. In the first of these, the question tackled is once the system to be controlled has been decided upon, how can a mathematical representation of the system be obtained?' Within the chapter techniques such as least squares model fitting and maximum likelihood estimation are discussed along with recursive methods which are mostly suitable for use on slowly time varying systems. Once a suitable model of the system has been obtained the next question to be faced is 'what objective do we wish to aim for in the design of a controller for the system. In Chapter 8 the basis for an optimal controller is considered. The primary dependance rests on the function that is to be optimized by maximizing)or minimizing if appropriate) its value over a period of time. Choice of functions, methods of optimisation and their implementation procedures are all covered.

Only too frequently the system under consideration does not simply consist of one input and one output, but has several inputs affecting each output and any one input will actually affect several outputs. The features and approaches apparent in the control of multivariable systems are put forward in Chapter 9, in which the topic is introduced by means of a case study from the aerospace industry. Subsequently the design of robust multivariable controllers is considered via a framework of various possible feedback configurations.

Over the last 15-20 years the need for controllers, which can adapt themselves to changes in both system and environment, has become more widespread. Use of an adaptive control scheme can often result in improved control action over a period of time and/or a controller which remains in operation despite significant alterations or system modifications. Alternative adaptive control methods with descriptions of the presented types of system for which they are suitable, are in Chapter 10. Chapter 11 meanwhile is concerned with a description of the use of Computer Aided Design (CAD) as a tool in the development and study of control systems. It is in fact preferable for the types of controllers considered in the preceeding three chapters to be investigated in terms of simulation and further CAD work before actual implementation takes place.

The next three chapters, 12, 13 and 14 are concerned primarily with various aspects of applying digital controllers in an industrial setting. In Chapter 12 reliability, fault tolerance and detection and types of error encountered with microprocessor based controllers are discussed. Further, two case studies are included in order to highlight the methods described for error minimization. The idea of microcomputer or computer implementation is continued in Chapter 13 in which software design procedures are considered along with hardware and interfacing techniques. Chapter 14 concentrates on applications of digital control in the electrical supply industry. This brings together many of the theoretical ideas put forward in some of the earlier chapters, for example the use of multiloop feedforward and

adaptive control. The main chapters are concluded with Chapter 15 in
which the employment of an expert system architecture around an
adaptive control shell is discussed. Examples are given of the
overall scheme in operation and features in the development of the
expert system are highlighted.

In conjunction with the main set of chapters just reviewed, a set
of case studies are presented in Chapters 16 to 21, and are intended
to demonstrate the implementation of both the more conventional and
modern control ideas. These chapters should therefore be regarded as
a supplement to the main text and an aid to understanding the
techniques presented. In all cases it is felt that the case study
chapter headings are self explanatory and the intention here is merely
to point to where they most appropriately tie in with the main text,
which is as follows:

a)   Chapter 16, Case Study I: follows Chapter 3.

b)   Chapter 17, Case Study II: follows Chapter 4 and 6.

c)   Chapter 18, Case Study III: follows Chapter 4 and 5.

d)   Chapter 19, Case Study IV: follows Chapter 8.

e)   Chapter 20, Case STudy V: follows Chapter 10.

f)   Chapter 21, Case Study VI: follows Chapter 11.

Finally it must be emphasised that in addition to the case studies,
almost all chapters, even the most theoretical, contain practical or
worked examples as a tutorial aid.

In conclusion the editors would like to thank all of the authors
for their contributions and their promptness in forwarding the chapters
for final preparations. The editors would like to give a special
mention to P.A. Witting for his organisation of the course in recent
years and his predecessors as Chairman of the organising committee
Dr. G.K. Steel, Dr. D. Sandoz, Dr. A.J. Morris and Prof. J.P. Stuart,
who nurtured the vacation school through its formative years. Also on
behalf of all of the authors the editors would like to thank those at
Peter Peregrinus and in the IEE who were responsible for the
production of this book, in particular John St. Aubyn at the former and
Andrew Wilson and Sarah Morrall of the latter.

May 1986   K. Warwick
                                                  D. Rees

Chapter 1

Introduction to digital control

P.A. Witting and Dr C.G. Proudfoot

1.1 INTRODUCTION

This introduction to 'Industrial Digital Control Systems' serves two purposes. Firstly it gives a history of computers in Control and discusses the advantages of using them. Secondly it gives an overview of the digital control systems theory to be presented in the later chapters.

In the 1950s and 1960s a Control Engineer had to master analogue technology as the major tool for computations in control systems implementation, and also for simulation purposes. Due to rapid developments in digital computers, however, the scene has changed dramatically. As well as being used for control system implementation computers are increasingly being used for design and analysis.

Because of these developments the approach to analysis, design and implementation of control systems is changing drastically. Originally it was only a matter of translating the earlier analogue designs into the new technology. As such, digital, or computer-controlled systems can be viewed as an approximation of analogue control systems. This approach does not always achieve the best performance however, because the full potential of computer control is not used. As an alternative approach it is possible to make full use of modern digital design and analysis methods, as discussed in the following chapters of this book, and briefly outlined in section 5 of this chapter.

Control engineering went through a considerable period of transition in the 1960s and 1970s with the development of 'modern' control theory. In essence the movement looked at control engineering from a somewhat different viewpoint to that of conventional frequency response/transfer function methods. The drive was for a better understanding of control systems and an improved quality of control. To achieve this much use was made of digital computers for the processing of data and this led to a move away from the Laplace transform/calcus approach and towards those techniques that could be readily programmed onto a computer.

The sheer unfamiliarity of the resulting mathematics (rather than its intrinsic difficulty) drove a wedge between the theoreticians and most practising control engineers. This rift was (and still is) rationalised by asserting that 'modern control is impractical'.   Originally there may have been some truth in this, the early theories could not cope with the imperfections found in real plant,and the analogue controllers were not suited to their implementation. However,the advent of inexpensive and virtually limitless computing power means that it is possible to implement even quite complex calculations cheaply and quickly.   The techniques of 'modern' control may now be put into practice. It is the purpose of the remaining two major sections of the introduction to show how the various aspects of digital control are related to each other and to show how they relate to modern industrial practice.

Figure 1.1 provides a 'road map' of the book.   It will be seen that there are three classes of material: introductory,   theoretical and applications.

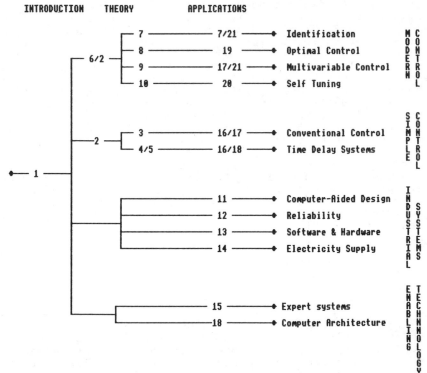

Figure 1.1 A road map of the material presented

These are applied to the four 'threads' of the school namely 'modern control' (the largest single one), 'conventional (simple?) control' where digital computers can offer performance improvements (36), 'industrial applications' and 'enabling technology'. In what follows some of the principal issues in digital control systems are identified and put in their context.

1.2 COMPUTERS IN CONTROL

A Computer controlled system can be represented as in Fig 1.2.

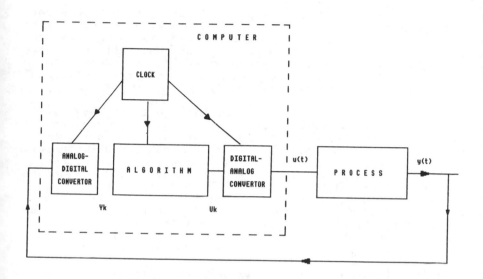

Figure 1.2 A Basic Computer Controlled System.

The output from the process y(t) is a continuous time signal. This output is sampled and converted into a digital signal using an analogue to digital (A-D) converter, the conversion being done at a sampling interval of T seconds. This sampled measurement is then processed using the algorithm and an output value is calculated. This output value is converted to an analogue signal using a digital to analogue (D-A) converter. The output of the D-A converter is usually (in process control, at least) held constant between conversions. It is seen in Fig 1.2 that the computer contains a clock for synchronising all processing and input/output operations. This is an essential feature of any computer control system.

The idea of using digital computers in control started to emerge in the 1950s. At that time, however, the computers used were slow, expensive and unreliable and were generally restricted for use as data loggers and for performing calculations for management information purposes. As reliability improved computers were used more and more and and by 1959 were being used for supervisory and sequence control.   The major tasks of the computer were to find optimal operating conditions, to perform scheduling and production planning and to give reports.   In these cases the ordinary analogue control equipment was still required.   In 1962 however, a radical departure from this approach was made by Imperial Chemical Industries (ICI) Ltd in the UK.   A Ferranti Argus Computer was installed at Burnaze Works (now Fleetwood) to measure 224 variables and manipulate 129 valves directly.   The era of Direct Digital Control (DDC) had started.

The growth of DDC was rapid.   In 1963 340 DDC computers were reported (37) and by 1969 this had grown to over 5000 (38).   In 1970 there was an IEEE special issue on computer control, with survey papers covering applications in the Electric Power Industry, the Paper and Pulp Industry, the Cement Industry and the Chemical Industry.   Digital computers were, indeed, here to stay.

Computer control was spreading but it was still expensive and, to some extent, unreliable, as one central processor was being used for a variety of purposes - management information, supervisory control and DDC control. In the earlier years of computer controlled processes the problem of unreliability was tackled by having a second computer sharing the same signals and running the same software.   These standby systems would be linked via a watchdog mechanism so that if one system failed the other would automatically take over.   In later systems analogue computer/auto/manual stations were used to maintain plant operability in the case of plant a computer shutdown.

The inherent unreliability of being restricted to a central processor changed with the availability of cheap (and more reliable) computing power in the form of the microprocessor.   Distributed control systems grew up with one master processor supervising a number of slaves, each of which could operate independently of the others.   The slave processor could be a simple microprocessor-based version of an analogue controller or a more powerful system looking after a particular area on the plant.   From the point of view of the distributed control approach it makes no difference.   Most manufacturers of industrial control computers now supply a range of different instruments and sytems all of which will have the ability to communicate with a central computer but operate independently of it.

In recent years the concept of distributed control using digital computers has grown rapidly with the further development of communications networks.   These allow all aspects of Factory computing to share a common data-base and communicate with each other, enabling, for example, decisions on factory scheduling to be communicated directly to local process controllers.

While the general power of computers has grown the

control techniques that are generally used have advanced little, the only commonly available control algorithm being the discrete equivalent of the analogue PI/PID controllers. There are, of course, advantages in using a digital computer, as discussed in section 1.3. It is often necessary, however, to take more care when implementing a digital controller rather than its analogue counterpart, particularly if it is operating at a fairly slow sample rate. Some of these problems are discussed in section 1.4.

1.3 ADVANTAGES OF DIGITAL CONTROL

Returning now to the control algorithm itself there are many real benefits in using a digital controller over its analog counterpart for process control. These benefits can be grouped in three main areas.

i.   Flexibility
ii.   Multiplicity of Function
iii. The ability to make use of advanced design and analysis techniques.

Each of these areas will be discussed in this section.

1.3.1 Flexibility

Assuming that suitable programming tools are available it is generally easier to change the specification of a digital control loop than with its analogue counterpart. This flexibility is as true of single loop controllers, where parameters can be specified with precision over a wide range, as of a larger system incorporating a whole control scheme, were it is possible easily to modify the function of the scheme. With a digital control scheme if it is desired to change a signal being used it is often only a case of respecifying what the signal should be, whereas substantial rewiring could be required with an analogue scheme. This ease of specification is particularly noticeable in the area of relays and sequence control where Programmable Logic Controllers (PLCs) have, in many areas, totally replaced analogue relay systems. (It should be stressed, however, that PLCs are not considered reliable enough, even in the 1980s to be used for duties where extreme reliability is required. Such applications include emergency shutdown systems and systems handling dangerous materials. In these cases either multiple PLCs are used, with one checking the others, or analogue relays are used).

1.3.2 Multiplicity of Function

A digital controller can be used for many tasks other than the straight implementation of a control algorithm. Consider for example Fig 1.3 which shows elements in a typical single-input/single-output control loop.

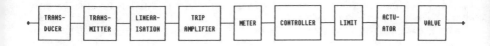

(a) Analogue

(b) Digital

Figure 1.3   Elements in a typical control loop

It is seen that in the analogue case besides the essential hardware consisting of the transducer, transmitter, actuator and valve there are a further five pieces of hardware. A linearising amplifier, a trip amplifier for alarms, the controller itself and a limiting amplifier. All of these latter functions are typically replaced using a modern microprocessor-based instrument. In addition, of course, the microprocessor-based instrument will have the flexibility and precision outlined in section 1.3.1 and the general reliability of the loop will be increased as it contains fewer instruments and less wiring. The reliability factor is particularly important as today's high technology process plant may contain many hundreds of control loops.

The multiplicity of function aspect is seen also in the way that one digital computer can, in general, handle a number of different control loops. Before a multi-loop controller (or computer) is used in this way, however, the safety aspects of losing single loop integrity should be considered.

1.3.3 Advanced Techniques

A third area for advantage for digital control systems in industry, and the one to which much of this book is addressed, is their ability to implement modern advanced control techniques and take advantage of a host of design and analysis tools not otherwise available. To date, however, the response of industry to these developments has been fairly slow. Even manufacturers of commercial control systems have in general used spare processing power for obtaining and displaying information rather than for increasing the capability of the controllers.

The exploitation of advanced control engineering technology has been mainly carried out by industrial and educational research departments. The tools are, however, sufficiently well developed to warrant serious consideration for systems where there is interaction or significant time delay, or for high order systems where conventional PI/PID techniques cannot be effectively utilised. It is often the case that these difficult control problems in industry have the highest potential financial returns. The tools that are outlined in section 1.6 of this chapter and detailed throughout the rest of the book can, if used correctly, offer a usefull and systematic approach to Control Engineering that can enable process plant to operate to a tighter specification leading to better quality product for the customer and increased efficiency, and therefore increased profit, for the producer.

1.4 IMPLEMENTATION PROBLEMS IN DIGITAL CONTROL

The main problems in implementing digital control are related to the fact that the signal is sampled and quantised. These are discussed in turn.

1.4.1 Sampling (39,40)

Selection of the sampling period in discrete systems is a fundamental problem. The proper choice depends on the properties of the signal, the reconstruction method, and the purpose of the system. Shannon's sampling theory states that the frequency of sampling should be chosen to be twice the highest frequency of the signal being sampled. This a very simple rule in the ideal case. In process control systems, however, a rational choice of sampling rate should be based on an understanding of its influence on the performance of the control system. It seems reasonable that the highest frequency of interest should be closely related to the bandwidth of the closed loop system. The selection of sampling rate can then be based on bandwidth, or, equivalently on the rise time of the closed loop system. It is important, of course, to pre-filter the signal being sampled in accordance with Shannon's sampling theorem so as to remove all unwanted high frequencies, and thus prevent the possibility of aliasing.

1.4.2 Quantisation (40)

It is not intended here to give a detailed analysis of the effects of quantisation only to make the reader aware that, in using the techniqus described in the later chapters of this book, quantisation is an important consideration. Generally quantisation effects arise from two sources.

i.)   A-D and D-A converters
ii.) Round off in arithmetic operations

With modern 16-bit or 32-bit processors using floating point arithmetic round-off is not generally a problem for simple arithmetic. It can however cause problems when using some of the modern identification and controller design techniques discussed later. For this reason such routines should be coded using numerically robust techniques such as UDU factorisation (41).

Quantisation at the A-D convertor can also cause problems, especially in industry where it is common to find transmitters with a far wider range than is required for normal operating purposes. Before installing a digital control system, therefore, it is essential to calculate the effect of such quantisation particularly if derivative action is to be used, and to take into account the expected normal operating range.

1.5 MODERN DIGITAL CONTROL

The techniques for designing modern digital control systems discussed in this book involve, in some way, using a model of the process to be controlled. This model is used either as the basis for extracting more information from the process or for predicting the result of taking certain actions and thence being able to choose the values of the plant inputs to give particular plant outputs. If a model of the process can be obtained directly (for example, if the equations and paramenters that govern the dynamics of the (linear) process are known) then the model can be used to drive the outputs of the system to desired values in a specified way. This is called pole-placement. It is often the case however, especially in the process industries, that the model of the process is non-linear or unknown, or (usually) both.

If the process is unknown then it must be identified in some way using data collected from the plant. Modern digital processing techniques can be used for this identification, involving two aspects: identification of model structure (order, time delays, interactions, etc) and the estimation of the model parameters. Once the structure of the model and its parameters are known then the model can be used as the basis for designing a control system.

If the process is nonlinear then the model used for designing the controller is a linearisation of the nonlinear model (or plant) about a particular operating point. If the controller is required to operate over a wide range then it may be necessary to repeat the design exercise at several points over the required range and then to use the flexibility inherent in a digital controller to switch controller parameters depending on operating point (for example, gain scheduling). Alternatively it may be possible, at least if the process non-linearity is not too severe, to use the techniques of adaptive control discussed in chapter 10.

In the remainder of this chapter an outline of different ways of representing discrete systems and designing controllers for them is given.

1.6 ASPECTS OF DIGITAL CONTROL THEORY

1.6.1 Simple Extensions of Continuous Control

It is intuitively satisfying to say that if a digital system has a high sampling rate then it approximates to a continuous system. Under these assumptions one may design controllers by relying on the vast store of classical control methods. The justification for using digital control in these circumstances must be that the practical limitations of analogue controllers (eg drift, inability to implement pure delay etc) are overcome, that the implementation is cheaper or that supervisory control and communications are more easily implemented.

However, the use of high sampling rates is wasteful of computer power and can lead to problems of arithmetic precision etc (9,30). One is therefore driven to find methods of design which take account of the sampling process and which may, therefore be used at lower sampling rates.

1.6.2 Modelling of Discrete Systems

The first step in the design process is to develop a model of the plant and other continuous components. It is found adequate in most instances to characterise the digital-input and digital-output processes as switch closures of negligible duration with all such closures occurring together (Fig.1.4).

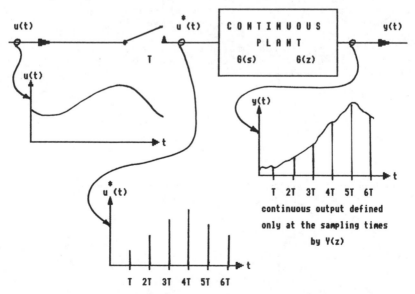

Figure 1.4 A Basic Sampled-Data System

Further attention may be confined to the state of all variables at these sample times, even though the continuous elements will have finite outputs between the samples as shown in figure 1.4.

1.6.2.1 Transfer function approach (1-4,6-9,11,16-22)   When these assumptions are valid one may develop a transfer function calculus of discrete systems which parallels the classical Laplace approach. For a system of order n the output at the kth sample time may be defined in terms of the input sequence (uo, ul.....) and the output sequence (yo, yl....)

$$y_k + a_1 y_{k-1} + a_2 y_{k-2} ----- a_n y_{k-n}$$

$$= b_0 u_k + b_1 u_{k-1} ---- b_n u_{k-n} \qquad (1.1)$$

or

$$y_n = \{ b_0 u_k + b_1 u_{k-1} ---- b_n u_{k-n} \} -$$

$$\{ a_1 y_{k-1} + a_2 y_{k-2} ---- a_n y_{k-n} \} \qquad (1.2)$$

That is, a difference equation relating the current output to the n previous outputs, the current input and the n previous inputs. Put another way, the current output is expressed as a weighted sum of the previous outputs and the current-plus-previous inputs. The modelling problem is one of determining how many such terms are required, the identification problem is one of associating values with each of the weighting terms.

The above is a 'time domain' description since it relates the output at a particular time to values of the variable occuring at other time instants. As with classical control there are advantages in using a transfer function approach. In the case of digital systems the transform variable is 'z' (rather than the Laplace 's').

The term y_{k-1} which appears in equation 1.1 simply represents the input waveform delayed by a single sampling interval, T. Likewise y_{k-2} is delayed by 2T etc. Thus in terms of the Laplace transform equation 1.1 could be re-written.

$$Y(s) + a_1 Y(s) e^{-sT} + a_2 Y(s) e^{-2sT} ---- a_n Y(s) e^{-nsT}$$

$$= b_0 U(s) + b_1 U(s) e^{-sT} ---- b_n U(s) e^{-nsT} \qquad (1.3)$$

since the Laplace variable only appears in the exponentials a z- transform notation is introduced.

$$z^{-1} = e^{-sT}$$

$$(1.4)$$

and equation (1.3) becomes

$$Y(z) + a_1 Y(z) z^{-1} + a_2 Y(z) z^{-2} ---- a_n Y(z) z^{-n}$$

$$= b_0 U(z) + b_1 U(z) z^{-1} --- b_n U(z) z^{-n} \qquad (1.5)$$

and collecting together like terms it is possible to obtain the familiar transfer function equation, but with 'z' as the variable rather than 's'

$$\frac{Y}{U}(z) = \frac{b_0 + b_1 z^{-1} ------ b_n z^{-n}}{1 + a_1 z^{-1} + a_2 z^{-2} ---- a_n z^{-n}} \qquad (1.6)$$

This is the system pulse transfer function. This transfer function is equally applicable to a continuous process defined by samples and to a purely discrete process such as a computer program.

1.6.2.2 State Space Approach (3-9,11)   An alternative approach is to use the concept of 'status'. The system state vector being a sufficient definition of the plant 'status' at any given time. For instance, the second order system of fig.1.5 is fully defined by the two parameters $[h_1, h_2]$ or by the two parameters $[h_1, \dot{h}_1]$
The term $\dot{h}_1$ being the first derivative of h_1. The notation [-----] is special case of a matrix, having only one column and n rows (n=2 in this case). Mathematically this is called a 'column vector' hence the term 'state vector' which is usually given the symbol $\underline{X}$.

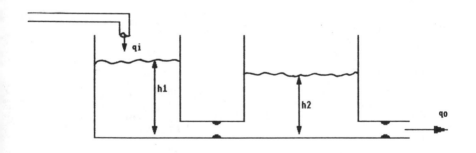

Figure 1.5   A Simple 2nd Order System.

This is the method favoured by 'modern'control and in the discrete case a plant is modelled by defining how the state at the next sample time $\underline{X}_{k+1}$ is determined from the current state $\underline{X}_k$ and the current input U_k

$$\underline{X}_{k+1} = \Phi\underline{X}_k + A\underline{U}_k \qquad (1.7)$$

and the current output of the system. $\underline{Y}$ is:

$$\underline{Y}_k = C\underline{X}_k + D\underline{U}_k \qquad (1.8)$$

The   modelling process requires the determination of Φ, Δ,C,D.   This representation of a system easily emcompasses m input,   p output systems of order n where m, n and p are not necessarily the same.

Φ is nxn and represents how the 'current state' $\underline{X}_k$ influences the next state $\underline{X}_{k+1}$.   It is a close relative of the   system impulse response.

Δ is n x m and controls how the forcing function influences the 'next state'

C is p x n and shows how the output(measured) variables, $\underline{Y}$, are related to the chosen 'state' variables in the 'state vector'

D is p x m and is essentially a 'feed forward' term showing how the output is influenced <u>directly</u> by the forcing function.

These matrices may be deduced directly from discrete Identification procedures or from the continous-time state space model.

$$\dot{\underline{X}} = A\underline{X} + B\underline{U}$$

$$\underline{Y} = C\underline{X} + D\underline{U}$$ (1.9)

Both the transfer function and state models of a system may be shown to equivalent and it is relatively easy to transform between the two (22).

1.6.3 System Design

System   design consists of devising a scheme whereby a system's   input is manipulated so that an adequate speed of response and steady-state accuracy is achieved.   This also involves   consideration of system stability,   integrity   and sensitivity.

<u>1.6.3.1 Classical Methods</u>   (8,9,11) The classical   transfer function   approach   to   design involves the   association   of performance   characteristics   with   particular   system   pole positions (or other equivalent measures).   Feedback is then applied   from   measurements   of   the   current   and   previous systems   outputs y_k , y_{k-1} etc to   modify   these   pole positions.

Similarly   the   current and past values of the system   error e_k , e_{k-1} etc may be combined in the forward path to   obtain similar effects

<u>1.6.3.2 State-Space design methods</u> (9,11)   In   the   state formulation   the   role of the roots of the   characteristics polynomial   in   classical   design   is   transferred   to   the eigenvalues of the matrix.   By measuring the current   value of   the   system state and   using this to provide   the   input

forcing function it becomes possible to manipulate the system eigenvalues. That is the system is forced by a 'forcing function made up of two terms $^\text{'}\underline{U}_k$ the externally imposed forcing function, plus $^\text{'}\underline{U}_k$ an additional forcing function derived from the current value of the 'state'

$$\text{ie} \qquad {}^\text{'}\underline{U}_k = F\underline{X}_k \qquad\qquad (1.10)$$

$$\text{and} \qquad \underline{U}_k = {}^\text{i}\underline{U}_k + {}^\text{'}\underline{U}_k$$

$$= {}^\text{i}\underline{U}_k + F\underline{X}_k \qquad\qquad (1.11)$$

thus the system equations are modified

$$\underline{X}_{k+1} = \Phi\underline{X}_k + B\underline{U}_k$$

$$= \Phi\underline{X}_k + B\{{}^\text{i}\underline{U}_k + F\underline{X}_k\}$$

$$= \{\Phi + BF\}\underline{X}_k + B\{{}^\text{i}\underline{U}_k\}$$

$$= \Phi'\underline{X}_k + B\{{}^\text{i}\underline{U}_k\} \qquad\qquad (1.12)$$

and F is chosen by a variety of methods, so that the new system matrix Φ', has the required eigenvalues and the desired transient and steady state perormance.

It is not always possible to vary particular eigenvalues via the given inputs $\underline{U}$ and the realisation that this may occur leads to the concept of CONTROLLABILITY. Likewise the modes associated with a particular eigenvalue may not be visible through the available outputs, $\underline{Y}$, thus giving, rise to the concept of OBSERVABILITY. A mode must be both controllable and observable if feedback is to be effective in changing it.

The above derivation assumes that the state $\underline{X}$ is measurable directly. This is usually untrue and only the system outputs $\underline{Y}$ are available to be operated upon. Provided that the system is observable, however, it is possible to design a subsidiary dynamic system called an OBSERVER, which will reconstruct $\underline{X}_k$ from $\underline{Y}_k$ and U_k (7,10,23). Using this estimate of $\underline{X}_k$, $(\hat{\underline{X}}_k)$, the feedback control may be implemented (29)

1.6.4 Optimal Control (7,12,13,19,22,24-26)

Whilst the pole placement techniques described above may provide adequate designs there is often strong economic pressure to produce a 'best' or optimal design. The criteria applied may include minimising energy useage, minimising elapsed time or some other combination. Often the criteria are chosen as much for their mathematical tractability as for their reality.

Two main approaches are used to achieve an optimal control scheme, Bellman's "Dynamic Programming" or the "Calculus of Variations". In general these approaches lead to open-loop solutions. Sometimes, however it is possible to achieve a feedback formulation of the solution and such solutions ought to be more tolerant of plant variations than their open-loop counterparts they are however bound to be sub-optimal in such circumstances.

Although optimal controllers are mathematically deducible for continuous-time control schemes they cannot generally be implemented except via the agency of a digital computer and so their derivation in the discrete time domain is more realistic in most cases. It is the reducing cost of this computing power combined with increasing pressure for efficiency which prompts the inclusion of optimal control in this book.

1.6.5 Self-tuning And Adaptive Control

The techniques described so far all assume, implicitly, that the plant has been perfectly modelled. An imperfection in this area will result in non-ideal performance and the methods vary in their sensitivity to these inaccuracies. The Smith Predictor control scheme (14) for instance can become totally unstable in the face of modelling errors. There exists some work by Owens et which is attempting to deal with the general problem of design in the face of modelling errors.

An alternative approach is to configure the control system so that it can adapt its control coefficients so as to achieve the closest possible match to some performance criterion. In this approach an on-line identification technique is used to recursively estimate the parameters of a model (of assumed structure) and these model parameters are then used, either implicitly or explicitly, to calculate the parameters of a controller. There are several approaches to the design of the controller itself (33,42-49), each relying heavily on the concepts of state-space and optimal control design discussed above.

Self-tuning or adaptive control has caught the imagination of industrial users and there have been many industrial applications (49,50). The approach is not yet fully established, however, partly because of its relative complexity, and partly because problems have been encountered in the general robustness of the approach in a full industrial environment These problems are, however being overcome. There are real benefits to be obtained in using self-tuning control, particularly if used as a periodic tuner rather than in a full adaptive context, where the major problem is in ensuring that the data used in the estimator is good data.

The technique of self-tuning control is discussed in

detail in chapter 10. Needless to say such control schemes require computer systems for their implementation, though these need not always be very powerful (36).

1.7. Summary

This introduction has set out the basic themes of the book and without going into detail has explained the relevance of and relationship between the major topic areas. In addition the relationship of the various theoretical approaches to current industrial practice has also been discussed.

REFERENCES

1.   Raven, F H, 1978,'Automatic control engineering', McGraw-Hill

2.   Shinners, S M, 1964,   'Control systems design', Wiley

3.   Power, H M., Simpson, R., 1978, 'Introduction to dynamics and control' McGraw-Hill

4.   Takahashi, Y., Rabins, M J., Auslander, D M.,1972, 'Control and dynamic systems' Addison Wesley,

5.   Rosenbrock, H H, 1974,'Computer-aided design of control systems' Academic Press

6.   Owens, D H, 1978,'Feedback and multivariable systems', Peter Peregrinus

7.   Owens, D H, 1981, 'Multivariable and optimal systems', Academic Press

8.   Katz. P, 1981, 'Digital control using microprocessors' Prentice-Hall

9.   Franklin, G F., Powell, D., 1980,'Digital control of Dynamic Systems' Addison-Wesley

10. Munro, N (Editor), 1979, 'Modern approaches to control systems design' Peter Pergrinus

11. Leigh, J R, 1984, 'Applied Digital Control' Prentice-Hall

12. Anderson, B D O., Moore, J B., 1971,'Linear optimal control' Prentice-Hall

13. Bryson, A E., Ho, Y C, 1969,'Applied optimal control'

Ginn & Co

14.   Marshall, J E,   1979, 'Control of time-delay systems'
       Peter Peregrinus

15.   Porter, B., Crossley, T R, 1972,'Modal control theory
       and applications' Taylor & Francis

16.   Wellstead, P, 1979, 'Introduction to physical
       systems modelling' Academic Press

17.   Eykhoff, P, 1974, 'System identification' Wiley

18.   Jury, E I, 1964, 'Theory and application of
       z-transform method' Wiley

19.   Lindorff, D P, 1965, 'Theory of sampled-data control
       systems' Wiley

20.   Bishop, A B, 1975, 'Introduction to discrete
       linear controls' Academic Press

21.   Cadzow, J A, 1973, 'Discrete-time systems' Prentice-
       Hall

22. ˙ Kuo,   B C,   1970,   'Discrete   data   control
       systems',Prentice-   Hall

23.   Luenberger, D G, 1966, 'Observers for multivariable
       systems', IEEE Trans AC-11, 2,190-197.

24.   Dorf, R C, 1964, 'Time domain analysis and design of
       control systems' Addison-Wesley

25.   Kucera, V, 1979,'Discrete linear control', Wiley

26.   Bellman, R., Dreyfus, S E, 1962, 'Applied Dynamic
       Programming' Princeton Univ. Press

27.   Munro, N, 1979, 'Pole Assignment', Proc. IEE 126,6,
       549-554.

28.   Jury, E I 1958, 'Sampled-data control systems', Wiley

29.   Cadzow,J A., Martens,H R., 1970, 'Discrete time and
       computer control systems' Prentice;Hall

30.   Edwards, J B., Owens, D H, 1977,'First order models
       for multivariable process control',Proc.IE124,11,1083-
       1088.

31.   Owens, D H., 1981, 'On the effect of feedback
       nonlinearities in first order multivariable   control'.
       IEE International Conference on 'Control and Its
       applications' University of Warwick

32. Owens, D H., Chotai, A. 1981, 'Controller Design for Unknown FMultivariable Systems using Monotone Modelling Errors', <u>Univ</u> Sheffield. <u>Dept</u> <u>Control</u> <u>Eng.</u>, <u>Research</u> <u>Report</u> <u>no</u> <u>144</u>

33. Clarke, D W., Gawthrop, P J , 1979,'Self-tuning control"<u>Proc.</u> <u>IEE</u> <u>126,6,</u>633-640.

34. Billings, S A., Harris, C J, 1981, 'Self-tuning and Adaptive Control: theory and applications, Peter Peregrinus

35. Thomas, H W., Sandoz, D J., Thompson M. 1983 'New Desaturation Strategy for Digital PID Controllers' <u>IEE</u> <u>Proc,</u> <u>Vol</u> <u>130,</u> <u>pt</u> <u>D,</u> <u>No</u> <u>4</u>, 189-192.

36. Dexter, A L, 1983, 'Self tuning control Algorithm for Single-Chip Microcomputer Implementation' IEE Proc, Vol 130, pt D, No 5, <u>255-260</u>

37. Ryan, F., 1963, 'Industries pulse:340 Digital Control Computers where they are', <u>Control</u> <u>Engineering</u> <u>Sept.</u> <u>1963</u>

38. Lapidus, G., 1969, 'A look at minicomputing applications', <u>Control</u> <u>Engineering,</u> <u>Nov</u> <u>1969</u>

39. Brown and Glazier, 1976, Telecommunications, Science Paperbacks

40. Astrom & Wittenmark, 1984,'Computer Controller Systems'Prentice Hall.

41. Bierman, G.J., 1977,'Factorisation methods for discrete sequential estimation', Academic Press

42. Astrom, K.J. and Wittenmark, 1973, 'On Self Tuning Regulators' <u>Automatica,</u> <u>Vol.</u> <u>9</u>, 185-199

43. Clarke, D.W. and P.J. Gawthrop, 1975, 'Self Tuning Controllers', <u>Proc.</u> <u>IEE,</u> <u>Vol.</u> <u>122,</u> <u>Pt</u> <u>D,</u> <u>No.</u> <u>9</u>,424-434.

44. Grimble, M.J., 1984, 'Implicit and Explicit LQG Self tuning Controllers', <u>Automatica,</u> <u>Vol.</u> <u>20,</u> <u>No.</u> 661-670

45. Clarke, D.W., Kanjilil, P.P. and Mohtadi, C., 1984, 'Generalised LQG Approach to Self tuning Control', <u>Oxford</u> <u>University</u> <u>Engineering</u> <u>Laboratory,</u> <u>Report</u> <u>No.</u> <u>1536/84</u>

46. Clarke, D.W., 1982, 'Model Following and Pole Placement Self-Tuners', <u>Optimal</u> <u>Control</u> <u>Methods</u> <u>and</u> <u>Applications,</u> <u>Vol.3,</u> <u>No.4</u>

47. Wellstead, P.E., Prager, D. and Zanker, D., 1979, 'Pole Assignment Self-tuning Regulator, Proc. IEE, Vol. 126, Pt.D, No. 8. 781-787.

48. Clarke, D.W., Mohtadi, C. and Tuffs, P.G., 1984, 'Generalised Predictive Control, Pts. 1 and 2, Oxford University Engineering Laboratory, Reports No. 1555/84 and 1557/84

49. Parks, P.C., Schaufelberger, W., Unbehauen, H. and Schmid, C.H.R., 1980, 'Application of Adaptive Control Systems', in 'Methods and Applications of Adaptive Control', Ed. Unbehauen, H. Springer-Verlag

50. Unbehauen, H. and Schmid, C.H.R., 1980, 'Applications of Adaptive Control in Process Control', in 'Application of Adaptive Control', Eds., Narendra, K.S. and Monopoli, R.V., Academic Press

Chapter 2

Digital signals and systems

Dr G.K. Steel

2.1 INTRODUCTION

The process of analogue to digital conversion applied
to a time varying signal involves the representation of the
signal as a sequence of ordinate values spaced out in time.
Such ordinate values each define the signal at one instant
of time and can be regarded as impulse values. The
sequence is then represented as a train of impulses
separated in time by the sample interval.

Sample sequences of this type are conveniently
described by their z-transform which is an alternative
statement of their Laplace transform. In the following
notes transform theory is used to indicate a number of
results which show the properties of sampled signals.
Firstly there is the problem of the ambiguity involved in
identifying a signal from the sample sequence representing
it. This ambiguity arises due to the fact that the
behaviour of the signal between sample instants is
undefined. Secondly there is the need to describe the
effect of processing the sample values in a computational
algorithm, as for instance in a digital control algorithm,
where we may wish to identify the effect on the frequency
response or the time response of the system.

2.2 LAPLACE TRANSFORM OF SAMPLED SIGNALS

When a signal $r(t)$ is sampled at intervals T the
sample at time $t = nT$ is represented as $r(nT).\delta(nT)$ where
$\delta(nT)$ is a Dirac pulse at $t = nT$. With $r(t)$ defined over
$0 < t < \infty$ the sample sequence is

$$r^*(t) = \sum_{n=0}^{\infty} r(nT)\delta(nT) \quad \dots\dots\dots\dots\dots\dots (2.1)$$

and the Laplace transform of this may be written in one of
three forms, Ragazzini and Franklin (1) :

(a) Sequence form

$$R^*(s) = \sum_{n=0}^{\infty} r(nT).e^{-snT} \quad \dots\dots\dots\dots\dots (2.2)$$

(b)   Spectral form

$$R^*(s) = \frac{1}{T} \sum_{n=-\infty}^{\infty} R(s+nj\omega_0) \quad \ldots\ldots\ldots\ldots \quad (2.3)$$

$$\omega_0 = \text{sampling frequency} = \frac{2\pi}{T}$$

(c)   Closed form

$$R^*(s) = \frac{1}{2\pi j} \int_{c-j\infty}^{c=j\infty} \frac{R(p)}{1-e^{-sT}e^{pT}} \, dp \quad \ldots\ldots \quad (2.4)$$

The significance of each of these forms will be examined in detail.

2.2.1 Sequence Form

This is the easiest form to interpret as it follows directly by taking the Laplace transform of each separate term in equation (2.1).

For example if $r(t) = e^{at}$ we have

$$R^*(s) = \sum_{n=0}^{\infty} e^{anT}e^{-snT} \quad \ldots\ldots\ldots\ldots \quad (2.5)$$

here the element e^{-snT} is seen as a delay operator which places the sample e^{at} at time $t = nT$.

2.2.2 Spectral Form

This is particularly useful in examining the ambiguity due to the loss of information about the signal between sample instants.

In interpreting equation (2.3), $R(s)$ is the Laplace transform of $r(t)$, the continuous time signal, and $R^*(s)$ is obtained as an infinite series of terms derived form $R(s)$ with s changed to $s+nj\omega_0$.

For example when $r(t) = e^{at}$, $R(s) = 1/(s-a)$ then

$$R^*(s) = \frac{1}{T}\left[\frac{1}{(s-a)} + \frac{1}{(s-a+j\omega_0)} + \frac{1}{(s-a-j\omega_0)} + \frac{1}{(s-a+2j\omega_0)} \right.$$
$$\left. + \frac{1}{(s-a-2j\omega_0)} + \ldots\right] \quad \ldots\ldots \quad (2.6)$$

The continuous signal $r(t)$ is characterised by $R(s)$ having a pole at $s=a$ in the s-plane. Each term in $R^*(s)$ also has a pole at $s=a+nj\omega_0$ which appears in the s-plane

as a point displaced vertically from s=a by $nj\omega_0$ as
indicated in Fig.2.1.

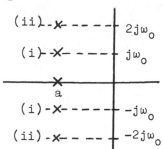

Fig. 2.1. Poles of R*(s)

The conjugate pair of poles (i) define a mode of the
form $e^{at}\cos(\omega_0 t)$. It is important to recognise that such a
signal if sampled at intervals T is capable of giving the
same sample sequence as that derived from e^{at}, the primary
signal. This can be observed in Fig. 2.2.

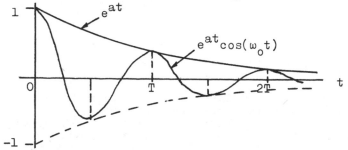

Fig. 2.2. Components of r*(t)

Similarly the pole pair (ii) define a mode with
frequency $2\omega_0$ which oscillates through two cycles in each
interval T and can synchronise with its peaks at the
sampling instants.

Hence if we examine the poles of R*(s) confusion
exists as to the true identity of r(t). In this example we
must disregard all poles except that at s=a.

With more complex functions r(t), the set of poles of
R(s) are duplicated with shift $nj\omega_0$ up the s-plane. It is
particularly significant that when r(t) contains a damped
oscillatory mode of frequency ω a conjugate pair of poles
appears in R(s) so that R*(s) has poles as indicated in
Fig. 2.3.

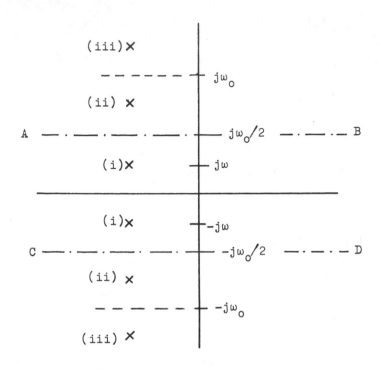

Fig. 2.3. Poles of R*(s)

Here the poles (i) identify r(t) while the pairs (ii),
(iii) etc. represent alternative modes which, if sampled,
could generate the same sample sequence. In this case we
can identify r(t) by disregarding all poles outside the
segment bounded by lines AB, CD. However if the signal
frequency ω is greater than $\omega_o/2$ the poles (ii) will fall

inside AB - CD while (i) move outside and the signal will
appear to contain a mode of lower frequency than the
original ω. We are unable to remove this ambiguity in
identifying the true signal frequency. The overlaying of
modes (ii) and (i) in this case is described as "aliasing".
Ambiguity can be avoided by observing the Shannon sampling
theorem which in these terms can be stated as:

If r(t) is to be recovered from the pulse sequence
r*(t) without ambiguity then the sampling frequency must be
at least twice the highest frequency of any oscillatory
mode in r(t).

The signal r(t) is identified by disregarding the
poles of R*(s) which lie outside the primary segment
AB.- CD in the s-plane i.e. $j\omega_o/2$ either side of the real
axis.

2.2.3 Closed Form

Both the sequence form and the spectral form give $R*(s)$ as an infinite series. An equivalent closed form expression is obtained by contour integration using

$$R*(s) = \frac{1}{2\pi j} \int_{c-j\infty}^{c+j\infty} \frac{R(p)}{1-e^{-sT}e^{pT}} \, dp.$$

The contour of integration is defined to enclose all the poles of $R(p)$. Poles due to $(1-e^{-sT}e^{pT})$ are placed outside the contour by a proper choice of s.

For example with $r(t) = e^{at}$ we have $R(p) = 1/(p-a)$. On substituting in the integral we evaluate the result by taking the residue at $p=a$ to get

$$R*(s) = \frac{1}{1-Ae^{-sT}} \qquad \text{with } A = e^{aT} \quad \dots\dots\dots(2.7)$$

This form is useful when we wish to evaluate $R*(s)$ for a particular value of s.

2.3 Z-TRANSFORM

The sequence form and closed form expressions for $R*(s)$ explicitly involve e^{sT} and may be simplified by substituting $z = e^{sT}$. The corresponding functions of z are defined to be the z-transforms of the signal $r(t)$ but they retain their significance as alternative forms of the Laplace transform of the sample sequence. Denoting $R*(z)$ as the z-transform we have:

Sequence form

$$R*(z) = \sum_{n=0}^{\infty} r(nT)z^{-n} \qquad \dots\dots\dots\dots(2.8)$$

Closed form

$$R*(z) = \frac{1}{2\pi j} \int_{c-j\infty}^{c+j\infty} \frac{R(p)}{1-z^{-1}e^{pT}} \, dp \quad \dots\dots\dots(2.9)$$

For example with $r(t) = e^{at}$ the sequence form is

$$R*(z) = \sum_{n=0}^{\infty} e^{anT}z^{-n} \qquad \dots\dots\dots\dots(2.10)$$

$$= 1+e^{aT}z^{-1}+e^{2aT}z^{-2}+e^{3aT}z^{-3}+\dots$$

which explicitly identifies the ordinates of $r(t)$ positioned in the sequence in accordance with the power of z^{-1} associated with each. The closed form gives

$$R^*(z) = \frac{1}{1-z^{-1}A} = \frac{z}{z-A} \quad \text{with } A = e^{aT} ...(2.11)$$

This form can be used to evaluate $R^*(z)$ for any value of z and indicates specifically that there is a pole at $z = A$ and a zero at the origin in the z-plane.

2.4   RELATIONSHIP BETWEEN THE S-PLANE AND Z-PLANE

We have noted that the z-transform of a signal is the same as the Laplace transform of the sample sequence with a change of variable $z = e^{sT}$. This means that all points in the s-plane have corresponding points in the z-plane. A point $s = \alpha + j\beta$ maps to $z = e^{\alpha T}e^{j\beta T}$, i.e. z is defined by a vector of length $e^{\alpha T}$ at an angle βT. The corresponding points shown in Fig.2.4 reveal the significance of this together with the important result that points inside the unit circle in the z-plane correspond to points within the primary segment of the s-plane and in the left hand half. The right hand half of the primary segment maps into the remainder of the z-plane outside the unit circle.

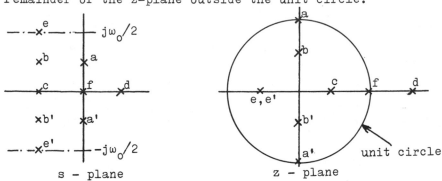

Fig 2.4.   z-plane/s-plane relationship

Poles of $R^*(s)$ in the left-hand half of the s-plane indicate modes which decrease with time while those in the right-hand half indicate modes which increase. Corresponding poles of $R^*(z)$ inside the unit circle represent decreasing time functions while those outside the unit circle represent increasing time functions. The unit circle is a critical boundary in the z-plane as is the imaginary axis in the s-plane.

A further feature of the mapping from the s-plane to the z-plane is that if a point $s = \alpha + j\beta$ is displaced vertically by an amount $jn\omega_o$ to $s = \alpha + j(\beta + n\omega_o)$ the value of z is unchanged. This means that segments of width ω_o above

and below the primary segment map to overlay points in the
s-plane derived from the primary segment. This feature
accounts for the usefulness of the z-transform since the
infinite set of poles in the s-plane, in Fig. 2.3 for
example, is reduced to a finite number of poles in the
z-plane.

2.5. Z-TRANSFORM INVERSION

Inversion of a z-transform involves taking $R^*(z)$ and
computing the ordinate values of the signal which it
represents. There are three basic methods available to do
this.

2.5.1 Polynomial Division

This is used to convert a closed form expression into
the sequence form from which the ordinate values are self
evident.

Given,

$$R^*(z) = \frac{a_0 + a_1 z^{-1} + a_2 z^{-2} + \ldots . a_p z^{-p}}{b_0 + b_1 z^{-1} + b_2 z^{-2} + \ldots . b_q z^{-q}} \ldots (2.12)$$

we divide numerator by denominator as polynomials in
ascending powers of z^{-1}. For example,

$$R^*(z) = \frac{Kz}{z-A} = \frac{K}{1 - z^{-1}A}$$

and on dividing out we get

$$R^*(z) = K(1 + A z^{-1} + A^2 z^{-2} \ldots . A^n z^{-n}) \ldots (2.13)$$

so that the ordinate at $t = nT$ is KA^n. The method is
simple to apply but tedious since all values must be
calculated up to any particular sample which is required;
unless the sequence structure can be recognised as in the
above simple example.

2.5.2 Partial Fraction Expansion

When $R^*(z)$ has simple poles we can expand in the form

$$R^*(z) = \frac{K_1 z}{z - A_1} + \frac{K_2 z}{z - A_2} + \ldots \frac{K_m z}{z - A_m} \ldots (2.14)$$

where $z = A_r$ is a pole position and there are m such poles.

This expansion is achieved by the usual method of
partial fraction determination except that we must first
form

$$\frac{R^*(z)}{z} = \frac{K_1}{z-A_1} + \frac{K_2}{z-A_1} + \ldots \frac{K_m}{z-A_m} \quad \ldots\ldots\ldots(2.15)$$

where $K_r = \underset{z \to A_r}{\text{Lim}} \left[(z-A_r) \frac{R^*(z)}{z} \right]$

The form required in equation (2.14) is produced on multiplying equation (2.15) by z. Inversion of $R^*(z)$ is then a matter of recognising each of the terms in equation (2.14) as contributing an ordinate $K_r A_r^n$ at t=nT as in equation (2.13) and summing the ordinates together.
For example with

$$R^*(z) = \frac{1+0.5 \ z^{-1}}{(1+0.2 \ z^{-1}) \ (1+0.4 \ z^{-1})}$$

write, $\frac{R^*(z)}{z} = \frac{z+0.5}{(z+0.2) \ (z+0.4)}$

$$= (\frac{1.5}{z+0.2}) - (\frac{0.5}{z+0.4})$$

then $R^*(z) = \frac{1.5z}{(z+0.2)} - \frac{0.5z}{(z+0.4)}$

and $\quad r(nT) = 1.5(-0.2)^n - 0.5(-0.4)^n \quad \ldots\ldots\ldots(2.16)$

This form allows any sample value to be calculated directly.

2.5.3 Contour Integration

$$r(nT) = \frac{1}{2\pi j} \oint z^{n-1} R^*(z)dz \quad \ldots\ldots\ldots\ldots(2.17)$$

where the contour integration is taken to enclose all the poles of $R^*(z)$.

In the example used above we get

$$r(nT) = \frac{1}{2\pi j} \oint z^{n-1} \frac{(z^2+0.5z)}{(z+0.2) \ (z+0.4)} dz.$$

The residue at z = -0.2 is,

$$\underset{z \to -0.2}{\text{Lim}} \left[z^{n-1} \frac{(z^2+0.5z)}{(z+0.4)} \right] = 1.5(-0.2)^n$$

and at $z = -0.4$

$$\operatorname*{Lim}_{z \to -0.4} \left[z^{n-1} \frac{(z^2 + 0.5z)}{(z+0.2)} \right] = -0.5(-0.4)^n.$$

The overall $r(nT)$ is the sum of these and again we have equation (2.16)

2.6 TRANSFER FUNCTIONS

A sampled signal may be processed by using the numerical values of successive samples in a computational algorithm. Each new sample leads to a result which is entered into the output sequence of the algorithm. The z-transform may be used to describe this input-output relationship in terms of a transfer function. It is assumed that the computing time is negligible so that the output samples appear at the same instants as the input samples.

For example we may compute the approximate integral of an input $r(t)$ using samples $r^*(nT)$ in the recurrence relationship:

$$i_{n+1} = i_n + T \, r_n \quad\dots\dots\dots\dots\dots\dots\dots(2.18)$$

where i_n is the integral value at $t=nT$ and r_n is $r(nT)$. The z-transform of the sequence generated by i_n may be defined as $I^*(z) = \sum i_n \, z^{-n}$ and the repeated application of equation (2.18) then gives rise to the relationship

$$z \, I^*(z) = I^*(z) + T \, R^*(z)$$

from which we form

$$\frac{I^*(z)}{R^*(z)} = \frac{T}{(z-1)} = W^*(z) \quad\dots\dots\dots\dots\dots(2.19)$$

Here $W^*(z)$ has the properties of a pulse-transfer function. Specifically $W^*(z)$ represents the z-transform of the output sequence following the application of a single unit input sample at $n=0$.

In equation (2.19) $W^*(z)$ is given in its closed-form. We may convert this to the sequence-form and obtain,

$$W^*(z) = T \, (z^{-1} + z^{-2} + z^{-3} \dots\dots) \quad\dots\dots\dots(2.20)$$

which shows a constant output value T for all samples. This is as would be expected by integrating a single input sample $r(0)=1$.

The process of computing the output sequence of a
particular computational algorithm for a given input
sequence can of course be carried out by numerical
processing with repeated application of the recurrence
relationship. A wider view of the dynamics of the output
sequence can be obtained by z-transform analysis. We
obtain the transform of the output $Q^*(z)$ from
$Q^*(z) = W^*(z)R^*(z)$, where $R^*(z)$ is the transform of the
input sequence. Inversion of $Q^*(z)$ as a closed-form
expression using contour integration leads to a formal
expression for the elements in the output as a function of
the element number n.

The use of the z-transform to obtain the pulse-
transfer function representing the input-output property of
a recurrence relationship is a powerful technique for
studying the dynamic behaviour of the recurrence
relationship, in particular its stability.

We observe that $W^*(z)$ represents the z-transform of the
output sequence generated by applying a single unit input
sample at t=0. The dynamic behaviour of the output is
characterised by the modal structure of this response.
Hence the poles of $W^*(z)$ are of particular interest as each
value of z for which $W^*(z)=\infty$ relates to an identifyable
mode. The poles are found by examining $W^*(z)$ in the closed-
form. For example equation (2.19) indicates a pole at z=1.
Such pole values may be identified with points in the
z-plane and it is useful to associate the modal characteristic with the positioning of points in the plane.

One useful way of establishing the modal form is to
exploit the relationship between the z-plane and the s-plane
This relationship was described in section 2.4.

If a point in the z-plane is mapped into the s-plane
the mode it represents can be identified as the sequence
which results from sampling a time function of the form
characterised by the position of the point in the s-plane.
Fig. 2.5 indicates how this relationship may be visualised.
Each of the pole positions given in the z-plane is mapped
into the corresponding position in the s-plane. The points
in the s-plane describe a continuous mode signal in the time
domain. The sequence of sample values associated with the
mode is obtained by sampling the continuous signal as shown.

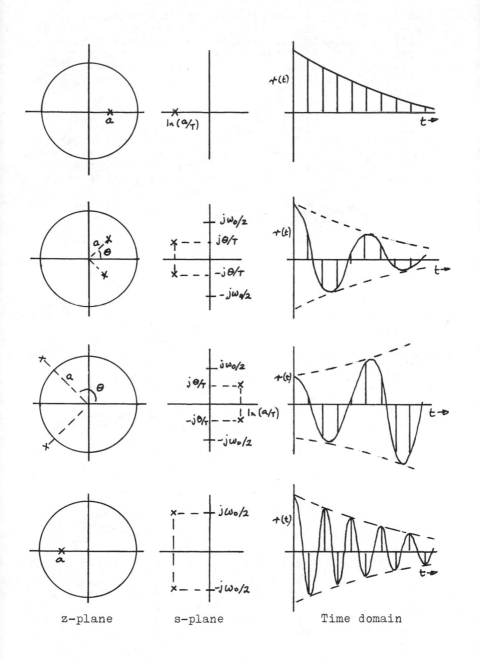

Fig. 2.5 Poles in the z-plane

Significantly it is evident that the sequence is divergent if the pole position in the z-plane is outside the unit circle, the corresponding s-plane point is then in the right hand half of the plane. Thus any recurrence relationship for which the pulse-transfer function W*(z) has poles outside the unit circle in the z-plane will be unstable.

If we apply this reasoning to the function given in equation (2.19) representing the integration algorithm we observe a pole at z=1. The corresponding point in the s-plane is the origin s=0. The mode associated with a pole at the origin in the s-plane is a constant step function and this mode is evidently present in the impulse response equation (2.20) where all samples have equal value.

2.7 ZEROS OF THE PULSE-TRANSFER FUNCTION

A typical pulse-transfer function takes the form

$$W^*(z) = \frac{N(z)}{D(z)} \quad \ldots\ldots\ldots\ldots\ldots\ldots\ldots(2.21)$$

where N and D are polynomials in z. The values of z which make D(z)=0 identify the poles of W*(z) and the significance of these values in characterising the modal components in the impulse response corresponding to W*(z) has been outlined in Section 2.6. The values of z which make N(z)=0 are the zeros of W*(z) and these have a different significance which may be understood as follows.

Firstly it is recognised that in the sequence form of a signal transform the multiplication of a sample value by z^{-n} delays the sample by n intervals. If we take a signal transform R*(z) and form $z^{-n} R^*(z) = G^*(z)$ this composite transform represents a signal g(t) of the same form as r(t) but delayed n sample intervals from t=0.

For example the signal $r(t) = e^{-at}$ has transform $z/(z-A)$ where $A = e^{-aT}$. If we now form a function $\frac{(z - B)}{(z - A)}$ which has the same pole at z=A but introduces a zero at z=B, the significance of this can be identified as follows; writing $\frac{(z - B)}{(z - A)} = \frac{z}{(z - A)} - Bz^{-1} \cdot \frac{z}{(z - A)}$

we observe that the first term is the transform of r(t), the second term is the transform of r(t) delayed by one sample interval and given a weighting factor B. By including the zero at z=B we have introduced another signal component of the same modal form but changed in amplitude and time origin.

In this example $N(z) = (z - B)$ and it is easy to extend the argument to a higher order polynomial which simply includes further delayed and weighted components of the same modal form.

As a particular case the denominator $D(z)$ may reduce to $D(z)=z^n$ so that $W^*(z)$ has poles at the origin of the z-plane, $z=0$. The resulting form of $W^*(z)$ is $z^{-n}N(z)$ which contains a finite number of terms, each representing one impulse with an appropriate delay. The coefficients in the polynomial $N(z)$ simply determine the sample values in the sequence.

2.8   FINAL VALUE THEOREM

It is often useful to be able to ascertain the limiting value of a sequence $r(nT)$ as n tends to infinity, i.e. the steady-state value. This can be done by operating on the transform $R^*(z)$ using the theorem;

$$\lim_{n \to \infty} r(nT) = \lim_{z \to 1} \left[(z-1) R^*(z) \right] \quad \ldots(2.22)$$

For example the time function $r(t) = 1 - e^{-at}$ gives $r(nT) = 1 - e^{-anT}$ and for a positive this tends to the value 1 as n tends to infinity.

The z-transform $R^*(z) = \dfrac{z}{(z-1)} - \dfrac{z}{(z-A)} = \dfrac{z(2z-(A+1))}{(z-1)(z-A)}$

$$\text{with } A=e^{-aT}$$

Applying the final value theorem to this we have,

$$\lim_{z \to 1} \left[(z-1) \cdot \frac{z(2z-(A+1))}{(z-1)(z-A)} \right] = 1$$

It is important to note that the above limiting value is obtained for any value of A. However we observe that if A is greater than 1 the value of a is negative and the exponential e^{-at} grows to infinity as time increases. The result from the final value theorem is incorrect in this case.

For the correct result to be obtained using the final value theorem there must be no unstable modes present i.e. all the poles of $R^*(z)$ must lie inside the unit circle in the z-plane.

2.9   CLOSED-LOOP SYSTEMS

A typical closed-loop digital control system is shown in Fig. 2.6.

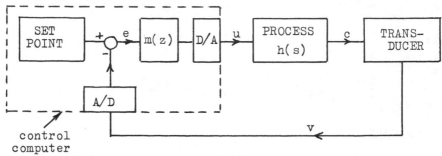

e = error   u = control signal
c = controlled variable   v = measured variable
m(z) = control function

Fig. 2.6   Control system

The system relies on a real-time clock to initiate the sequence of operations in the control computer at the required sampling intervals. At the sampling time the analogue to digital conversion of the measured value is first initiated and on completion a digital value is stored in memory. The control algorithm is then run to evaluate the current error and process this in accordance with the recurrence relationship defining the control function. Finally the digital value of the control signal is passed to the process interface through a digital to analogue converter.

Three aspects of this processing sequence are of particular significance,

    1. processing time
    2. analogue output clamping
    3. quantisation

2.9.1 Processing Time

The time taken from the start of analogue to digital conversion to the final output from the digital to analogue converter may be significant when the process reaction is fast. This can be taken into account by including a delay element with transfer function $e^{-s\delta}$ where δ is the time delay. In analysing the control system Fig. 2.6 it is then convenient to associate this delay with the process rather than the computer taking the process transfer function to be $e^{-s\delta} h(s)$. The system is then represented as shown in Fig. 2.7.

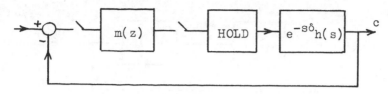

Fig. 2.7   Block diagram

Here the sampling switches are regarded as being synchronised. In determining the pulse-transfer function of the element $e^{-s\delta} h(s)$ it is necessary to use the modified z-transform (1) to take account of the time delay.

2.9.2   Analogue Output Clamping

The need to hold the analogue value, obtained from the analogue to digital converter, at the process input point until the next value is presented can be represented by the "hold" block shown in Fig. 2.7.

A single control signal sample u(0) creates an input to the process of the form,

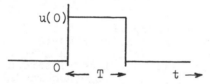

Fig. 2.8   Process input

The pulse-transfer function of the combined hold and process blocks is the z-transform of the process output response to this input with u=1.

We note that the time function Fig. 2.8 can be regarded as a step function u(0) applied at time t=0, followed by a step -u(0) applied at time t=T. If the process transfer function is h(s), the output transform following the first step is $u(0)\frac{h(s)}{s}$. The second step reverses this with a delay of one sample interval. If F(z) is the z-transform of $\frac{h(s)}{s}$ the overall transfer function is;

$$H(z) = (1 - z^{-1})F(z) \quad \ldots\ldots\ldots\ldots\ldots(2.23)$$

The control system Fig. 2.6 can now be represented by the simplified block diagram of Fig. 2.9.

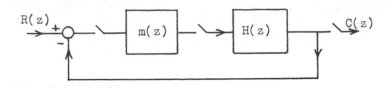

Fig. 2.9   Simplified block diagram

2.9.3   Quantisation

The analogue to digital conversion process inherently involves representing a continuous amplitude variation by a finite number of discrete levels. The levels are separated by the amplitude corresponding to the least significant bit of the converter.

A similar quantisation effect is also present in and digital processing algorithm where numerical coefficients are represented in finite word length. In addition the results of computation are rounded off to a finite word length before being passed to the digital to analogue converter.

In the foregoing analysis of sampled signals and pulse-transfer functions using z-transforms it is assumed that all variables are represented exactly. Quantisation (Kuo (5)) and round off (Franklin and Powell (4)) errors are non-linear effects which cannot be simply accommodated in z-transform theory.

The effects of these errors on system performance can be significant however. Where a digital processor is used to realise a required control function the errors have the effect of making the pole and zero positioning in the z-plane imprecise (Rabiner and Gold (3)). Also significant steady-state error may persist and limit-cycle oscillations develop (5).

2.10   SYSTEM ANALYSIS

Once the closed-loop system of Fig. 2.6 has been represented in the simplified block diagram Fig. 2.9 we may analyse the closed-loop behaviour.

The closed-loop pulse-transfer function is given by,

$$\frac{C(z)}{R(z)} = Q(z) = \frac{m(z)H(z)}{1 + m(z)H(z)} \quad \dots\dots\dots(2.24)$$

If $m(z).H(z)$ is expressed as a polynomial ratio
$\frac{N(z)}{D(z)}$ then,

$$Q(z) = \frac{N(z)}{N(z) + D(z)}\cdots\cdots\cdots(2.25)$$

The poles of this transfer function define the natural modes of the closed-loop. The denominator polynomial $N(z) + D(z)$ is identified as the closed-loop characteristic polynomial and the values of z for which $N(z) + D(z) = 0$ identify the modes.

Stability requires that all the zeros of $N(z) + D(z)$ must lie inside the unit circle of the z-plane.

2.11   SYSTEM DESIGN

In systems such as the single-loop control system Fig. 2.6 the design problem generally evolves around the choice of the control function $m(z)$ in order to impart a satisfactory form to the closed-loop transfer function $Q(z)$. This choice is constrained by the form of the function $H(z)$ representing the fixed process elements.

A wide variety of design procedures is avaliable, these fall into two categories,

(a)   direct synthesis procedures

(b)   iterative design procedures

The direct synthesis procedures (Franklin and Powell (4)) (Kuo (5)) assume that the control function $m(z)$ is not restricted in any way by hardware or software limitations and can be allowed to take any form demanded by the nature of the fixed process elements and the specification of the required system performance. This design approach has found wider application in digital control systems than has the equivalent technique used with continuous systems. In the digital controller the realisation of the required $m(z)$ may involve no more than programming a special purpose software procedure. With continuous systems the limitation was the complications involved in designing special purpose analogue controllers.

Complete flexibility in the choice of control function is not always available however. For example in using a standard process control computer the choice of function may be restricted to using a standard algorithm with variable parameters; in many cases a three-term control algorithm. The design procedure must then examine the effect of the choice of controller parameters on the system performance and lead the designer to make an appropriate final choice. Iterative design techniques lend themselves to this approach. These are based on similar techniques evolved for continuous system design using root-locus or frequency response graphs to characterise the response (4)(5).

REFERENCES

1. Ragazzini,J.R., and Franklin,G.F.,1958, 'Sampled-data control systems',McGraw-Hill,New York.

2. Kuo,B.C.,1963,'Analysis and synthesis of sampled-data control systems',Prentice-Hall,Englewood Cliffs, N.J.

3. Rabiner,L.R. and Gold,B.,1975,'Theory and application of digital signal processing', Prentice-Hall, Englewood Cliffs,N.J.

4. Franklin,G.F. and Powell,J.D.,1980,'Digital control of dynamic systems',Addison-Wesley,Reading,Mass.

5. Kuo,B.C.,1980,'Digital control systems', Holt,Rinehart and Wilson,New York.

Chapter 3

Digital controllers for process control applications

P.A. Witting

This chapter is concerned with the realisation of simple controllers by digital means. Controllers may be designed ab initio in the digital domain and this approach has much to commend it. However, for those well versed in analogue control it is arguably easier to carry out the design using the familiar continuous domain techniques and then to simulate the controller thus designed very simply in the digital domain. This technique works very well for systems where the sampling rate can be made fast compared to the basic system time constants and in particular for process control systems which are usually very slow. They also usually have plants which are rather poorly defined thus rendering some of the more sophisticated design techniques difficult to apply.

However, one needs a good reason to move over to digital control. Many reasons may be cited including:-

STABILITY: Typical analogue controllers are prone to drift, especially when long time constants are involved. Digital realisations totally eliminate this problem.

COST: With current technology digital controllers are often less costly than their analogue counterparts.

COMMUNICATIONS: It is very difficult to interface analogue controllers to supervisory systems. This is much simpler with digital systems.

ALGORITHMS: Analogue controllers are very tied to those compensators which can be realised fairly simply by integrators etc. This limitation does not dominate the choice of digital compensators.

ADAPTION: One usually has to choose a compromise controller to cover a number of rather different operating regimes. It is very simple to allow a digital

controller to switch between algorithms
to suit the prevailing circumstances.

TIME DELAY: Analogue controllers can only approximate a pure time delay. Digital cont-
rollers can achieve time delays exactly
and this opens up a whole range of
possibilities for the control of
systems with delay elements (i.e. most
process control systems).

   The remainder of this chapter is concerned with the
means of mechanising digital controllers from their
counterparts. This will be illustrated by consideration of
the 3-term controller because of its widespread use in
process control. However, the method is equally applicable
to other types of controller this will be mentioned briefly.
The latter part of the chapter is concerned with
implementation aspects of digital controllers.

3.1   A BRIEF REVIEW OF 3-TERM CONTROLLERS

   Process control is characterised by systems which are
relatively slow and complex and which in many cases include
an element of pure time delay. Even where a pure time delay
element is not present the complexity of the system, which
will typically contain several 1st order subsystems, will
often result in a process reaction curve which has the
appearance of a pure delay. Further it is rare for process
control systems to include an integration term, a situation
which, when combined with a pure time delay results in very
low controller gain settings and consequently large steady-
state offsets. The above considerations lead to a very
difficult control engineering problem especially when they
are allied to the considerable difficulty of deriving a
process model of sufficient accuracy. For this reason
process control engineers have traditionally fallen back on
approximate models of their processes, these models usually
take one of the following forms.

$$\frac{Ke^{-s\tau}}{(1 + sT_1)} \tag{3.1}$$

$$\frac{Ke^{-s\tau}}{(1 + sT_1)(1 + T_2)} \tag{3.2}$$

   The parameters τ and T_1 in (3.1) and to a lesser extent,
τ, T_1 and T_2 in (3.2) may be derived from fairly simple tests
on a plant. The crudity of such models however precludes
the use of the more sophisticated design techniques which
have been formulated in recent years. It is therefore still

common practice to utilise the simple 2-and 3-term controllers that have found applications in process control for 40 years or more.

3.1.1 The Proportional (Single-Term) Controller

The simplest method of exercising control over a plant is to incorporate a "proportional controller" as shown below.

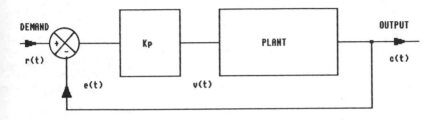

Fig 3.1   Simple Proportional Control System

Such a controller produces an actuating signal $v(t)$ which is proportional to the error signal present at its input.

i.e. $v(t) = K_p e(t)$

Assuming that the plant has a finite steady state gain, K as shown in equations 1 and 2, then a simple proportionally controlled plant will ALWAYS exhibit a steady-state error between input and output.

$$\text{steady-state error} = e(t)\bigg|_{t \to \infty} \frac{1}{1 + K_p K} \qquad (3.3)$$

This error can, of course be made negligibly small by increasing the gain of the proportional controller, unfortunately, as the gain is increased the performance of the closed loop system becomes more oscillatory and takes longer to settle down after being disturbed. Eventually, a value of gain is reached at which the system never settles, but is unstable. Unfortunately, most process plants have a considerable amount of pure time delay and this severely restricts the value of gain, $K_p K$, which can be used. This is illustrated by the graph below which shows the critical gain for a plant having the type of transfer function shown in equation 3.1 above. It will be observed that, as the pure time delay becomes comparable with the 1st order time constant, T_1 , the maximum gain (Ku) becomes very limited. To provide a reasonably acceptable degree of stability it is only possible to use about HALF of this maximum gain.

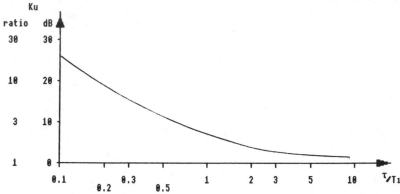

Fig 3.2 Gain Versus Normalised Integral Time

The   net   result of this is that it is   impractical   to rely on simple gain adjustments to provide the necessary accuracy and stability and an additional element must be added to the controller to provide the required performance.

The graph below relates the frequency of oscillation Wu that will be observed in a plant of this type to the relative amount of pure time delay.

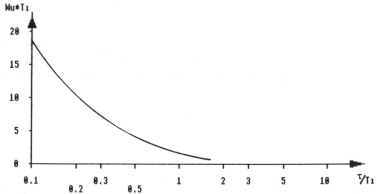

Fig   3.3   Oscillation   Frequency   Versus   Normalised Integral Time

3.1.2 Proportional - Plus - Integral (Two-Term) Controllers.

In order to avoid the difficulty outlined above it is necessary to change the nature of the control system. To understand this consider a simple temperature control system.

The plant is always subject to heat losses and so, in

order that a constant temperature be maintained, a constant heat input must be provided. To provide this the heat source (be it a steam valve or a thyristor controller) must continue to operate, thus requiring a finite input signal $v(t)$. With a proportional controller this implies that there will be a finite error signal to be amplified. This requirement may be removed by inserting an integration block (Laplace transform $1/s$) in cascade with the error signal. This provides a constant output in the absence of an input and a constant rate-of-change of output in the presence of a constant input as shown below.

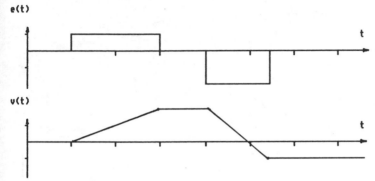

Fig 3.4   Input-Output Characteristic of an Integrator

If such a device is placed in a control system as shown below, the presence of a positive error will cause the controller output to increase thus increasing the heat output. Eventually an equilibrium condition is attained in which the error is zero while the output of the integrator is just sufficient to provide the necessary heat input. With this arrangement the steady-state error is totally eliminated, irrespective of the value of the gain, Kp. The gain, however, does have an effect on the system performance. If Kp is increased the system will be found to respond more rapidly and to become less stable.

In fact, the addition of the integrator has a considerable destabilising effect and it is for this reason that proportional action is combined with the intergrator as shown below.

The P + I control has the equation:

$$v(t) = Kp \left[e(t) + \frac{1}{Ti} \int_0^t e(t)\ dt \right] \qquad (3.4)$$

or

$$\frac{V(s)}{E(s)} = Kp \left[1 + \frac{1}{sTi} \right] = \left[\frac{1 + sTi}{s} \right] \frac{Kp}{Ti} \qquad (3.5)$$

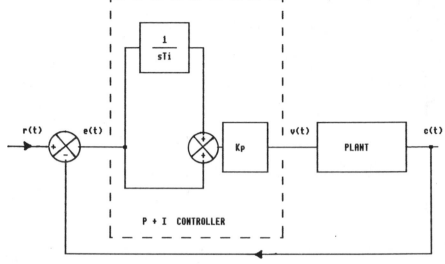

Fig 3.5   Plant With P + I Controller

The term Ti is called the "integral time constant".   It will be seen that the resulting transfer function is a little more complex than the simple integrator.   The 1/s term on which the elimination of the steady-state error rests is still present but a zero has been added and this considerably enhances the stability of the system.

3.1.3   Proportional-Plus-Integral-Plus Derivative   (3-Term) Controllers.

The three-term (PID) is an enhancement of the two term controller described above.   An additional contribution to the actuation signal v(t) is computed, based on the rate at which the error is changing.   Clearly this term has no effect on the steady-state performance but, for constant demand r(t), it always tends to reduce the actuation signal and so combats the tendency for the output to overshoot when recovering from a disturbance.   The reverse is true when the input demand is changed - an effect known as set point kick.   The net result of adding this extra term is a further improvement in system stability.

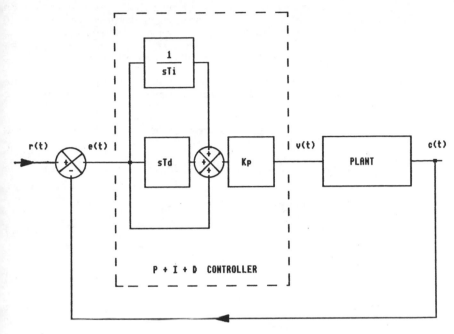

Fig 3.6   Plant with P + I + D Controller

The PID controller has the equation

$$v(t) = Kp \left[e(t) + \frac{1}{Ti} \int_0^t e(t) \, dt + Td\frac{de(t)}{dt} \right] \qquad (3.6)$$

$$\frac{V(s)}{E(s)} = Kp \left[1 + \frac{1}{sTi} + sTd \right] = KpTd \left[\frac{s^2 + s/Td + 1/TiTd}{s} \right] \qquad (3.7)$$

Again it will be appreciated that the error removing integration term, 1/s is present. The term Td is called the derivative time constant.

3.1.4 Setting One-Two-and Three-Term Controllers

Clearly the controllers discussed so far may be designed using the normal techniques of control systems design. However, because of the uncertainty which frequently surrounds the exact details of the plant transfer function, approximate methods are commonly used. The most famous (though not necessarily the best) of these methods is that due to Ziegler and Nichols(1).

This   method was determined after many   experiments   on typical   process-type plants and has been in use for some 40 years.   The essence of the method is very simple.

(a)   set up the plant with a proportional controller

(b)   adjust the proportional gain constant, Kp,   until the closed loop system just oscillates.   The gain used is termed Ku.

(c)   measure the period of the resulting   oscillations   and call this time Tu.

The   "best"   parameters   for   the   controller   may   then   be determined as:

Proportional Controller

Kp = 0.5Ku   (3.8)

Proportional-plus-Integral Controller:

Kp= 0.45Ku   (3.9)

Ti = 0.83Tu

Proportional-plus-Integral-Derivative Controller:

Kp = 0.6Ku $\longrightarrow$ Ku   (3.10)

Ti = 0.5Tu

Td = 0.125T

This method of determining the values of Tu and Ku has a   number   of   disadvantages,   not   the   least   being   the possibility of damage to the plant and also the   difficulty, in practice,   of determining when sustained oscillation   has been   obtained.   As   an alternative to their "closed   loop" method Zeigler and Nichols proposed an "open loop" method in which the step response (reaction curve) of the plant   alone is determined.

The   input to the plant is subjected to a unit   change and   the   reaction curve recorded (see   below).   From   this curve the quantities R' (unit reaction rate) and L   (lagtime) are   measured.   From   these   the   values of Ku and   Tu are estimated using formulae provided by Zeigler and Nichols.

Tu = 4L

$$Ku = \frac{2}{R'L}$$ (3.11)

Where it is impractical to subject the plant to a unit step change in input, the values measured on the graph must be scaled appropriately.

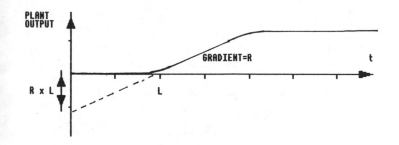

Fig 3.7   Typical Reaction Curve

3.2   DIGITAL REALISATION OF 3-TERM CONTROLLERS

A conventional   3-term controller has the   following transfer function:-

$$V(s) = E(s) \left[1 + \frac{1}{sTi} + sTd \right] Kp \qquad (3.12)$$

Kp = proportional gain

Ti = integral time constant

Td = derivative time constant

or in time terms

$$v(t) = Kp \left[e(t) + \frac{1}{Ti} \int_0^t e(t)dt + Td\frac{de(t)}{dt} \right] \qquad (3.13)$$

Remembering that microprocessors can only sample the signals at discrete points in time it will be necessary to approximate each of the terms in (3.13) using the sampled values of e(t).

The first (proportional) term is trivial

The derivative term may be approximated by what is called the "backward difference".

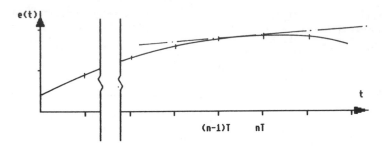

Fig 3.8   Calculation of Derivative Term

$$\left. \frac{de(t)}{dt} \right|_{t=nT} \doteq \frac{e(nT) - e([n-1]T)}{T} = \frac{e_n - e_{n-1}}{T} \qquad (3.14)$$

where T = sampling period

The integral term is evaluated by approximating the area under the e(t) curve.   This is most simply carried out using rectangular integration as shown below.

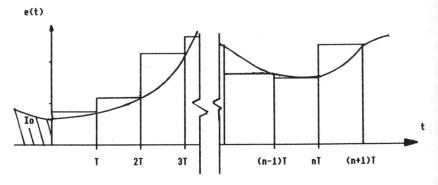

Fig 3.9   Calculation of Integral Term

$$I(nT) = I_n = \int_{t=-\infty}^{t=nT} e(t)dt$$

$$= \int_{-\infty}^{0} e(t)dt + \int_{0}^{nT} e(t)dt$$

$$= I_0 + \int_0^{nT} e(t)\,dt$$

$$= I_0 + T \sum_{k=1}^{n} e(kT) \qquad\qquad (3.15)$$

notice that $I_n = I_{n-1} + Te(nT)$

Bringing all these expressions together results in the difference equation/algorithm for 3 term control.

$$v(nT) = Kp \left[e(nT) + Td\,\frac{e(nT) - e([n-1]T)}{T} + \frac{I_{n-1}}{Ti} + \frac{T}{Ti}\,e(nT) \right]$$

using $v(nT) = v_n$ and $e(nT) = e_n$

$$v_n = Kp \left[e_n\left(1 + \frac{Td}{T} + \frac{T}{Ti}\right) - e_{n-1}\frac{Td}{T} + \frac{I_{n-1}}{Ti} \right] \qquad (3.16)$$

Equation (3.16) is simple enough to implement digitally, but suffers from one particular disadvantage which is manifest when the system it is controlling is switched from manual to automatic control. The initial output of equation (3.16) will simply be

$$v_n = Kp\left(1 + \frac{Td}{T} + \frac{T}{Ti}\right)e_n \qquad (3.17)$$

Since the controller has no knowledge of the previous sample values. It is quite unlikely that this output value will coincide with that previously available under manual control. As a result the transfer of control will cause a "bump" which may seriously disturb the plant operation. This can only be overcome by laboriously aligning the manual and computer outputs or by adding complexity to the controller so that it will automatically "track" the manual controller. The problem arises mainly from the need for an "initial condition" on the integral, I_{n-1}, and the solution adopted for digital 3-term controllers is similar to that used with conventional analogue controllers, namely to externalise the integration so that the controller only has to calculate the change in output required. The external integrator being a part of both the manual and automatic systems.

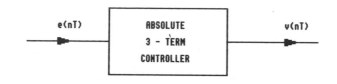

(a)   Conventional Controller

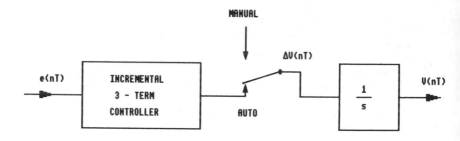

(b)   Incremental Controller

Fig.   3.10

The external integration may take the form of an electronic or pneumatic integrator but frequently the type of actuating element is changed so that incremental algorithms are use with actuators which, by their very nature contain integral action. For example, a pneumatic valve might be replaced by a motor driven valve. This brings in its train the additional advantage that, if there is a power failure, the plant will continue to operate at the last, known good, settings. Thus ensuring, for a time at least, safe operation in most circumstances. The digital incremental 3-term controller may be readily derived from equation (3.17). This is a simple and easily programmed algorithm which merely requires the storage of 3 coefficients and 3 sample values.

$$\Delta v_n = v_n - v_{n-1}$$

$$v_n = Kp \left[e_n \left(1 + \frac{Td}{T} + \frac{T}{Ti} \right) - e_{n-1} \frac{Td}{T} + \frac{I_{n-1}}{Ti} \right] \qquad (3.18)$$

and

$$v_{n-1} = Kp \left[e_{n-1}(1 + \frac{Td}{T} + \frac{T}{Ti}) - e_{n-2}\frac{Td}{T} + \frac{I_{n-2}}{Ti} \right] \qquad (3.19)$$

but

$$I_{n-1} = I_{n-2} + Te_{n-1}$$

thus

$$\Delta v_n = Kp \left[e_n (1 + \frac{Td}{T} + \frac{T}{Ti}) - e_{n-1} (1 + \frac{2Td}{T}) + e_{n-2}\frac{Td}{T} \right] (3.20)$$

C1 C2 C3

A pictorial representation of this algorithm is shown below:

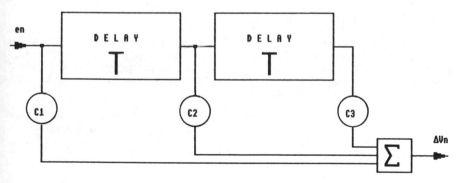

Fig 3.11   Pictorial Representation of PID Algorithm

3.2.1   Choice of Parameter Values For 3-Term Control

The "traditional" Zeigler-Nichols criterion gives the settings for Ti and Td defined by equation 3.10.

Ti = 0.5Tu

Td = 0.125Tu

$\qquad (3.21a)$

It has been suggested by Lee (2) and reported by Roberts and Dallard (3) that the sampling period for the system should be chosen in the range.

$$0.0625Tu < T < 0.125Tu$$

(3.21b)

Typically $T = 0.1Tu$

Using these values in the difference equation (3.20) leads to the control algorithm.

$$\Delta v_n = Kp \left[2.45e_n - 3.5e_{n-1} + 1.25e_{n-2} \right]$$

(3.22)

or in Z-transform terms:

$$\frac{\Delta V}{E}(z) = Kp \left[2.45 - 3.5z^{-1} + 1.25z^{-2} \right]$$

(3.23)

for the incremental algorithm, and:

$$\frac{V}{E}(z) = Kp \left[\frac{2.45 - 3.5z^{-1} + 1.25z^{-2}}{1 - z^{-1}} \right]$$

(3.24)

for the original algorithm.

The fact that, apart from the gain constant, this algorithm is the same for all loops could provide worthwhile economies in the programming of multi-loop microprocessor controllers. The primary difference between loops will be the sampling frequency and the gain Kp.

A further advantage of this method is that there is only a single "tuning" parameter. Roberts and Dallard report that the performance of this controller is not much inferior to that of a controller in which all 3 parameters are tuned.

Graphs to aid in the selection of Ti, Td and Kp when these parameters are kept independent are given in Lopez et al (4).

The Zeigler-Nichols method referred to above can be further refined by taking account of the self regulation present in the plant. This modification is due to Cohen and Coon (5). One particular disadvantage of the Zeigler-Nichols method is that the fact that it tends to give rather oscillatory responses. ($\zeta = 0.2$)

3.3 IMPLEMENTATION OF OTHER ALGORITHMS

The simple method described in section 3.2 is, of course, not confined to the mechanisation of a 3-term controller, it may be applied to other algorithms.

Consider the simple compensator

$$\frac{V}{E}(s) = K \frac{1 + s\tau}{1 + s\alpha\tau} \qquad (3.25)$$

$$\alpha > 1 \text{ for lag compensators}$$

$$\alpha < 1 \text{ for lead compensators}$$

in differentiated form this is

$$v(t) + \alpha\tau \frac{dv(t)}{dt} = Ke(t) + K\tau \frac{de(t)}{dt} \qquad (3.26)$$

or, using the previous representation of differentials as finite differences we have

$$V_n + \alpha\tau \left[\frac{v_n - v_{n-1}}{T} \right] = Ke_n + K\tau \left[\frac{e_n - e_{n-1}}{T} \right]$$

ie

$$v_n \left[1 + \frac{\alpha\tau}{T} \right] = K \left[1 + \frac{\tau}{T} \right] e_n - K\frac{\tau}{T}e_{n-1} + \frac{\alpha\tau}{T}v_{n-1}$$

$$v_n = e_n K \left[\frac{T + \tau}{T + \alpha\tau} \right] - e_{n-1} \left[\frac{K\tau}{1 + \alpha\tau} \right] + v_{n-1} \left[\frac{\alpha\tau}{1 + \alpha\tau} \right] \qquad (3.27)$$

This method may readily be extended to compensators with higher order terms in s. For instance

$$s^2 E(s) \equiv \frac{d^2 e(t)}{dt^2}$$

$$\frac{d^2 e(t)}{dt^2} = \frac{\left[\dfrac{de(t)}{dt}\right]_{t=nT} - \left[\dfrac{de(t)}{dt}\right]_{t=(n-1)T}}{T}$$

$$= \frac{\left[\dfrac{e_n - e_{n-1}}{T}\right] - \left[\dfrac{e_{n-1} - e_{n-2}}{T}\right]}{T}$$

$$= \frac{e_n - 2e_{n-1} + e_{n-2}}{T} \tag{3.28}$$

thus $\displaystyle \int \frac{d^2 e(t)}{dt^2} \equiv \left[\frac{1 - 2z^{-1} + z^{-2}}{T}\right] E(z) \tag{3.29}$

And the process may be repeated for higher orders.

An alternative method for arriving at a Z-transfer function is by simple s → z substitution.

strictly $z = e^{sT}, \quad s = \dfrac{1}{T} Ln(z)$

However, this substitution does not lead to the required polynomial form of z transfer function. To achieve this an approximation to the natural logarithm is used:

$$Ln(z) \approx 2 \left[\frac{1 - z^{-1}}{1 + z^{-1}}\right]$$

hence

$$s \approx \frac{2}{T} \left[\frac{1 - z^{-1}}{1 + z^{-1}}\right] \tag{3.30}$$

and the substitution is used directly, so that the simple compensator is transformed:-

$$\frac{V}{E}(s) = K \left[\frac{1 + s\tau}{1 + s\alpha\tau}\right]$$

$$V(z) = E(z) \frac{1 + 2\frac{\tau}{T}\left[\frac{1 - z^{-1}}{1 + z^{-1}}\right]}{1 + 2\frac{\alpha\tau}{T}\left[\frac{1 - z^{-1}}{1 + z^{-1}}\right]} K \qquad (3.31)$$

$$= E(z) \left[\frac{\left[1 + 2\frac{\tau}{T}\right] + \left[1 - 2\frac{\tau}{T}\right]z^{-1}}{\left[1 + 2\frac{\alpha\tau}{T}\right] + \left[1 - 2\frac{\alpha\tau}{T}\right]z^{-1}}\right] K \qquad (3.32)$$

ie $\quad V(z)\left[\left[1 + 2\frac{\alpha\tau}{T}\right] + \left[1 - 2\frac{\alpha\tau}{T}\right]z^{-1}\right] =$

$$E(z) K \left[\left[1 + 2\frac{\tau}{T}\right] + \left[1 - 2\frac{\tau}{T}\right]z^{-1}\right] \qquad (3.33)$$

or

$$v_n\left[1 + 2\frac{\alpha\tau}{T}\right] + v_{n-1}\left[1 - 2\frac{\alpha\tau}{T}\right] =$$

$$e_n\left[1 + 2\frac{\tau}{T}\right]K + e_{n-1}\left[1 - 2\frac{\tau}{T}\right]K \qquad (3.34)$$

$$v_n = e_n\left[\frac{T + 2\tau}{T + 2\alpha\tau}\right]K + e_{n-1}\left[\frac{T - 2\tau}{T + 2\alpha\tau}\right]K - v_{n-1}\left[\frac{T - 2\alpha\tau}{T + 2\alpha\tau}\right] \qquad (3.35)$$

This transformation, known as the Bilinear z-Transform is only strictly valid for sT<<1. In fact if this limitation is violated the frequency axis "warped" according to the formula.

$$w' = \frac{2}{T}\,\mathrm{Tan}\left[\frac{wT}{2}\right] \qquad (3.36)$$

where w is the "designed" frequency and w' is the "achieved" frequency.

Thus the corner frequencies of compensators which violate the wT<<1 rule have to be pre-warped to account to this.

3.4   IMPLEMENTATION ASPECTS

Sections 3.2 and 3.3 have discussed how, in a simple-minded way, digital controllers may be implemented from analogue prototypes. Any digital controller however has to work within the restrictions imposed by digital computers, further it ought to take advantage of the opportunities offered by such computers.

The added features of such controllers might include:

. filtering

. desaturation (of 3-term controllers)

. linearisation

. adaption

whilst the problems include, amongst others:

. conflicting requirements, in the selection of sampling rate.

. arithmetic inaccuracy - leading to noise, deadbands and instability.

It is with these opportunities and problems that this section is concerned.

3.4.1 Refinements of the 3-Term Algorithm

The calculation of integral and derivative action outlined earlier was fairly crude and improved accuracy is possible. For instance, trapezoidal integration or Simpson's rule integration is possible.

Because Simpson's rule requires more computation than trapezoidal integration the latter is more common. Using this techniques we have

$$I_n = I_{n-1} + \left[\frac{e_n + e_{n-1}}{2} \right] T \qquad (3.37)$$

which gives rise to the simple 3-term algorithm

$$v_n = Kp \left[e_n \left[1 + \frac{Td}{T} + \frac{T}{2Ti} \right] - e_{n-1} \left[\frac{Td}{T} - \frac{T}{2Ti} \right] + \frac{I_{n-1}}{Ti} \right] \qquad (3.38)$$

which is the same as the previously given form but with slightly modified coefficients.

The corresponding incremental algorithms is:

$$\Delta v_n = Kp \left[e_n \left[1 + \frac{Td}{T} + \frac{T}{2Ti} \right] \right.$$

$$\left. - e_{n-1} \left[1 + \frac{2Td}{T} - \frac{T}{2Ti} \right] + e_{n-2} \frac{Td}{T} \right] \qquad (3.39)$$

The calculation of the derivative action term is subject to a number of problems due to noise. The action of taking the derivative of a signal will always tend to emphasise any high frequency noise terms in the signal. A number of methods may be employed to combat this effect.

One method is simply to filter the error term with a low-pass filter of the form:-

$$\frac{e'}{e} = \frac{1}{1 + s\tau}$$

where e is raw error

and e' is the smoothed error

$\qquad (3.40)$

Where τ is chosen so that the bandwidth of the filter includes all significant frequencies in the desired error signal but excludes "noise" frequencies.

This process may be reduced to a differential equation.

$$e' + \tau \frac{de'}{dt} = e$$

thus

$$e'_n + \frac{\tau}{T} \left[e'_n - e'_{n-1} \right] = e_n$$

ie

$$e'_n = e_n \frac{T}{T + \tau} + e'_{n-1} \frac{\tau}{T + \tau}$$

or for $T \ll \tau$

$$e'_n = e_n \frac{T}{\tau} + e'_{n-1} \qquad (3.41)$$

This expression may be included in the 3-term algorithm at the expense of a slight increase in computation time and storage requirement.

A slightly simpler smoothing effect may be obtained by using a running average technique.

$$e'_n = \frac{e_n + e_{n-1} \ - \ - \ - \ e_{n-m}}{m \ + \ 1} \tag{3.42}$$

Bilberro (8) reports also that the 4 point central difference estimate of de/dt has been used with success.

$$\frac{de}{dt}\Bigg|_{t=nT} \approx \frac{e_n - e_{n-3} + 3e_{n-1} - 3e_{n-2}}{6T} \tag{3.43}$$

The error signal is formed as the difference between set point and measured variable. Noise corruption will only be present on the measured variable and it may be more sensible to apply the smoothing to this signal rather than to the derivative directly (3.12a). Sudden changes in set point will of course, give rise to very large values of de/dt which will, however, only persist for one calculation. This phenomenon, known as "set point kick" or "derivative Kick" is present in both analogue and digital controllers. However, the latter do provide a simple means for eliminating its worst effects. The filter scheme in (3.12b) below will provide some smoothing and protection but the scheme in (3.12c) below, which provides for different filter time constants in the set point and measured variable paths may prove more effective. Filtering the set point with a long time constant is equivalent to forcing the set point changes to be ramped in.

In addition, or insted of linear filtering, it is perfectly possible for a digital controller to sense when the set point has been changed, and, in these circumstances, to feed forward a compensating signal which will balance the set point kick (3.12d below). This is, in fact equivalent to taking the derivative of the feedback signal only (filtered or otherwise) and this is the method normally employed for implementation.

Another severe problem in analogue controllers is that of "integral wind up" caused by persistent offset between measured and set values. The effect of this is to lengthen the system settling time.

Analogue controllers are limited in what they may achieve to aleviate this problem. Typically, the integral term is clamped to a limiting value or is reset to zero. Both of these techniques reduce the period of "controller paralysis" but suffer from the disadvantage that the integral term is set to an inappropriate value once the control output comes back within the dynamic range of the actuator being driven. Their actions may be mimicked quite easily in digital controllers provided that the P, I and D terms are separately calculated and hence are available for direct manipulation. While such an approach may work it does not overcome the problem of misalignment in the integral term - an effect comparable to the auto/manual

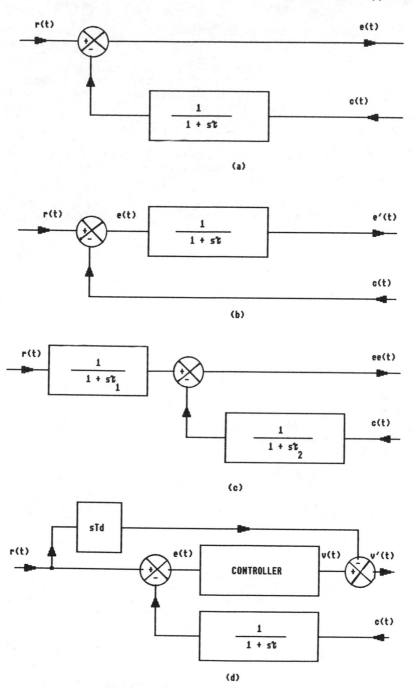

Fig 3.12 Various Noise Reduction Techniques for PID Controllers

changeover "bump" often experienced with non-incremental controllers.

Fortunately it is simple to include mechanisms for the exact adjustments of the integral term into digital controllers. Such techniques require the controller to monitor the plant performance and to detect the point of desaturation. At this time an appropriate value for the integral term is computed and used as the starting point for control action. Such calculations would be impossible for analogue controllers and they represent another instance of a qualitative improvement in controller performance which is attributable directly to the digital implementation.

In Thomas, Sandoz and Thompson (10) it is shown that for dominantly first order plants the required setting of the integral term is:

$$I = \frac{1}{K_c K_p} Y_D + \frac{1}{K_c} L \qquad\qquad (3.44)$$

where

Kp = controller gain

Kc = plant gain

Y_D = value of controlled variable at time of desaturation

L = load disturbance

All of the above parameters being readily estimated either on-line or off-line.

The net effect of this setting for I is to allow the plant to perform thereonwards as if it had emerged from a steady state with its output at Y_D. Thus a well behaved linear, non-saturating, step response is obtained after desaturation.

3.5   Other Controller Exhancements

The above discussion was related specifically to 3-term controllers, although the ideas of signal smoothing etc. could well be applied to any controller. Other features which might be incorporated in any controller (including 3-term controllers) include linearisation and adaption and these will be briefly outlined.

3.5.1   Linearisation

Most processes control systems include significant non-linearities whose presence may cause degraded performance. Typical of such non-liniarities are those occuring in flow measurement (square root) and temperature measurement.

By applying an inverse non-linear law to the smoothed, measured input a significant reduction in the effect of the non-linearity may be achieved.

There are a number of ways in which the correction may be applied.

(i)   The correction law may be expressed mathematically using whatever functions are required. This method is often rather wasteful of computer time and is totally impossible if the relevant functions are not available.

(ii)   A lookup table may be constructed. This simply represents the x - y co-ordinates of the inverse function. It is very wasteful of computer memory but is the quickest to execute.

(iii)   A lookup table may be constructed with very few points, intermediate values being determined by interpolation. This represents a compromise solution.

3.5.2   Adaption

A full adaptive control scheme is outside the scope of "simple controllers" and will be discussed elsewhere. However, simple schemes in which the control algorithm or the control parameters are switched according to the state of one or two easily measured plant parameters are readily implemented with digital controllers.

Such schemes are termed "programmed adaption" and can provide very worthwhile improvements in plant performance. One problem that must be solved in such schemes is that of providing "bumpless" transfer from one control algorithm to the next.

3.6   Sample Rate Selection

The sampling rate employed by a digital controller is an important parameter. The choice is governed by a number of considerations.

(i)   An excessively high sampling rate imposes a heavy burden on the computer.

(ii)   A high sampling rate makes the simple design methods more exact.

(iii)   A high sampling rate agravates the problems caused by derivative action with noisy signals.

(iv)   The sampling rate must satisfy Shannon's law.

(v)   Excessively high sampling rates agravate the problems of finite-precision arithmetic.

Clearly rule (iv) imposes a lower limit on the sampling rate, all of the other rules (except ii) provide some "fuzzy" guidance as to the upper limit.

Rule (iii) may be explained thus:

$$\frac{de}{dt} \approx \frac{e_n - e_{n-1}}{T} \qquad\qquad (3.45)$$

but the measured values of e are in error due to noise by an amount $\leqslant \varepsilon$ (say). Thus the calculated value of de/dt is

$$\frac{de}{dt} = \frac{(e_n \pm \varepsilon) - (e_{n-1} \pm \varepsilon)}{T} \qquad\qquad (3.46)$$

$$= \text{true derivative} \pm \frac{2\varepsilon}{T}$$

Clearly the error term, $2\varepsilon/T$ gets larger as the sampling period shortens.

Rule (v) may be illustrated by reference to the difference equation of the low pass filter given by equation 3.41.

$$e'_n = e_n \cdot \frac{T}{\tau} + e'_{n-1} \qquad\qquad (3.47)$$

T = sample time

τ = filter time constant

Clearly we have to represent the coefficients one and T/τ. Assuming that we require at least 5% precision on the representation of the relative sizes of the coefficients (which, of course, govern the filter action) this implies a dynamic range in the number systems of

$$1 : \frac{T}{20\,\tau}$$

Hence a careless choice of $T \approx \tau/100$ leads to a dynamic range of 2000:1 requiring 12 bit accuracy (including sign bit). In general rapid sampling implies high arithmetic accuracy, and this matter has been given a more extensive treatment in the literature than is possible in this chapter. (12-18)

3.7 Considerations of Computational Accuracy

Many microprocessors currently in use employ a word length of 8 bits. It is, therefore pertinent to consider whether the limited accuracy of calculation available in these devices will significantly disturb the performance of a 3-term algorithm.

The incremental 3-term algorithm may be illustrated diagrammatically as shown below:

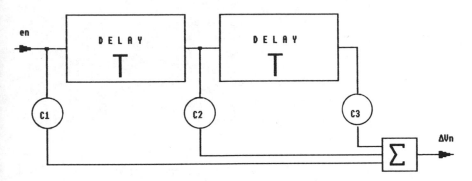

Figure 3.13   Pictorial Representation of 3-Term Algorithm

Where c_1 = 2.45, c_2 = -3.5, c_3 = 1.25 if the previously given recommendations are followed and if Kp = 1.

Because digital representation of quantities usually involves a truncation error, the coefficients will each be subject to a maximum error of Δc. Likewise the measured value of the error signal will be subject to an error δ.

The maximum value of Δv occurs if

$$e_{n-2} = e_n = -e_{n-1} = e_{max} \tag{3.48}$$

thus:

$$\Delta v_{max} = e_{max}(2.45 + 3.5 + 1.25) = 7.2e_{max} \tag{3.49}$$

The actually realised value of Δv, however, could be in error due to coefficient rounding etc.

ie $\Delta v_{actual} = (e_{max} \pm \delta)(C_1 - C_2 + C_3 \pm 3\Delta C)$

$$\doteqdot 7.2e_{max} \pm \delta (C_1 - C_2 + C_3) \pm e_{max} 3\Delta C \tag{3.50}$$

Where the first term represents the desired output from an exact algorithm, the second term represents the error due to input quantisation and the third term represents the error due to coefficient trunction, assuming the worst case situation.

Thus maximum error = $(C_1 - C_2 + C_3) + e_{max} (3\Delta C)$

under these circumstances:

$$\%error = 100 \left[\frac{7.2\delta + e_{max}\ 3\Delta C}{7.2e_{max}} \right]$$

$$= \frac{\delta}{e_{max}} 100 + \frac{300}{7.2} \Delta C \qquad (3.51)$$

At less than maximum output, say e

$$\%error = \frac{\delta}{e} 100 + \frac{300}{7.2} \Delta C \qquad (3.52)$$

It might be realistic to assume that the output actuator has a resolution of about 3% and thus sensible to require that the error be contained within this band.

Allocating the error equally between each cause gives (at full scale, emax) and using 3.51.

$$100 \frac{\delta}{e_{max}} = 1.5$$

and

$$\frac{300}{7.2}\Delta C = 1.5$$
$$(3.53)$$

Consequently:

$$\Delta C = 0.036 \qquad\qquad \frac{\delta}{e_{max}} = 0.015 = \frac{1}{2^6} \quad (3.54a)$$

Thus the resolution on the input is 1 part in 2^6 which implies a word length of 7 bits including sign. (This would increase to 10 bits if the accuracy is to be maintained down to e = emax/10

Similarly the coefficient resolution must be

$$\frac{\Delta C}{C_{max}} = \frac{0.036}{3.5} \qquad (3.54b)$$

Or a resolution of 1 part in 97 (1 part in 2^7) which implies a word length of 8 bits (including sign) for coefficient storage.

The above, admittedly rough and ready, calculation shows that an 8 bit microprocessor may be exected to give a reasonable performance as a 3 term controller without special precautions.

There is, however, one source of error which has been overlooked. When the 8 bit value of a coefficient is multiplied by the 8-bit value of a signal, a 16 bit number is produced. Truncating this to 8 bits results in a maximum error of $1/2^7$.

Since this occurs 3 times in the 3-term algorithm, there is a possible error of $= 2.4\%$ in the answer. To reduce this error to about 0.3% (i.e. 1/10 of the overall allowed error) and thus retain the validity of the previous calculations would require the product to be retained to about 11 bits; in practice it would seem easiest to store the input variables and the coefficients to single precision (8 bits) but to do all arithmetic as double precision (16 bits).

A phenomenon which has not been mentioned is that of quantisation-error-induced deadband. This is essentially a steady-state effect which results in the output settling to a different value depending on the direction from which it is approached. This is discussed by Blackman (11) for feedback controllers and in section 3.7.1 below.

It should be pointed out that where fixed-point calculations are used problems of coefficient scaling can occur and many commercially available systems use either hardware or softward floating point arithmetic.

3.7.1 Offset Errors Caused by Computational Inaccuracies

As a result of the above mentioned sources of computational error, and the fact that variables can only be stored to the accuracy determined by the computer word length, it is possible for a digital control system to exhibit a steady-stage offset error in excess of that predicted by conventional linear control theory.

This may be explained qualitatively as follows. Consider a controller operating in steady-state conditions, the plant has stabilised and neither the input, the error signal, nor the output change between sampling instants. A change in operating conditions (such as an increase in load) now occurs and results in the error changing by ε. Ideally the output will change to correct this error but it is possible because of the finite accuracy of computation, storage and representation of the output that the actual binary number presented at the output of the controller will not change. Thus the increased offset, ε, remains uncorrected.

Consider the controller represented by the z-transform

$$\frac{h}{e}(z) = \lambda D(z) = \frac{\lambda \displaystyle\sum_{n=0}^{N} b_n z^{-n}}{\displaystyle\sum_{m=1}^{M} a_m z^{-m}} \qquad M \geqslant N \qquad (3.55)$$

Where λ is a scaling factor and $a_0 = 1$

N.B.   *The variable 'h' is equivalent to the 'v'
         used previously. The change being made in
         the interests of consistency with Knowles
         (19).*

The difference equation relating the controller output,
h, to the error signal, e, is given by:-

$$h(k) = \sum_{n=0}^{N} b_n \lambda e(k-n) \;-\; \sum_{m=1}^{M} a_m h(k-m) \qquad (3.56)$$

or, using $f = \lambda e$

$$h(k) = \sum_{n=0}^{N} b_n f(k-n) \;-\; \sum_{m=1}^{M} a_m h(k-m) \qquad (3.57)$$

In fact the controller output may be in error due to
computational inaccuracies. The inaccurate output is:

$$h'(k) = e(k) + \Delta(k) \qquad (3.58)$$

The sampled error signal, e'(k) will be in error because of the finite representation used:

$$e'(k) = e(k) + r_q(k) \qquad (3.59)$$

$$e(k) = \text{true error signal}$$

$$r_q(k) = \text{quantisation error}$$

Likewise the scaling of the error signal by:

$$f(k) = \lambda e'(k) \qquad (3.60)$$

will result in another error, $r_\lambda(k)$, such that the stored representation of f(k) is f'(k), given by:

$$f'(k) = \lambda e'(k) + r_\lambda(k) \qquad (3.61)$$

This erroneous value of f'(k) is used to compute the value of h'(k) using the formula:

$$h'(k) = \left[\sum_{n=0}^{N} b_n f'(k-n)\right] - \left[\sum_{m=1}^{M} a_m h'(k-m)\right] + R_D(k) \qquad (3.62)$$

Where R_D represents the errors accumulated in the calculation of the product terms of the equation.

Note that multiplications by one or zero do not incur an error.

Substituting (3.61) & (3.59) in (3.62) gives:

$$h'(k) = \left[\sum_{n=0}^{N} b_n \left[\lambda e(k-n) + \lambda r_q(k-n) + r_\lambda(k-n)\right]\right]$$

$$- \left[\sum_{m=1}^{M} a_m h'(k-m)\right] + R_D(k) \qquad (3.63)$$

Once a steady-state has been reached, the terms:

$$h(k) \quad = h(k-1) \quad = h$$

$$\Delta(k) \quad = \Delta(k-1) \quad = \Delta$$

$$h'(k) = h'(k-1) = h'$$

$$e(k) \quad = e(k-1) \quad = e$$

$$e'(k) = e'(k-1) = e'$$

$$r_q(k) = r_q(k-1) = r_q$$

$$r_\lambda(k) = r_\lambda(k-1) = r_\lambda$$

$$R_D(K) = R_D(k-1) = R_D$$

$$\varepsilon(k) \quad = \varepsilon(k-1) \quad = \varepsilon$$

Thus, using these steady-state values in (3.63) we have

$$h' = \left[\sum_{n=0}^{N} b_n (\lambda e + \lambda r_q + r_\lambda)\right] - \left[\sum_{m=1}^{M} [a_m h']\right] + R_D \tag{3.64}$$

Assume that the actual error signal, e, is now purturbed so thatthere is a new error signal e_2:

$$e_2 = e + \varepsilon \tag{3.65}$$

and a new steady-state computer output, h' given by:

$$h'_2 = \left[\sum_{n=0}^{N} b_n (\lambda [e + \varepsilon] + \lambda r_{q2} + r_{\lambda 2})\right] - \left[\sum_{m=1}^{M} a_m h'_2\right] + R_{D2} \tag{3.66}$$

where the symbols have the same meaning as before and the subscript "2" represents the values obtained under the new conditions. Now assume that ε is so small that no change in output results. Then:

$$h'_2 = h'$$ (3.67)

and we have from (3.64) & (3.66)

$$\left[\sum_{n=0}^{N} b_n [\lambda e + \lambda r_q + r_\lambda] \right] - \left[\sum_{m=1}^{M} a_m h' \right] + R_D$$

$$= \left[\sum_{n=0}^{N} b_n [\lambda e + \lambda \varepsilon + \lambda r_{q2} + r_{\lambda 2}] \right] - \left[\sum_{m=1}^{M} a_m h' \right] + R_{D2}$$ (3.68)

Which gives:

$$\sum_{n=0}^{N} b_n [\lambda (r_q - r_{q2}) + (r_\lambda - r_{\lambda 2})] + (R_D - R_{D2})$$

$$= \sum_{m=1}^{M} b_n \lambda \varepsilon$$ (3.69)

or

$$\varepsilon = (r_q - r_{q2}) + \frac{r_\lambda - r_{\lambda 2}}{\lambda} + \frac{R_D - R_{D2}}{\lambda \sum_{n=0}^{N} b_n}$$ (3.70)

Now r_q, r_{q2}, r_λ, $r_{\lambda 2}$, all result from a single operation thus their maximum value is:

$$q \quad \text{for truncation}$$

or

$$q/2 \text{ for rounding}$$

Thus the worst possible values for $(r_q - r_{q2})$ and $(r_\lambda - r_{\lambda 2})$ are

$$| r_\lambda - r_{\lambda 2} | \leqslant 2q$$ truncation   (3.71)

$$| r_q - r_{q2} | \leqslant 2q$$

Likewise R_d and R_{D2} both result from the errors in calculating the difference equation (3.64). Assume that amon the coefficients a_m, b_n, there are μ whose value is neither one nor zero. Then we have:

$$| R_D | \leqslant \mu q$$ truncation   (3.72)

$$| R_{D2} | \leqslant \mu q$$

Thus the maximum value of $(R_D - R_{D2})$ is $2\mu q$.

Thus it is posible to put a maximum bound on ϵ :

ie $$| \epsilon | \leqslant 2q[1 + 1/\lambda] + q \dfrac{2\mu}{\lambda \sum\limits_{n=0}^{N} b_n}$$ (3.73)

This represents the worst case with the error signal change fromone side of the "dead band" to the other. So it may be concluded that:

Dead band width $$\leqslant \pm q \left[(1 + 1/\lambda) + \dfrac{\mu}{\lambda \sum\limits_{n=0}^{N} b_n} \right]$$ (3.7

This is identical to the formula given by Knowles(19) excep that he considers the second term negligible. The symbols have been kept the same as in (19) for convenience of cross referencing.

3.8 CONCLUSIONS

This chapter has discussed how digital techniques may be used to implement relatively simple controllers.   It has been shown that, while such implementations may not have any control   theoretic   advantages   over   their   analogue counterparts, the digital implementation does allow for various enhancements. These include the linearisation of transducer characteristics, the implementation of simple adaption and desaturation; so that the resulting system has significant   practical   advantage   over   the   analogue equivalent.

A number of methods for transferring analogue designs to the digital domain have been demostrated.   These include the   method   of   finite   differences   and   bilinear transformation.

Various aspects of importance in practical designs have been   discussed,   including   both   the   selection   of   an appropriate sampling interval and the problems of finite-precision arithmetic.

References

1.   Zeigler J G & Nichols N B   "Optimum Settings for Automatic Controllers". Taylor Instrument Company Bulletin TDS - 10A100.

2.   Lee.   "Direct Digital Control,"1965 Society for Instrument Technology Symposium 22nd April.

3.   Roberts and Dallard, 1974, "Discrete PID Controller a   Single Tuning Parameter" Measurement and Control Vol 7   T97-T101.

4.   Lopez, Murril and Smith, 1969, "Tuning PI and PID Digital   Controllers",   Instruments   and   Control Systems, February 1969 89-95.

5.   Cohen & Coon "Theoretical Considerations of Retarded Control",Taylor Instrument Company Bulletin TDS-10A102.

6.   Jury E I "Sampled Data Control Systems", Wiley

7.   Ragazzini J R & Franklin G, 1958, "Sampled Data Control Systems", MacGraw-Hill.

8.   Bibero R J, 1977, "Microprocessors in Instruments and Control", Ch 5, Wiley New York.

9.   Fensome D A, 1983,"Understanding 3-Term Controllers", Electronics and Power Sept 1983, 647.

10.   Thomas H W Sandox D J Thompson M, 1983, "New
        Desaturation Strategy for Digital PID Controllers"
        IEE Proc. Vol. 130 pt. D. No. 4, 189-192.

11.   Blackman R B "Data smoothing and Prediction", 75-81,
        Addison Wesley.

12.   Pertram J E "The Effect of Quantisation in Sampled-
        Feedback Systems", 1958, Trans. AIEE, 77 pt.2, 177.

13.   Slaughter J B, 1964, "Quantisation Errors in Digital
        Control Systems" Trans IEEE PTGAC, 1964, 70.

14.   Tsypkin Ya Z, 1960, "An Estimate of the Influnece of
        Amplitude Quantisation of Processes in Digital
        Automatic Control Systems, "Automatica i
        Telemkhanika, 1960, 3.

15.   Knowles J B, 1965, "The Effect of a Finite Word-
        Length Computer in a Sampled-Data Feeback System"
        Proc. IEE, 112, 1197.

16.   Knowles J B and Edwards R, 1965, "Finite Word-Length
        Effects in Multi-Rate Direct Digital Control
        Systems", Proc. IEE, 112, 2376.

17.   Kaneko T and Liu B 1968, "Round-Off Error of Floating
        -Point Digital Filters", 6th Annual Allerton
        Conference.

18.   Knowles J B, and Edwards R, 1966, "Computational
        Error Effects in a Direct Digital Control Systems",
        Automatica, 4 7.

19.   Knowles J B, 1980, "A Simplified Analysis of
        Computational Errors in a Feedback System
        Incorporating a Digital Computer" in "The Design
        and Analysis of Sampled Data Control Systems" - an
        IERE Seminar held at the Royal Institution 19 Feb
        1980.

Chapter 4

Design of digital controllers

Dr D. Rees

4.1   INTRODUCTION

During recent decades the design procedures for analog controllers have been well formulated and a large body of knowledge accumulated. This methodology, based on conventional design techniques of the root locus and Bode plot, or the process reaction curve methods of Ziegler - Nicholas (or Cohen and Coon) may be applied to designing digital controllers. The procedure would be to first design the analog form of the controller or compensator to meet a particular performance specification. Having done this the analog form can be transformed to a discrete representation by means of the the z transform or directly using difference equations.

This approach has already been introduced in previous chapters, where the most common type of digital controller was considered, namely, the discrete form of the three term algorithm. Undoubtedly, this is the most popular and widely used control algorithm and virtually all microprocessor based systems commercially available today use this form of controller. The enhancements introduced in recent years have been in auto-tuning of the three-term algorithm using on-line plant identification and computation of an optimum setting for the 2 or 3 term control parameters. The reason for the popularity of this type of controller arises from the wealth of knowledge that currently exists within process industries, where its operation is well understood and the procedure for its tuning clearly outlined. Also the controller has proved remarkably robust in controlling difficult systems.

The alternative approach is to design controllers directly in the discrete domain, based on the time domain specification of a closed-loop system response. The controlled plant is represented by either a discrete model, as in the case of certain industrial processes where continuous dynamics is inappropriate, or by a discretised model, which is a continuous system observed, analysed and controlled at discrete intervals of time. Since the time response is the ultimate objective of the design, then this approach which we shall now consider provides a direct path to the design of controllers. The features of direct

digital designs are that sample rates are   generally   lower
than   for   three-term   equivalents;   the behaviour   of   the
closed-loop   system   is   less dependent   upon   the   sampling
interval;   that   design is directly "performanced based"   ;
and   for systems with significant delays,   the design of   a
suitable   controller   is straight forward and can achieve   a
better   response than is possible from a 3-term controller.
The   reason why sampling rates are generally higher for   the
discretised   form   of the 3 term controller is the   need   to
approximate   integral   and   derivative   terms   accurately.
However,   this   is   not   a   significant   factor   as   current
microprocessor   controllers can operate at orders of   50-100
samples/second, which is adequate for many processes.

4.2   PROCESS MODELS

The starting point of all designs procedures is   based
on a model of a process.   The techniques of Ziegler-Nichols,
and variants of it are based on fitting a model to the open-
loop   reaction   curve,   which represents the measured   plant
dynamics.   Most common plants may be approximated to simple
linear   models   which   are   "best fits" of   the   process
reaction   curve   data.   The models normally fall   into   the
following three categories:

First Order lag plus time-delay

$$G(s) = \frac{Ke^{-\theta s}}{1+s\tau} \tag{4.1}$$

Cascaded lag plus time-delay

$$G(s) = \frac{Ke^{-\theta s}}{(1+\tau_1 s)(1+\tau_2 s)} \tag{4.2}$$

Underdamped second-order lag plus time-delay

$$G(s) = \frac{Ke^{-\theta s}}{\tau^2 s^2 + 2\tau\xi s + 1} \tag{4.3}$$

It is difficult to fit models higher than second-order
to the process reaction curve.   This in most cases is not a
disadvantage as normally the response is dominated by one or
two   time   constants and the smaller time constants   can   be
accounted for in the time delay.

Since   we are designing a discrete controller then   it
is necessary to evaluate the pulse transform of the   process
and   this   must   take   into   account   the   sample   and hold
operation   of the D/A converter.   This arrangement is shown
in Fig.4.1.

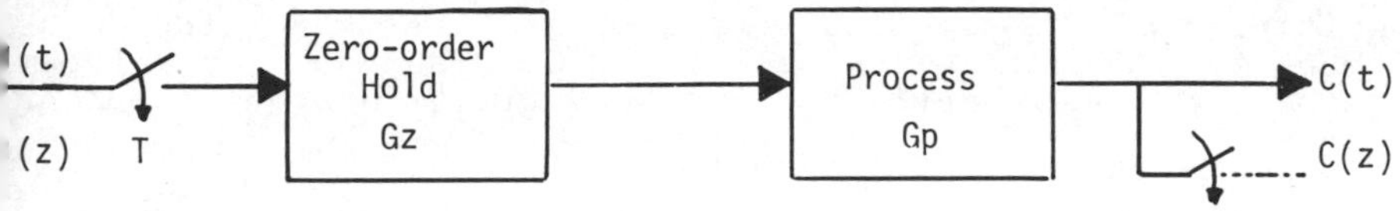

Fig.4.1 Block diagram of an open loop system with hold device.

If the process is described by a first-order lag plus time delay, then the pulse transfer function $G(z) = G_z G_p(z)$ is

$$G(z) = \mathcal{Z}\left[\frac{1-e^{Ts}}{s} \frac{Ke^{-\theta s}}{1+s\tau}\right] = (1-z^{-1})\mathcal{Z}\left[\frac{Ke^{-\theta s}}{s(1+s\tau)}\right] \tag{4.4}$$

Assuming that delay θ is an integral number of sample periods, say $\theta = kT$, for a sample period of T, then

$$G(z) = K(1-z^{-1})z^{-k}\mathcal{Z}\left[\frac{1}{s(1+s\tau)}\right]$$

Since

$$\mathcal{Z}\left[\frac{1}{s(1+s\tau)}\right] = \frac{(1-e^{-T/\tau})z^{-1}}{(1-z^{-1})(1-z^{-1}e^{-T/\tau})}$$

then

$$G(z) = K\frac{(1-e^{-T/\tau})z^{-1}}{(1-e^{-T/\tau}\,z^{-1}}z^{-k} \tag{4.5}$$

The second-order lag plus dead time process model of eqn.4.2 is an improved representation of many processes. In this case, the pulse transform relating $C(z)$ to $U(z)$ is

$$\frac{C(z)}{U(z)} = \frac{Kz^{-k}(b_1+b_2z^{-1})z^{-1}}{(1-e^{-T/\tau_1}z^{-1})(1-e^{-T/\tau_2}z^{-1})} \tag{4.6}$$

where

$$b_1 = 1 + \frac{\tau_1 e^{-T/\tau_1} - \tau_2 e^{-T/\tau_2}}{\tau_2 - \tau_1}$$

$$b_2 = e^{-T(1/\tau_1 + 1/\tau_2)} \frac{+\tau_1 e^{-T/\tau_2} - \tau_2 e^{-T/\tau_1}}{\tau_2 - \tau_1}$$

This again assumes that the delay is an integer multiple of the sampling period. If this is not the case then the dead time can be written as a sum of an integer number of sample periods minus some fraction of a sampling period.

$$\theta = kT - \beta T$$

where β is in the range 0 to 1.

For the first order model plus dead time of eqn.4.1 the pulse transform when $\theta \neq kT$ is

$$\frac{C(z)}{U(z)} = \frac{K(c_1 + c_2 z^{-1}) z^{-k-1}}{1 - (e^{-T/\tau}) z^{-1}} \tag{4.7}$$

$$c_1 = 1 - e^{-\beta T/\tau}$$

$$c_2 = e^{-\beta T/\tau} - e^{-T/\tau}$$

The process model can be obtained from step or frequency response tests. Once the form of the model is known however, an alternative approach would be to estimate the parameters of the discrete model directly using techniques such as least squares identification. This is a particularly attractive approach for digital control as the techniques used for identification produce a discrete model.

4.3   GENERAL SYNTHESIS METHOD

Generalised digital control loops can be represented by plant transfer functions with discrete controllers and zero order holds (Fig.4.2).

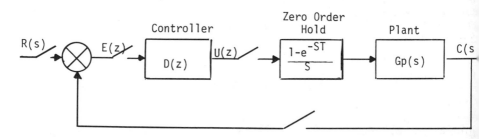

Fig.4.2. A continuous process under digital control.

Designing a controller with the direct synthesis method consists of determining the controller transfer function D(z) that is required to produce a specific closed-loop response. The overall, closed loop transfer function K(z) is chosen so that the system has a desirable transient response for a specific input, with a time constant which is

appropriate to the response time of the plant. The direct
synthesis approach assumes that the plant can be represented
by low-order models. For the system shown in Fig.4.2.

$$\frac{C}{R}(s) = \frac{D(z)G(z)}{1+D(z)G(z)} = K(z)$$ (4.8)

Then, if G(z) is known and K(z) is specified, the
required compensator D(z) is given by

$$D(z) = \frac{1}{G(z)} \frac{K(z)}{1-K(z)}$$ (4.9)

Equation 4.9 can be readily implemented, but care must
be taken in choosing K(z). The design calls for a D(z)
which will cancel the plant effects and add whatever is
required to give the desired K(z). This does mean that we
have to place constraints on K(z) so that we don't try and
achieve the impossible.

4.3.1 Constraint of causality.

D(z) will be of the form:

$$D(z) = \frac{b_0+b_1z^{-1}+b_2z^{-2}}{1-a_1z^{-1}-a_2z^{-1}}$$ (4.10)

For D(z) to be realisable then the numerator must be
of lower order or of the same order of z as the denominator.
Otherwise the controller would require future values of the
error to calculate the current value of output. This means
that for a plant which has a transportation delay z^{-k} the
desired closed loop transfer function K(z) must also include
the same delay. This fact can be readily seen from eqn.4.9.

4.3.2 Constraint of stability.

The roots of the characteristic equation of the
closed-loop system are the roots of the equation

$$1 + D(z)G(z) = 0$$ (4.11)

Now suppose there is a common factor in DG(z) as would
occur if D(z) were called upon to cancel a pole or zero of
G(z). Since a common factor remains a factor of the
characteristic polynomial, and if this factor is outside the
unit circle then the system is unstable. To avoid this,
then D(z) should not contain the poles or zeros on, or
outside the unit circle, and this is satisfied by meeting
the constraints:

    1 - K(z)   must contain as zeros all the poles of
                G(z) that are outside the unit circle.

and $K(z)$ must contain as zeros all the zeros of $G(z)$ that are outside the unit circle.

4.4 DAHLIN DESIGN

Dahlin's method is a particular case of the general synthesis method where the plant $G(z)$ is assumed to be modelled by a first or second order transfer function, and the desired closed-loop transfer function $K(s)$ is a first order lag of the form

$$K(s) = \frac{\lambda e^{-\theta s}}{s+\lambda}$$

or in the z domain (with ZOH included-eqn.4.5)

$$K(z) = \frac{\left[1-e^{-\lambda T}\right] z^{-k-1}}{1-e^{-\lambda T}z^{-1}} \tag{4.12}$$

The reciprocal of the time constant (λ) can be used as a tuning parameter with large values giving increasingly tight control. Clearly, a realistic closed loop response time must be chosen which takes account of the bandwidth limitations of actuators etc. and the need to prevent high frequency noise problems. It is normal to chose a closed loop time constant which is 2 to 3 times as fast as the open loop value. Using the direct synthesis relationship of eqn.4.9, the controller pulse transform is of the form

$$D(z) = \frac{1}{G(z)} \frac{\left[1-e^{-\lambda T}\right] z^{-k-1}}{\left[1-e^{-\lambda T}z^{-1}-(1-e^{-\lambda T})z^{-k-1}\right]} \tag{4.13}$$

4.4.1 Features of the Dahlin Controller

To illustrate an important feature of this controller the relationship of eqn.4.13 can be re-written as follows

$$D(z) = \frac{1}{G(z)(1-z^{-1})P(z)} \frac{\left[1-e^{-\lambda T}\right]z^{-k-1}}{} \tag{4.14}$$

where

$$P(z) = 1-\beta z^{-1}+\beta z^{-2}(1-z^{-(k+1)+1}+\cdots+\beta z^{-m}(1-z^{-(k+1)+m})+\cdots +\beta z^{-k}$$

$$\beta = 1-e^{-\lambda T}$$

and m is an integer in the range 2 to k.

Since $(1-z^{-1})$ is a factor of the denominator polynomial of $D(z)$ then the controller has integral action which will ensure that there will be zero steady-state error between the plant output and the desired setpoint $R(t)$. This integral action is a direct consequence of specifying a unity feedback structure, together with requiring a closed-loop response which approaches the setpoint exponentially, since this ensures zero error between the closed loop output and $R(t)$.

Another feature of the direct synthesis method is that the controller attempts to cancel the plant transfer function, provided that there are no poles or zeros, on or outside the unit circle. If this cancellation is exact the combined forward path transfer function will be

$$D(z)G(z) = \frac{1-e^{-\lambda T}}{(1-z^{-1}}$$

(4.15)

Consequently, the combination of plant and controller is simply an integrator and a gain term, resulting in a closed loop transfer function whose rise time depends on the gain. This can be compared with the continuous case for a plant with two simple lags T_1 and T_2; where setting the controller derivative time to T_2 and the integral time to T_1 results in a forward path transfer function

$$G(s) = \frac{K}{s}$$

(4.16)

provided $T_1 >> T_2$, so that the closed loop bandwidth is directly determined by the gain K. This is illustrated by the Bode diagram of Fig.4.3.

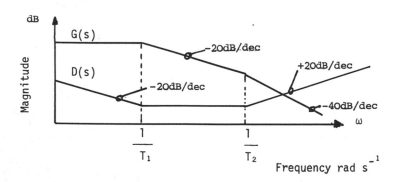

Fig.4.3 Continuous Equivalent of Dahlin Design.

For a plant modelled by a first order lag where the delay is not an integral number of sample periods (eqn.4.7), then using eqn.4.9 the controller pulse transform becomes

$$D(z) = \frac{\left[1-e^{-\lambda T}\right]\left[1-e^{-T/\tau}z^{-1}\right]}{K\left[1-e^{-\lambda T}z^{-1}-(1-e^{-\lambda T})z^{-k-1}\right]\left[c_1+c_2z^{-1}\right]} \qquad (4.17)$$

and for a second order lag plus dead-time process (eqn.4.6)

$$D(z) = \frac{(1-e^{-\lambda T})(1-e^{-T/\tau_1}z^{-1})(1-e^{-T/\tau_2}z^{-1})}{K\left[1-e^{-\lambda T}z^{-1}-(1-e^{-\lambda T})z^{-k-1}\right]\left[b_1+b_2z^{-1}\right]} \qquad (4.18)$$

4.4.2 Application of Dahlin Controller

To demonstrate the Dahlin design procedure an implementation is considered for the temperature process (Fig.4.4) which is covered in case study 1.

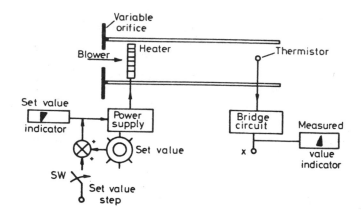

Fig.4.4 Schematic of the Temperature Process.

Air drawn through a variable orifice by a centrifugal blower, is driven past a heater grid and through a length of tubing to the atmosphere again. The process consists of heating the air flowing in the tube to some desired temperature. The detecting element consists of a bead thermistor fitted to the end of a probe inserted into the airstream 28 cms from the heater.

On this process, a step response test was made (Fig.4.5) and a first and second order model was evaluated (Table 4.1). These results taken from Irwin and Thompson paper clearly only apply to a specific orifice setting.

Also in Table 4.1 are the pulse transfer functions for the continuous models. Recall that this is the z transform of the product of the zero order hold and process transfer function.

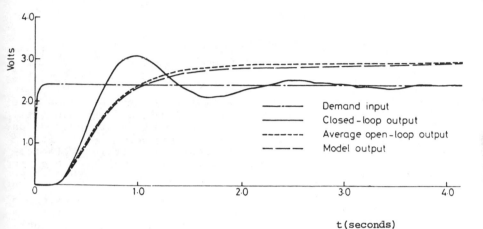

Fig.4.5 Step responses for Process and Second
Order Model.

	Continuous transfer function $G_p(s)$	Pulse transfer function for $T = 0.1s$ $G(z) = GzGp(z)$
First Order Model	$\dfrac{1.24e^{-0.3s}}{1+0.53s}$	$\dfrac{0.213z^{-4}}{1-0.828z^{-1}}$
Second Order Model.	$\dfrac{1.24e^{-0.3s}}{(0.377s+1)(0.132s+1)}$	$\dfrac{0.0896(1+0.709z^{-1})z^{-4}}{(1-0.767z^{-1})(1-0.469z^{-1})}$

Table 4.1 First and Second Order Models of a
Temperature Process

Using the second order model the Dahlin Controller equation (eqn.4.18) for a closed loop response with a time constant of 0.15s ($\lambda = 6.667$), becomes

$$D(z) = \frac{5.413(1-0.767z^{-1})(1-0.469z^{-1})}{(1-0.513z^{-1}-0.485z^{-4})(1+0.709z^{-1})} \qquad (4.19)$$

which in polynomial form is

$$D(z) = \frac{5.413(1-1.235z^{-1}+0.3596z^{-2})}{1+0.1961z^{-1}-0.364z^{-2}-0.485z^{-4}-0.345z^{-5}} \qquad (4.20)$$

With this controller, the closed-loop step response is shown in Fig.4.6, and although this would be judged entirely satisfactory according to the design requirements placed on K(z) the actuation signal is clearly unacceptable. It is both highly oscillatory and has negative excursions. (Note:- the results of Fig.4.6 were obtained by simulation due to the need for negative actuation which clearly could not be obtained from the actual process.)

The question of the actuation signal was not considered in the design method, where it was assumed that a satisfactory overall transfer function was the only requirement. In designing a control system however, the behaviour of the actuation signal is also very important, since this will often produce valve movement (or a similar piece of equipment) with practical constraints.

The negative values of actuation can be overcome by clamping the U(n) signal to zero when negative, but this results in a response with overshoot (Fig.4.7).The cause of the oscillations, and the negative excursions arising from them are due to the poles that lie in the vicinity of z = 1.

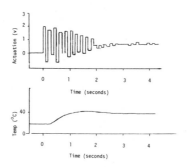
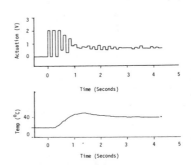

Negative actuations clamped at zero.

Fig.4.6 and Fig.4.7 System responses using Dahlin controller.

For the controller of eqn.4.19, it can be seen that the pole at z = -0.709 contribute to this oscillation. This is the pole that was introduced to cancel the plant zero.

Many plants have sampled-data transfer functions with zeros
near, or even outside the unit circle, resulting in poles
which are oscillatory or even unstable.

This can be overcome by removing the ringing poles and
adjusting the d.c. gain of the controller to be 'the same'.
This is done by substituting z = 1 into the factor
$(1+0.709z^{-1})$ which gives

$$D(z) = \frac{3.17(1.0.767z^{-1})(1-0.4688z^{-1})}{(1-0.5132z^{-1}-0.486z^{-4})} \qquad (4.21)$$

The resulting performances (Fig.4.8) provides a
considerable improvement in the manipulated variable,
without greatly degrading the system response.

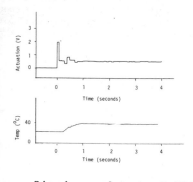

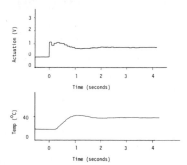

Ringing pole at -0.709   All ringing poles
removed.   removed

Fig.4.8 and Fig.4.9 System responses using Dahlin
controller.

A further improvement can be obtained by removing all
poles which lie in or around z = 1.   For example eqn.4.19
can be written in terms of its factors as

$$D(z) = \frac{5.413(1-0.767z^{-1})(1-0.4688z^{-1})}{(1-z^{-1})(1+0.73z^{-1})(1-0.244z^{-1}+0.664z^{-2})(1+0.709z^{-1})}$$
$$(4.22)$$

Setting z = 1 in all the factors with the exception of
the integrator term $(1 - z^{-1})$, then

$$D(z) = \frac{1.289(1-0.767z^{-1})(1-0.4688z^{-1})}{(1-z^{-1})} \qquad (4.23)$$

This controller results in the responses of Fig.4.9,
where it can be seen that the ringing in the manipulated
variable has been completely removed at the expense of an
increase in overshoot.   However, the degradation is
surprisingly small compared to the improvements made to the

actuation signal. An interesting feature of the eqn.4.23 is
that it now has the form of a PID controller. The general
z-transform representation of the common PID controller
is

$$D(z) = Kc \left[1 + \frac{T/Ti}{1-z^{-1}} + \frac{T_d}{T}(1-z^{-1}) \right] \qquad (4.24)$$

which can be written as

$$D(z) = Kc \left[\frac{\left[1 + \frac{T}{Ti} + \frac{Td}{T} \right] - \left[1 + 2\frac{Td}{T} \right] z^{-1} + \frac{Td}{T} z^{-2}}{(1-z^{-1})} \right] \qquad (4.25)$$

and it can be readily seen that it has an identical
structure to eqn.4.23 and the controller parameters in terms
of proportional gain, integral action time and derivative
action time can be evaluated by equating coefficients.

4.5 KALMAN DESIGN

As we have seen, the Dahlin algorithm is based on the
specification of the output in response to a setpoint change
without any constraints placed on the manipulated variable.
An alternative approach originally proposed by Kalman is to
design a digital controller with restrictions placed on both
the manipulated and controlled variables. For example, the
specifications for a step change in set point might be for
the response to settle at the final value within a specific
number of sampling periods, with the actuation signal
assuming only a specific number of values before reaching
the final value.

This approach, allows the designer to take account of
load changes and to specify that the error sequence should
be zero after a specified number of sample instants or that
the error should reduce in a specified manner. It can be
shown that in the case of a second-order system a minimum of
two values of manipulated variable are required before
the setpoint can be reached. Similarly for a third order
system, 3 values etc. Taking account of these
restrictions we can write expressions for the
controlled, and manipulated signal, as follows:-

$$C(z) = \sum_{n=0}^{\infty} C_n z^{-n} \qquad (4.26)$$

$$U(z) = \sum_{n=0}^{\infty} U_n z^{-n} \qquad (4.27)$$

It has already been seen that a general second-order
process has a pulse transfer function of the form
(eqn.4.6)

$$H(z) = \frac{C(z)}{U(z)} = \frac{Kz^{-k}(b_1 z^{-1} + b_2 z^{-2})}{(1-e^{-T/\tau_1} z^{-1})(1-e^{-T/\tau_2} z^{-1})} \qquad (4.28)$$

It then follows, for a step input

$$R(z) = \frac{1}{1-z^{-1}} \qquad (4.29)$$

that the relationships for the closed-loop transfer functions for the controlled and actuation variables are:

$$K(z) = \frac{C}{R}(z) = (1-z^{-1})^{-1}(C_{k+1} z^{-(k+1)} + z^{-(k+2)} + z^{-(k+3)} + ---)$$

$$(4.30)$$

and

$$Q(z) = \frac{U}{R}(z) = (1-z^{-1})^{-1}(u_0 + u_1 z^{-1} + u_2 z^{-2} ---) \qquad (4.31)$$

Specifying that the response settles to the final value within two sampling periods then eqn.4.30 and 4.31 become

$$K(z) = C_{k+1} z^{-(k+1)} + (1-C_{k+1}) z^{-(k+2)}$$

$$= p_1 z^{-(k+1)} + p_2 z^{-(k+2)} \qquad (4.32)$$

$$Q(z) = u_0 + (u_1 - u_0) z^{-1} + (u_2 - u_1) z^{-2}$$

$$= q_0 + q_1 z^{-1} + q_2 z^{-2} \qquad (4.33)$$

From the direct synthesis relationship of eqn.4.9 the controller and the process pulse transfer function expressed in terms of $K(z)$ and $Q(z)$ are respectively

$$D(z) = \frac{K(z) Q(z)}{1-K(z)K(z)} = \frac{Q(z)}{1-K(z)} \qquad (4.34)$$

$$H(z) = \frac{C(z)/R(z)}{U(z)/R(z)} = \frac{K(z)}{Q(z)} \qquad (4.35)$$

Equation 4.32 requires that the coefficients of the numerator in the pulse transform, $H(z)$ sum to unity (to achieve zero offset). To satisfy this condition $H(z)$ can be written as:-

$$H(z) = \frac{K(z)}{Q(z)} = \frac{\dfrac{b_1}{b_1+b_2}z^{-1} + \dfrac{b_2}{b_1+b_2}z^{-2}z^{-k}}{\dfrac{1}{K(b_1+b_2)}(1-e^{-T/\tau_1}z^{-1})(1-e^{-T/\tau_2}z^{-1})} \qquad (4.36)$$

It then follows that

$$K(z) = \frac{b_1}{b_1+b_2}z^{-(k+1)} + \frac{b_2}{b_1+b_2}z^{-(k+2)} \qquad (4.37)$$

$$Q(z) = \frac{1}{K(b_1+b_2)}(1-e^{-T/\tau_1}z^{-1})(1-e^{-T/\tau_2}z^{-1}) \qquad (4.38)$$

$$D(z) = \frac{1}{K(b_1+b_2)} \frac{(1-e^{-T/\tau_1}z^{-1})(1-e^{-T/\tau_2}z^{-1})}{1 - \dfrac{b_1}{b_1+b_2}z^{-(k+1)} + \dfrac{b_2}{b_1+b_2}z^{-(k+2)}} \qquad (4.39)$$

4.5.1 Application of Kalman Controller

The Kalman controller was applied to the temperature process previously considered, with the process modelled by the second order model. Initially a sampling time of 0.1 second was chosen, but this led to an actuation signal that gave an undamped oscillation (Fig.4.10 and Fig.4.11) so T was increased to 0.3 second.

The process transfer function for T = 0.3s is

$$H(z) = 0.448 \frac{(1+0.362z^{-1})z^{-2}}{(1-0.451z^{-1})(1-0.103z^{-1})} \qquad (4.40)$$

and using eqn.4.39 the Kalman controller is

$$D(z) = 1,637 \left[\frac{1-0.554z^{-1}+0.0464z^{-2}}{1-0.734z^{-2}-0.266z^{-3}} \right] \qquad (4.41)$$

which results in the responses of Fig.4.12. It can be seen that the temperature response is acceptable but the manipulated variable exhibits considerable ringing. Again, as in the case of the Dahlin controller, the ringing can be eliminated by removing all factors in the denominator of eqn.4.41 with the exception of the integrator term, as follows:

$$D(z) = 1.637 \frac{(1-0.553z^{-1}+0.0464z^{-2})}{(1-z^{-1})(1+z^{-1}+0.266z^{-2})} \qquad (4.42)$$

Setting z = 1 in the denominator quadratic term yields

$$D(z) = 0.72 \frac{(1-0.554z^{-1}+0.0464z^{-2})}{(1 - z^{-1})} \qquad (4.43)$$

and results in a ringing free controller as is demonstrated in the response of Fig.4.13. The structure of the controller is identical to that determined using Dahlin procedure, however it leads to a more lightly damped response. Again the algotithm is equivalent to the discrete form of the PID controller.

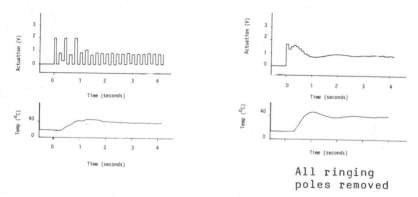

All ringing
poles removed

Fig.4.10 and Fig.4.11 System response using Kalman
controller (T=0.1s)

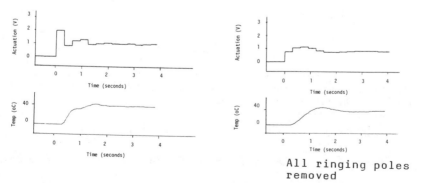

All ringing poles
removed

Fig.4.12 and Fig.4.13 System responses using Kalman
controller (T=0.3s)

4.6 PREDICTIVE CONTROLLER DESIGN

The previous section outlined a direct synthesis design method which was based on specifying the closed loop transfer function. In particular, the algorithms of Dahlin and Kalman were considered and it was shown that under certain conditions the algorithms reduce to the discrete forms of the standard PI and PID strategies. Arising from this equivalence, it can be seen that such direct design methods can be readily used to tune conventional PID controllers, an area of work which has been investigated by Chiu et al (10).

Another approach which falls into the direct design category, and is the theme of this section is predictive control. In its simplest form the implementation requires an estimate of the process output at the next sample instant, and based on this the actuation signal is chosen such that the output is made equal to the desired value. Clearly this approach relies upon a process model to 'predict' the future value of the controlled variable and uses this value as the input to the controller. Such an approach is intuitively appealing because of its simplicity, and can be readily extended to handle measurable disturbances using feedforward.

4.6.1 Direct-Single Step Design

It has already been seen from the previous section that a second order plant transfer function $G(z)$, without a pure time delay, relating the measured variable $C(z)$ to the actuation signal $U(z)$ is given by:

$$G(z) = \frac{C(z)}{U(z)} = \frac{K(b_1+b_2z^{-1})z^{-1}}{(1+a_1z^{-1}+a_2z^{-2})} \qquad (4.44)$$

$$a_1 = -(e^{-T/\tau_1}+e^{-T/\tau_2})$$

$$a_2 = e^{-T(1/\tau_1+1/\tau_2)}$$

b_1 and b_2 have been previously defined.

Assuming that K is included in the b_1 and b_2 coefficients, then eqn.4.44 may be written in terms of the input and output sequences as

$$C(z) (1+a_1z^{-1}+a_2z^{-2}) = U(z) (b_1+b_2z^{-1})z^{-1} \qquad (4.45)$$

which expressed in time-domain difference equation form is

$$C(n+1) = a_1C(n)-a_2C(n-1)+b_1U(n)+b_2U(n-1) \qquad (4.46)$$

Using this notation, C(n) represents the output at the current sample instant 'n' and consequently this equation represents the prediction of the system output, at sample instant 'n+1' under the control action U(n).

It can be readily seen that the simplest predictive control strategy is to choose the current actuation U(n) in such a way that the next output C(n+1) is equal to the desired setpoint, say W(n). It then follows that eqn.4.46 becomes

$$b_1U(n) + b_2U(n-1) = W(n) + a_1C(n) + a_2C(n-1) \tag{4.47}$$

$$U(n) = \frac{1}{b_1}\left[W(n) + a_1C(n) + a_1C(n-1) - b_2U(n-1)\right] \tag{4.48}$$

which gives the control action required at the current sampling instant in order that C(n+1) equals W(n).

Taking the z-transform of eqn.4.47 gives.

$$U(z)\ (b_1 + b_2z^{-1}) = W(z) + C(z)\ (a_1 + a_2z^{-1}) \tag{4.49}$$

which results in the block diagram form shown in Figure 4.14.

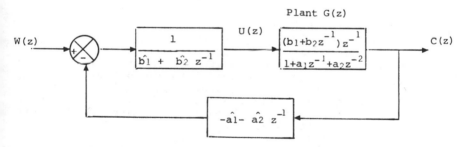

Fig.4.14 Direct Single Step Design (Deadbeat)

A controller based on this design procedure, which requires that the closed-loop response has a finite settling time, minimum rise time, and zero steady state error, is referred to as a deadbeat controller. It aims to transfer the plant output from its current operating value C(n) to the desired setpoint W(n) in a single sample instant, and keep it at that value. Such a design procedure is however unrealistic, since it is attempting to drive the plant to the desired value in a single sample instant and this assumes that there is no limit on the actuation signal. Also in the case of a second order process it is very likely that the system response will be oscillatory between sample values although the design will guarantee correct output values at sample instants. This assumes that the plant model accurately represents the process, with plant parameters being exact, which again is an unrealistic assumption.

4.6.2 Model Following Design

A better strategy to that of the dead-beat design is to arrange for the output to follow a setpoint with a predetermined transient response profile. For example, we can specify that, instead of reaching the setpoint W(n) in one sample time, the plant should instead, only get a fraction (α) of the way. By varying the weighting factor the response can be tailored from the deadbeat case ($\alpha = 1$) to a much slower one, but which nevertheless approaches infinitessimally close to W(n).

Thus, letting

$$C(n+1) = C(n) + \alpha(W(n)-C(n)) \tag{4.50}$$

which on taking z-transforms, yields

$$\frac{C(z)}{W(z)} = \frac{\alpha z^{-1}}{1-(1-\alpha)z^{-1}} = \frac{(1-\beta)z^{-1}}{1-\beta z^{-1}} \tag{4.51}$$

where $\beta = 1 - \alpha$

This transfer function is one that we have met before (eqn.4.12) and represents a first order lag with time constant λ where $\beta = e^{-\lambda T}$.

Consequently this simple design procedure is equivalent to specifying a first order "model following" response to the setpoint W(n). In this respect it is similar in design philosophy to the Dahlin design method although it results in a different structure with both feedback and feedforward components to the controller.

Expressing eqn.4.50 in terms of β, and evaluating the actuation signal U(n) for this model following design, the prediction equation (eqn.4.46) is re-written as

$$C(n)+(1-\beta)(W(n)-C(n))=-a_1C(n)-a_2C(n-1)+b_1U(n)+b_2U(n-1) \tag{4.52}$$

On rearranging,

$$U(n) = \frac{1}{b_1}\left[(a_1+\beta)C(n)+a_2C(n-1)-b_2U(n-1)+(1-\beta)W(n)\right] \tag{4.53}$$

which yields the pulse transform

$$U(z) = \frac{1}{b_1+b_2z^{-1}}\left[(1-\beta)W(z) + \left[(a_1+\beta) + a_2z^{-1}\right]C(z)\right] \tag{4.54}$$

and results in the block diagram form shown in Fig.4.15.

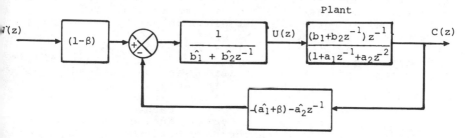

Fig.4.15 Model Following Design

To accommodate a more general closed loop characteristic, it is usual to form an "auxiliary output" $\emptyset(z)$ from the system output $C(z)$ as

$$\emptyset(z) = P(z) \ C(z) \tag{4.55}$$

where $P(z)$ is the inverse of the desired closed loop response. If a dead-beat controller is now designed to set $\emptyset(n+1)$ to the required setpoint $W(n)$, then the output

$$C(z) = \frac{\emptyset(z)}{P(z)} = \frac{W(z)}{P(z)} \tag{4.56}$$

and has a response $1/P(z)$ as required. The important features of $1/P(z)$ are that it should have a time constant which is reasonable in the light of the plant characteristics and have a steady-state gain of unity to ensure that steady state system output is equal to the setpoint value. For example the response specified by eqn.4.53 would be realised by choosing

$$P(z) = \frac{(1-\beta z^{-1})}{(1-\beta)} \tag{4.57}$$

Comparing the implementations resulting from predictive designs with those of Dahlin approach it can be seen that the plant zero $(b_1 + b_2 z^{-1})$ is once again cancelled by the controller, and consequently, the "ringing pole" problem still occurs. This can be removed by the same procedure as that adopted with the Dahlin algorithm. However, unlike the Dahlin algorithm (and Kalman) the absence of integral action in the forward path of the controller means that the predictive controller as designed cannot adequately deal with disturbances, but more about that later.

4.6.3 Control Weighting Design

The controller designs already given can be considered

as a simple minimisation of a performance criterion. In the
case of the dead-beat controller, the squared error between
the prediction $C(n+1)$ and $W(n)$ is minimised, by setting
$C(n+1) = W(n)$ or for model following control by minimising
the squared error between the prediction $C(n+1)$ and some
function of the setpoint. Since actuation changes clearly
are of importance, as well as model-following errors, then a
more general performance criterion can be defined which
takes these into account. In this way accuracy of model
following can be traded for large excursions in plant
actuations, and a suitable compromise reached.

Let us then consider a controller designed such that a
performance index (J) which includes both the squared error
and squared actuation, is minimised

$$J = (C(n+1)-W(n))^2 + \gamma U(n)^2 \qquad (4.58)$$

where λ is a weighting which determines the relative
importance of actuations against system error. Now $C(n+1)$
is dependent on $U(n)$ so minimising eqn.4.58 with respect to
$U(n)$ results in

$$\frac{dJ}{dU} = \frac{2(C(n+1)-W(n))\partial C(n+1)}{\partial U(n)} + 2\gamma U(n) = 0 \qquad (4.59)$$

From the prediction equation eqn.4.46

$$\frac{\partial C(n+1)}{\partial U(n)} = b_1$$

Therefore

$$C(n+1)-W(n) + \frac{\gamma}{b_1} U(n) = 0 \qquad (4.60)$$

and replacing $C(n+1)$ by its prediction formula

$$-a_1 C(n) - a_2 C(n-1) + b_1 U(n) + b_2 U(n-1) - W(n) + \frac{\gamma}{b_1} U(n) = 0 \qquad (4.61)$$

which yields for the actuation signal at the current sample
instant the difference equation

$$U(n) = \frac{1}{b_1 + \frac{\gamma}{b_1}} \left[W(n)+a_1 C(n)+a_2 C(n-1)+b_2 U(n-1) \right] \qquad (4.62)$$

and using z transform notation

$$U(z) = \frac{1}{(b_1 + \frac{\gamma}{b_1}) + b_2 z^{-1}} \left[W(z)+C(z)(a_1 + a_2 z^{-1}) \right] \qquad (4.63)$$

which results in the block diagram form shown in Fig.4.16.

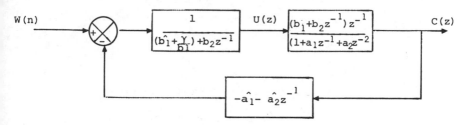

Fig.4.16 Design Using Control Weighting

The only difference between this and Fig.4.14 is the forward-path part of the controller, which represents the pole that cancels the plant zero. It can be seen that the pole position is now dependent on the value of γ.

As γ varies, this pole moves from a position where it cancels the zero exactly ($\gamma = 0$) to a point closer to the origin, with a consequent reduction in the forward path gain. This can be compared with the Dahlin solution, where in the case of the plant having a zero near to the unit circle no cancellation occurs (to avoid the oscillatory actuations) but the steady state gain of the controller is maintained. With the control weighting approach it can be seen that a compromise can be obtained by moving the ringing pole closer to the origin. The penalty paid for this is that the model following is less exact but it does give the advantage of significant reductions in the actuation signal.

4.6.4 Incremental Form of the Predictor

An unfortunate result of the loop structure which emerges from the predictive approch is that zero steady state error is only guaranteed in the absence of disturbances, and if the model coefficients are exact.

From Fig.4.14 and 4.15 it can be seen that the steady state conditions of the loop for the steady state value of the output to be equal to the setpoint, results in an actuation signal of

$$Uss = Wss \; \frac{(1 + \hat{a}_1 + \hat{a}_2)}{\hat{b}_1 + \hat{b}_2} \qquad (4.64)$$

This value of actuation is only produced if the estimated plant parameters are exact and if there is no load disturbance. In effect, the predictor given by eqn.4.46 is only accurate in the steady state if these conditions hold.

To overcome this problem, the predictor equation may be written in incremental form so that it may predict the absolute output C(n+1) from previous outputs and changes in actuations.

Thus,

$$C(n+1)-C(n)=-a_1(C(n)-C(n-1))-a_2(C(n-1)-C(n-2))+b_1\Delta U(n)+b_2\Delta U(n-1)$$

where $\Delta U(n) = U(n)-U(n-1)$ (4.65)

Rearranging,

$$C(n+1) = (1-a_1)C(n)+(a_1-a_2)C(n-1)+a_2C(n-2)+b_1\Delta U(n)+b_2\Delta U(n-1)$$ (4.66)

which   for   the   model   following   design   previously
considered and using the incremental form of the   predictor,
results in an actuation signal given by

$$U(n) = \frac{1}{b_1}\left[(1-\beta)W(n)+(a_1+\beta-1)C(n)+(a_2-a_1)C(n-1)-a_2C(n-2)+b_1U(n-1)\right.$$
$$\left. -b_2(U(n-1)-U(n-2))\right]$$

and in z transform notation gives   (4.67)

$$U(z) = \frac{1}{(1-z^{-1})(b_1+b_2z^{-1})}\left[(1-\beta)W(z) + (a_1+\beta-1)+(a_2-a_1)z^{-1}-a_2z^{-2})C(z)\right]$$
                                                                       (4.68)

which results in the block diagram of Fig.4.17

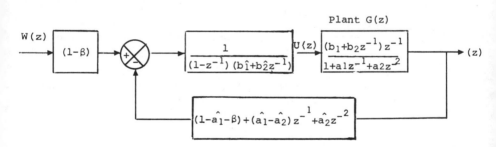

Fig.4.17 Model Following Design using Incremental
form of Controller

It can be seen that now an integrator term $(1-z^{-1})^{-1}$
has   been   introduced in the forward path,   and that in   the
steady state the feedback path reduces to

$$(1-\hat{a}_1-\beta) + (\hat{a}_1 - \hat{a}_2) + \hat{a}_2 = 1 - \beta$$ (4.69)

which   is   now   independent   of   the   accuracy   of   the
estimated plant parameters.   Consequently this   controller
can   reduce   steady-state   errors   caused   either   by   load
variations or by parameter inaccuracies, to zero.

Alternatively, the structure obtained by the predictive approach can be transformed, using simple block diagram manipulation, into a unity feedback structure. If this is done, then it becomes identical to the Dahlin design, complete with integral action. However, the predictive method is still valuable because of its simple conceptional basis, and that it can easily handle measurable disturbances.

4.6.5 Feedforward Compensation

One important feature of predictive control is that measurable disturbances can readily be included in the algorithms. Consider a disturbance V(t) which affects the plant, then the plant output can be described by

$$C(z) A(z) = B(z) U(z) + L (z) V (z)$$

where $L(z) = l_1 z^{-1} + l_2 z^{-2} + l_3 z^{-3} - - - -$ (4.70)

Which can be written for two significant disturbance samples.

$$C(n+1) = -a_1 C(n) - a_2 C(n-1) + b_1 U(n) + b_2 U(n-1) + l_1 V(n) + l_2 V(n-1) \quad (4.71)$$

In setting C(n+1) to the desired target value, it is a straight forward task to take account of the disturbance terms in calculating the current actuation U(n). The only restriction is that the plant output can be expressed in the form given in eqn.4.70 which implies that the denominator dynamics are common to both the actuations and the disturbances. Effectively this means that both actuations and disturbances act on the plant at the same point (Fig.4.18) and the success of this technique(Morris, A.J. et al (7)) indicates that this is a reasonable assumption in many practical cases.

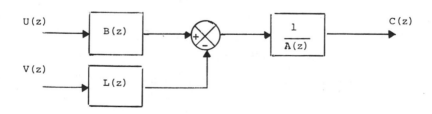

Fig.4.18 Disturbance structure for feedforward
        compensation.

4.7   CONCLUSIONS

The digital control design methods described have been

available   in various forms for almost 30 years,   but it   is
still   unusual   to see a commercial   process   controller   or
servo system which uses them.   To some extent this is due to
the   success   of   conventional   PID control,   which has   a
remarkable   "robustness" even on ill-defined processes,   and
for   which   the   majority   of   process   engineers   have   an
intuitive "feel".

Onc   "feature" of the digital approach is that   sample-
rates are, in general, slower than for the pseuuu-continuous
controller,   and,   in   the   early   days   of   direct   digital
control,   when computers were extremely expensive,   this was
seen   as an advantage,   since one computer could handle many
control   loops more easily.   However,   a slow sample   rate,
although   appropriate for the dynamics of the   plant,   means
that   disturbances   can be present for   quite   long   periods
before corrective action is taken, and this, together with a
psychological   feeling   that   the controller   is   not   doing
anything,   is   now seen as a disadvantage in many operator's
eyes.   To some extent the problem can be remedied by   using
simple   transformations   to convert the coefficients   for   a
faster   sample   rate (Astrom (8)) whilst still retaining the
essential simplicity of the predicitive design methodology.

A further advantage of the direct design method is that
they   can   readily   handle   delays,   resulting   in   a   more
satisfactory   response.   Some of the ideas covered in   this
chapter   are further developed using case studies in chapter
17 and 18.

REFERENCES

1.   Dahlin,E.B.,1968
        "Designing and Tuning Digital Controllers"
        Instruments and Control Systems, Vol.41,No.6.

2.   Kalman,R.E.,1954.
        Discussion following Bergen,AR and Ragazzini,J.R.
        "Sampled-data Processing Techniques for Feedback
        Control systems."Trans.AIEE, Nov.236-247.

3.   Franklin, G.F.and Powell, J.D., 1980
        "Digital control of Dynamic Systems" Addison-Wesley.

4.   Irwin,G.W. and Thompson,S, 1984,
        "Use of a microcomputer for classical control system
        design work"   Eurocon '84, Computers in Communication
        and Control.

5.   Chiw, Corripio, and Smith, 1973.
        "Digital Control Algorithms - Part 1 and 2 Dahlin and
        Kalman Algorithms". Instruments and Control Systems,
        Oct/Nov.

6.   Clarke,D.W.,1984.
        "Self-Tuning Controller Design and Implementation in
        Real-Time Computer Control", Ed.,S.Bennett & D.Linkens

Peregrinus.

7.   Morris, A.J. et al.,1981.
"Self-Tuning Control of some pilot plant process"
Microprocessor and Microsystems, Vol.5 No. 1.

8.   Astrom, K.J., 1979.
"Simple Self Tuners" Lund Report TFRT-7184.

9.   Thomas, H.W., 1984.
Design of Digital Controllers" Proceedings of the
IEE Vacation School on "Industrial Digital Control
Systems", Balliol College, Oxford.

10.   Chiw, Corripio, and Smith, 1973
Digital Control Algorithms, Part IV, Tuning PI and PID
Controllers. Instruments and Control Systems."

Chapter 5

Control of time delay systems

Dr J.E. Marshall

5.1 INTRODUCTION

This chapter is concerned with the effects on control of the presence of time-delay in a plant. An important design principle due to O.J.M. Smith is introduced and applied in this chapter. This principle is appropriate to continuous as well as discrete control. It is a principle that has already been introduced implicitly in the Dahlin algorithm met in an earlier chapter, and we compare these methods in section 5.4.2.

There is a close connection between digital control and time-delays. All computer based control depends on storage. Storage implies a delay. This is explored in section 5.2.

An essential feature of predictive control, to be introduced in this chapter, is the need for an explicit system model. An important practical consideration is what happens when the model differs either deliberately or by chance from the actual system. This phenomenon, which is common, is called mismatch. Depending on how performance is measured it is often the case, particularly when delays are not large, to change performance for the better by deliberate mismatch – rather akin to controller tuning – but an extension of that idea. When such model-tuning is appropriate is to be discussed.

Smith predictor methods are open to various criticisms. To what extent these are valid (if at all) depends on the application and the nature of the plant. If the plant is not self-regulating (i.e. open loop stable) more sophisticated methods are needed. For example the self-tuning algorithms to be treated in Chapter 10.

A particular strength of Smith's method is its use of a delay-free equivalent controller. This means that the design of a time-delay system may sometimes be reduced to an equivalent delay-free design. When such a reduction is not possible, such as for the case of delayed state-feedback there are alternatives, which will be described.

The reader may find the text-book (1) and a 1981 survey (2) useful as an introduction to more general properties of time-delay systems. The control of time-delay systems is still an active research field and some new work not described in the chapter may be found at the end of the references.

5.2   SIMULATION OF A PURE TIME-DELAY

We shall denote the pure time-delay with typical input $x(t)$ and output $x(t - \tau)$ by its transfer function $\exp(-s\tau)$. The delay τ is sometimes suffixed to distinguish it from other delays. For example, $\exp(-s\tau_0)$ is used (later) as a model of a plant delay, magnitude τ_0.

Consider the discrete digital simulation of a delay. We require for our present purposes a continuous (analogue) input, and a continuous (analogue) output. An obvious realisation, of which an improved form is suggested later, is shown in Fig. 5.1.

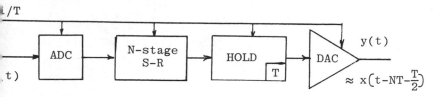

Fig. 5.1   Digital simulation of a delay

Crudely speaking this realisation produced a delay NT where N is the number of stages in the shift register (S - R) and T is the inter sample interval. 1/T is the clock rate. A careful analysis of the system gives the expression for output. It is, when $x(t)$ is the sinusoid $\sin \omega t$.

$$y(t) = \frac{\sin(\frac{\omega T}{2})}{(\frac{\omega T}{2})} \cdot \sin \omega (t - \frac{T}{2} - NT)$$

$$+ \frac{\sin\left[(\frac{2\pi}{T} - \omega)\frac{T}{2}\right]}{(\frac{2\pi}{T} - \omega)\frac{T}{2}} \cdot \sin\left[(\frac{2\pi}{T} - \omega)(t - \frac{T}{2} - NT)\right] \quad (5.1)$$

$$+ \text{ similar terms in } \frac{2\pi}{T} + \omega, \ \frac{4\pi}{T} \pm \omega, \ \frac{6\pi}{T} \pm \omega, \text{ etc.}$$

For T sufficiently small (to be checked, in practice) the higher terms may be neglected. The magnitude of the first neglected term (the second term in the expression) is not negligible unless the number of samples per cycle $k = (\frac{1}{T})/(\frac{\omega}{2\pi})$ is large. The magnitude of the first term (the fundamental term) and the first neglected term are given in the table. We denote them by A_0, A_{-1} respectively. $k = 2$ corresponds to the sampling theorem limit.

TABLE 5.1.   Sampling rate and frequency response

SAMPLES/CYCLE	k=2	3	4	5	6	7	8	9	10
A_0	.6367	.827	.900	.935	.955	.966	.974	.980	.983
A_{-1}	.6367	.413	.300	.234	.191	.161	.139	.122	.109

Note that A_{-1} is given by $A_0/(k - 1)$.

A_0 is given by $\dfrac{\sin(\pi/k)}{(\pi/k)}$, from which further
entries in the table may be deduced.   Provided that k is
sufficiently large, we see that $y(t) \approx x(t - NT - T/2)$.
This expression arises for other inputs also.   The delay
T/2, arising from the HOLD of duration T is common to all
practical digital control schemes and shows that all
discrete control schemes contain a delay.

A more efficient delay simulation uses the carousel
principle (1) in which N stores are addressed cyclically
with READ and WRITE instructions.   We shall see later that
small differences in delay affect performance so that sometimes the T/2 term assumes practical importance.

Many control schemes require the use of models in the
realisation of a controller.   It is the need to model delay
accurately that makes discrete control natural for time-
delay systems.   Later we look at systems which combine
continuous and discrete control.   Having shown the close
connections between digital systems and time-delay we will
classify some simple time-delay schemes.

5.3   SOME ELEMENTARY TIME-DELAY SYSTEMS (TDS)

Fig. 5.2 shows three simple TDS structures.   In each
case G(s) denotes the delay-free part, called the subplant,
and exp(-sτ) denotes the plant delay.

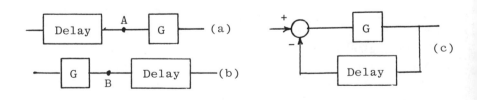

Fig. 5.2 (a) input (b) output (c) feedback, delays

In cases (a) and (b) there is an obvious "delay-free equivalent plant". In (c) there is not, Control schemes for (a)
and (b) have much in common. Case (c), an example of a
"delayed-state" system, needs different methods. In 5.2(a)
and 5.2(b) the points denoted A and B are INACCESSIBLE for
control purposes.

5.4 SMITH'S PRINCIPLE, AND METHOD

5.4.1 Introduction.

Smith's principle (3,4,1) leads to an example of prediction methods. It is applied to systems where the subplant and delay appear in series as in Fig. 5.2(a), 5.2(b).
Smith's method is this:- choose the controller C*(s) (see
Fig. 5.3(a)) so that the system transfer function is identical to that of Fig. 5.3(b). C(s) is

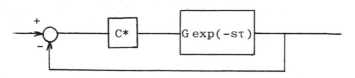

Fig. 5.3(a)   The C*(s) controller

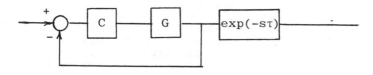

Fig. 5.3(b)   Externalised delay

Equating the respective transfer functions we find

$$\frac{C^*(s)G(s)\exp(-s\tau)}{1 + C^*(s)G(s)\exp(-s\tau)} = \frac{C(s)G(s)\exp(-s\tau)}{1 + C(s)G(s)} \quad . \tag{5.2}$$

From which it follows that

$$C^*(s) = C(s)/\left[1 + C(s)G(s)\left(1 - \exp(-s\tau)\right)\right]. \tag{5.3}$$

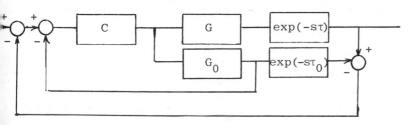

Fig. 5.4   SP scheme

Fig. 5.4 shows a realisation of Fig. 5.3(a) with the terms of C*(s) made explicit, for the delayed output case.

The G(s) and exp(-sτ) in the <u>upper</u> forward path are reversed for the corresponding delayed input case. Note the assumptions implicit in Smith's principle. "Design C*(s) so that the output of the time-delay system is a time-delayed replica of the output of the corresponding delay-free system." For this to be possible we require, at least:

(1)   There is no function stored in the plant delay.
(2)   There are zero initial conditions on G(s).
(3)   There is no disturbance to the plant.
(4)   The plant is stable.
(5)   The plant is well understood, i.e. the models of G(s) and of exp(-sτ) appearing in the lower forward path, are to be IDENTICAL in transfer function to those of the subplant and delay.

Let us denote corresponding models by $G_0(s)$, $\exp(-s\tau_0)$ and give the explicit transfer function of the system of Fig. 5.4. Denote input and output transforms by $\bar{x}(s)$, $\bar{y}(s)$ respectively then

$$\frac{\bar{y}(s)}{\bar{x}(s)} = \frac{C(s)G(s)\exp(-s\tau)}{1 + C(s)G_0(s) + C(s)\big(G(s)\exp(-s\tau) - G_0(s)\exp(-s\tau_0)\big)}.$$

$$(5.4)$$

The final bracketed denominator term is to be identically zero for Smith's principle to apply. This term is denoted the "mismatch term".(1)   The dynamics of the controlled system depend essentially on the magnitude of this mismatch term. In any real situation this term will not be zero. It is reasonable in the light of Smith's principle to require it to be small. Thoughtful consideration of Fig. 5.4 shows that the method is equivalent to making point B of Fig. 5.2(b) available for control. The signal at B is a "prediction" of the output. For this reason (and others equivalent to it) such controllers are called <u>predictor</u> controllers. The expression "Smith predictor" is used here, but there are other predictors, for example that of Fuller. (5)   This latter author uses the useful expression "copy co-ordinates" for those states of the model corresponding to certain inaccessible states of the system to be controlled. Of the various assumptions listed, the reasons for which are not all self-evident, it is important to explore (5) further. When system and model do not agree we say that there is "mismatch". As we shall show, mismatch may <u>sometimes</u> improve performance.

5.4.2   Dahlin re-visited.

Dahlin's method (Chapter 4) is a special case of the general synthesis method, whereby a desired closed-loop transfer function is pre-specified. This together with a choice of structure, unity negative feed-back, say, will give in turn the series controller. This design philosophy applies equally well to continuous control but is not usual there. Smith and Dahlin both recognise that if the plant contains an explicit <u>series</u> delay there is no point in expecting an instantaneous response at the output (closed loop or open loop). Further any attempt to make the output respond to input before the delay time has expired is to ask the impossible. Any attempt to do this must lead to controllers which are anticipative, and hence physically unrealistic. So if in the Dahlin algorithm we pre-specify the closed-loop transfer function with a delay equal at least to the series plant delay we have a situation like that of the Smith case. In the Smith case we design the "best" delay-free controller. In the Dahlin (and the general synthetic method) we pre-specify a desired delay-free closed-loop response.

We recall that Smith's principle applies to continuous as well as discrete controllers. Obviously if the desired closed-loop response is chosen in a certain way then the Smith controlled system and the Dahlin controlled system would have identical performance - and hence identical series controllers $D(z)$.

5.4.3 The effects of mismatch.(6,7,8,9)

By Smith's principle the output of the controller system is a time-delayed version of the delay-free output. At first sight it would appear that provided the output of the delay-free system is made satisfactory by a suitable choice of $C(s)$ then it follows that the output from the time-delay system will be.

Recall that the controller $C*(s)$ uses models of subplant and delay. It is a natural question to ask: to what extent is performance degraded by mismatch? The answer depends on how performance is quantified.

Let us take, for example, the "overshoot to a step input" as a criterion of performance. Suppose that $C(s)$ is designed to minimise overshoot of the delay-free system. Then the corresponding TDS design will have the same overshoot, as the output is just a time-shifted replica of the delay-free output. Now, if τ_0, the model delay, is treated <u>as a design variable</u>, and modified about its nominal value, the plant delay τ, it is found in many cases that overshoot is changed. In particular there are cases in which increasing τ_0 <u>reduces</u> overshoot. Similarly, changes can be made to parameters of $G_0(s)$ the subplant model, but this is usually less efficient as a way of modifying overshoot. It is certainly much more difficult to analyse. A technique "improvement by mismatch" has been described in the 1982 Proc. IEE Pt. D.(7).

Examples discussed in that paper, and others, will be given in the lecture. It is popularly believed that only increase of τ_0 improves performance. This is not true. There are cases where <u>decrease</u> of τ_0 improves things.

Another measure of performance that lends itself to analytical study is Integral of Squared Error (ISE) criterion. Recent results (10,11) have shown that it is possible to obtain closed form solutions for the ISE criterion for some simple time-delay schemes, including delayed feedback

schemes not designable by Smith's method. This leads to
parametric optimal control methods for TDS, called "Analytical design of TDS".

Another source of mismatch, which occurs is due to the
way in which the time delay is modelled. Before the digital
realisation was possible it was usual to use an analogue
realisation via the Padé approximant. This replaces the
transfer function $\exp(-s\tau_0)$ by a ratio of low order polynomials in S. O.J.M. Smith used such an approximant in his own
work. In the context of discrete control such methods are
not appropriate, and there are times when their use is misleading.

5.4.4   When is Smith's method not appropriate?(8)

There is a small but interesting class of problems, in
which Smith's method does not give the best answer. There
are cases where, roughly speaking, the delay is not very
large, and the plant is "sluggish". In such cases it is
found that the model delay should be made smaller than τ.
If τ is not large it may be the case that $\tau_0 = 0$ gives the
best performance.(8) Consider the transfer function
denominator

$$1 + C(s)G_0(s) + C(s)\big(G(s)\exp(-s\tau) - G_0(s)\exp(-s\tau_0)\big). \qquad (5.5$$

With $\tau_0 = 0$ the result is $1 + C(s)G(s)\exp(-s\tau)$ whether the
subplant is well modelled or not! The resulting closed-loo
transfer function is

$$\frac{C(s)G(s)\exp(-s\tau)}{1 + C(s)G(s)\exp(-s\tau)}. \qquad (5.6$$

This is the same as conventional unity negative feed-back.

Such cases as these are easily recognised. Recall thi
is only when τ is small.

5.4.5   Smith's method and disturbance.(12)

A frequent criticism of Smith's method is that it does
not give good results in the presence of disturbance. Note
that due to the delay all control schemes have inevitable
errors due to the obvious delay in the application of
control. The control input cannot affect the output at all
for a time interval τ after the onset of disturbance. It i
not surprising that Smith's method is not robust to disturbances. The absence of disturbance, and other signals was
assumed in the derivation of C*(s)! If there is disturbanc
it must be considered at the design stage, it is naive (13)
to expect automatic rejection of disturbance.

Let us distinguish cases where disturbance is infrequent, and where it is continuous. The latter case is deal
with in later lectures. We shall assume, in order to make
progress, that a disturbance d(t), of known form but unknow
magnitude, is incident on the plant at the output (see Fig.
5.5). Let its transform be kD(s), where k is an unknown
value. Express d(t) as two terms $d_I(t) + d_T(t)$, which we

call the <u>inevitable</u> part and the <u>tail</u>. $d_I(t)$ has time
duration τ. It cannot be reduced by control. (We assume
that feed-forward which assumes foreknowledge of the
incidence of disturbance is not on.)

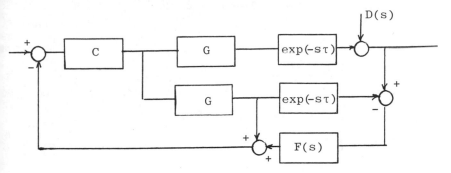

Fig. 5.5   SP scheme modified for disturbance rejection

Write $D(s) = k\big(D_I(s) + \exp(-s\tau)D_T(s)\big)$. Introduce a
feed-back element $F(s)$ into the scheme of Fig. 5.4. We
obtain Fig. 5.5 which also shows the disturbance input. We
require to choose $F(s)$ to minimise the effect of $D(t)$ at the
output. The response to the disturbance may, again, be
considered as two parts. The inevitable part that cannot be
controlled plus a tail. We attempt to reduce the tail as
much as possible. With realistic plant it is unlikely that
it can be reduced to zero.

When there is no mismatch (which we assume here) $\bar{y}(s)$
is independent of $F(s)$, as it appears in a product with the
"mismatch term" in the closed-loop transfer function. $\bar{y}_0(s)$
has two terms, the inevitable one, and the tail. After
subtracting the inevitable part we achieve the following
expression for the tail

$$y_T(s) = \exp(-s\tau)D_\tau(s) +$$

$$+ \frac{CGF\exp(-s\tau)D(s)}{1 + CG_0 + CF\big(G\exp(-s\tau) - G_0\exp(-s\tau_0)\big)} \qquad (5.7)$$

where we have dropped the obvious s-dependences for brevity.
$F(s)$ is chosen to make this as small as possible. When
plant and models match we obtain

$$F = \frac{1 + CG}{CG} \cdot \frac{D_T(s)}{D(s)} . \qquad (5.8)$$

This is a rational transfer function. $\frac{1 + CG}{CG}$ is the inverse
of the closed loop delay-free transfer function. The closer

that the realisation of F(s) approaches this value, the mor
negligible the tail. Results similar to this are found in
the stochastic case. Extensions of Smith's method by modification of the structure of Fig. 5.5, have been suggested.
In particular the work of Hang and his students is recommer
ded.(14,15,20)

5.5   DELAYED FEED-BACK

In the delayed feed-back case Fig. 5.2(c) there is no
helpful delay-free equivalent. There are few well established methods here. We note that there is not the accessi
bility problem however of Fig. 5.2(a) and Fig. 5.2(b). Let
us assume the structure given in Fig. 5.6, and further
assume that the loop is stable.

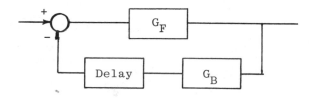

Fig. 5.6 Delayed feed-back

Note that this closed-loop system has an infinite number of
poles because $1 + G_F(s)G_B(s)\exp(-s\tau) = 0$ has an infinite
number of roots. If τ is small most of the roots will be
well away from the origin and the dominant poles may be
controlled via conventional techniques followed by "fine
tuning" to offset the approximation of neglecting the other
poles. If τ is very small, and in its absence the loop is
stable then conventional delay-free design may be used, as
elsewhere.

The example which follows, though of continuous type,
shows that the common belief that increasing delay makes
matters worse is not always true.

Consider the delayed feed-back system consisting of ar
integrator with delayed feed-back. The closed-loop transfe
function is $\dfrac{\frac{K}{S}}{1 + \frac{K}{S}\exp(-s\tau)}$. Define error as input minus

output. Minimise the ISE cost

$$J = \int_0^\infty \left(x(t) - y(t)\right)^2 dt. \qquad (5.9$$

It may be shown (10,16) that

$$J = \frac{\cos(k\tau)}{2k\big(1 - \sin(k\tau)\big)} - \tau. \tag{5.10}$$

Let k take its maximum practical value k_{max}. Then a little calculus gives $J_{min} = 0.685 J(0)$, when $\tau = \pi/6k_{max}$. The delay-free value $J(0)$ is $\frac{1}{2k_{max}}$.

A very early second order example (17) with a similar conclusion is referenced by Bateman.(18) In this case involving the control of a light aircraft it was possible to adjust a small delay in a transducer by mechanical means. This made the second order system more lively (increasing delay was equivalent to less damping). This improved the performance as judged by the pilot. It is also a case of sluggish plant, with a small series delay.

5.6 CONCLUSIONS

This chapter has been concerned with linear plants with delay, and simple design criteria. The methods given here extend in some cases to nonlinear plant, (1,15) and to optimal control schemes.(5,6,19) The feed-back delay case (state feed-back) is still an active research area, and practical methods are emerging. We have not discussed sub-plants which are unstable. Recent results (20) address this problem by means of simple modification to controllers of the Smith predictor type.

REFERENCES

1.   Marshall, J.E., 1979, 'The Control of Time-Delay Systems', IEE Control Engineering Series, Vol. 10, Peter Peregrinus.

2.   Chotai, A., Garland, B. and Marshall, J.E., 1981, 'A Survey of Time-Delay Systems Control Methods'. In Proceedings of International Conference on Control and its Applications, 316-322, IEE Conf. Pubn. 194.

3.   Smith, O.J.M., 1959, 'A Controller to Overcome Dead-Time', I.S.A.J., 6(2), 28-33.

4.   Smith, O.J.M., 1958, 'Feedback Control Systems', McGraw Hill.

5.   Fuller, A.T., 1968, 'Optimal Non-linear Control of Systems with Pure Delay', Int. J. Control, 8(2), 145-168.

6.   Hocken, D., Marshall, J.E. and Salehi, S.V., 1983, 'Time-Delay Control: Mismatch Problems'. Third IFAC Symposium on Control of Distributed Parameter Systems. Eds. Barbary, J-P., LeLetty, L., Pergamon.

7.   Marshall, J.E. and Salehi, S.V., 1982, 'Improvement of System Performance by the Use of Time-Delay Elements',

IEE Pt. D, 127, No.5, 177-181.

8. Hocken, R.D., Salehi, S.V. and Marshall, J.E., 1983, 'Time-Delay Mismatch and the Performance of Prediction Control Schemes', Int. J. Control, 38(2), 433-447.

9. Hocken, R.D. and Marshall, J.E., 1982, 'Mismatch and the Optimal Control of Linear Systems with Time-Delays' Opt. Control Applns. and Methods, 3, 211-219.

10. Walton, K. and Gorecki, H., 1984, 'On the Evaluation of Cost Functionals with Particular Emphasis on Time-Delay Systems', IMA J. of Maths. Control and Infn., I(3), 283-306.

11. Walton, K. and Marshall, J.E., 1983, 'Closed-Form Solution for Time-Delay Systems Cost Functionals', Int. J. Control, 39(5), 1063-1071.

12. Marshall, J.E., 1983, 'Prediction Principles for Time-Delay System Controller Synthesis', Acta Applicandae Mathematicae, 1, 189-120.

13. Nielsen, G., 1969, 'Control of Systems with Time-Delay' Fourth IFAC Congress, Warsaw, 25-38.

14. Hang, C.C. and Wong, F.S., 1979, 'Modified Smith Predictors for the Control of Processes with Dead-Time' Proc. ISA Annual Conf., Chicago, October.

15. Marshall, J.E., 1974, 'Extensions of O.J. Smith's Method to Digital and Other Systems', Int. J. Control, 19(5), 933-939.

16. Marshall, J.E. and Walton, K., 1984, 'Analytical Design of TDS Controllers', Proc. Fourth IMA Conference on Control Theory, Cambridge. (Preprints).

17. Editorial Staff, 1937, 'The Damping Effect of Time-Lag' Engineer, 163, 439.

18. Bateman, H., 1945, 'The Control of an Elastic Fluid', Bull. Am. Math. Soc., 51, 601-646.

19. Hocken, R.D. and Marshall, J.E., 1983, 'The Effects of Mismatch on an Optimal Control Scheme for Linear Systems with Control Time-Delays', Opt. Control Applns. and Methods, 4, 47-69.

20. De Paor, A.M., 1985, 'A Modified Smith Predictor and Controller for Unstable Processes with Time-Delay', Int. J. Control, 41, 1025-1036.

21. Hang, C.C. and Tham, Q.Y., 1982, 'Digital extensions of the modified Smith predictors', ISA Conf., Philadelphia October.

Chapter 6

State-space concepts

Dr. G.K. Steel

5.1 INTRODUCTION

The application of state-space concepts to control engineering developed in the decade from 1950 and gave rise to "modern control theory". Previous formulations of control theory relied on concepts related to transfer functions; the output/input relationship of a system expressed as a ratio of Laplace transforms of the input and output signals. Allied to this the powerful concept of frequency response, which had its roots in communication theory, formed the foundation of "classical control theory". In this the notion of a "signal" was basic and the control system was regarded as a signal processing element.

Modern control theory took a more detailed view of the internal structure of a system, and the many variables associated with it, so that the concept of "state" replaced that of the "signal" as the primary interest. This change of view was necessary to come to terms with problems of time optimal control and non-linear system stability. A consequence of this is a shift of view from the transfer function as an operator on signals to the matrix as an operator on states. The system response is seen as a progression of changes of state evolving through time.

5.2 FORMULATION OF STATE EQUATIONS

The state variables associated with a dynamic system are those variables which relate directly to the capability of the system to accumulate, or store a comodity, as in a reservoir. Most commonly in physical systems the comodity stored is energy, but it may be capital in an economic system or goods stocked in a production process. Specifically in mechanical systems the velocity of each mass and the extension of each spring are state variables; by analogy in electrical networks the current in each inductor and the voltage across each capacitor. Macfarlane (3) Kuo (5)). See examples of state variables in Fig. 6.1.

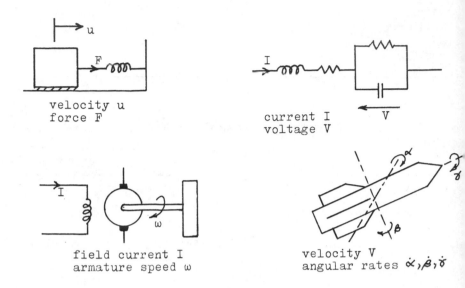

Fig. 6.1   State variables
            (a) Electromechanical systems

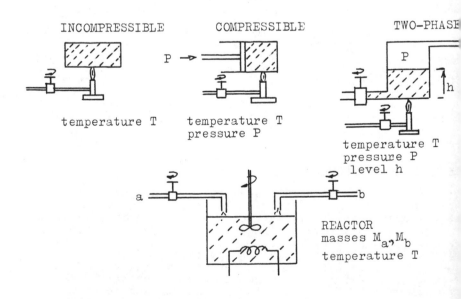

Fig. 6.1 State variables
          (b) Thermal systems

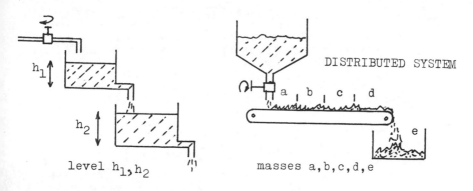

Fig. 6.1 State variables
(c) Mass transport systems

Natural physical constraints determine the rate at which the comodity can flow in or out of the storage element. The relationships which determine the rate of change of the state variables become the governing differential equations of the system. In general we have a set of n first-order simultaneous equations relating the n state variables in the form,

$$|\dot{X}| = f(\ |X|\ ,\ |U|\ ,t) \qquad \ldots\ldots\ldots\ldots (6.1)$$

where $|X|$ is the state vector matrix, a column of n state variables, $|U|$ is the input vector, a column of m input variables, and t is time. When the system is linear and not time varying these equations may be written,

$$|\dot{X}| = |A|\ |X| + |B|\ |U| \qquad \ldots\ldots\ldots\ldots (6.2)$$

where A and B are matrices with constant coefficients. Also a set of output variables $y_1, y_2, \ldots y_r$ can be defined as a linear combination of the state variables,

$$|Y| = |C|\ |X| \qquad \ldots\ldots\ldots\ldots\ldots (6.3)$$

Equation (6.1) or (6.2) are called 'state equations' and determine the rate of change of state which, in turn, constrains the evolution of the system state through time.

The state variable formulation has been basic to the techniques of system simulation; either using analogue or digital computers (Gordon (7)). The simulation is structured as a set of interconnected integrators. The integrators represent the storage elements and their outputs, being the integral value accumulated, are the state variables. The input to each integrator is the rate of change of its output so the state equations are set up arranging interconnections to form the function given in equation (6.1).

6.3 DERIVATION OF STATE EQUATIONS

The process of deriving the state equations for a given physical system involves translating fundamental physical laws into the required form of simultaneous first-order defferential equations. In some cases the analysis can be formalised to take advantage of the structural features of the equation.

6.3.1   First Principles

In the case of simple electrical and mechanical systems with lumped elements, a formal procedure can be evolved. With regard to electrical systems the state variables are conveniently defined to be the currents in the inductors and the voltages across the capacitors as these variables relate directly to the energy stored in the circuit. For example with the circuit Fig. 6.2.

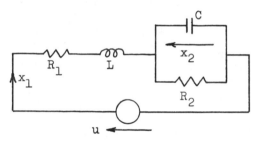

Fig. 6.2 Example

the current in the inductor x_1 and the voltage across the capacitor x_2 give rise to equations,

$$\dot{x}_1 = \frac{1}{L} \text{ (voltage across L)}$$

$$= \frac{1}{L} (-R_1 x_1 - x_2 + u)$$

$$\dot{x}_2 = \frac{1}{C} \text{ (current through C)} \qquad \ldots\ldots\ldots(6.4)$$

$$= \frac{1}{C} (x_1 - x_2/R_2)$$

or in matrix form,

$$|\dot{X}| = \begin{vmatrix} -R_1/L & -1/L \\ 1/C & -1/R_2 C \end{vmatrix} |X| + \begin{vmatrix} 1/L \\ 0 \end{vmatrix} u \ \ldots(6.5)$$

This technique can be further formalised by the application of network theory (5).

A similar procedure can be applied to simple mechanical systems with lumped elements. The state variables are the velocities of the masses and the forces on the springs. In terms of these variables the state equations relate force to rate of change of velocity for each mass and velocity to rate of change of force for each spring.

Although these techniques are powerful generalisations there are many systems which have a more complex structure and require specialised treatment. For example, the inverted pendulum

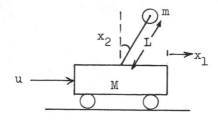

Fig. 6.3 Inverted pendulum

The state variables are chosen to be the position of the carriage x_1, and the velocities $x_3 = \dot{x}_1, x_4 = \dot{x}_2$. We then get,

$$\dot{x}_1 = x_3$$
$$\dot{x}_2 = x_4$$
$$\dot{x}_3 = -x_2 mg/M + u/M$$
$$\dot{x}_4 = x_2 g/L - u/ML$$

$$\ldots\ldots\ldots\ldots(6.6)$$

The latter two equations embrace the torque balance on the pendulum and the force balance on the carriage.

A further example is the stirred-tank reactor (Tou (4)) outlined in Fig. 6.4.

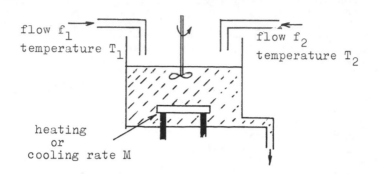

Fig. 6.4 Stirred-tank reactor

The component streams with flow rates f_1 and f_2 react to produce a third compound and involve exothermic and endothermic exchanges. With the mass in the tank remaining constant the state variables are x_1 = mass of component (1) x_2 = mass of product, x_3 = temperature. The state equations result from the application of the principles of heat and mass balance and take the form

$$\dot{x}_1 = f_1 - (f_1 + f_2 - e^a)x_1$$
$$\dot{x}_2 = (1 - x_1 - x_2)e^b - (f_1 + f_2)x_2$$
$$\dot{x}_3 = H_1 e^a x_1 + H_2(1 - x_1 - x_2)e^b + \qquad \dots\dots(6.7)$$
$$f_1(T_1 - x_3) + f_2(T_2 - x_3) + M$$

where $a = (\alpha_1 - \beta_1/x_3)$
$\qquad b = (\alpha_2 - \beta_2/x_3)$

and H_1, H_2, α_1, α_2, β_1, β_2 are constants.

These equations are fundamentally non-linear and cannot be expressed in matrix form as they stand. A linear form may be derived assuming small changes in the variables with a restricted range of applicability about the operating point.

6.3.2   High Order Differential Equations

A further possibility is to derive the state equations from a high order differential equation governing the process input-output relationship. For example, with an aircraft altitude control system (4) we have

$$\frac{d^4h}{dt^4} + 2\zeta\omega_0 \frac{d^3h}{dt^3} + \omega_0^2 \frac{d^2h}{dt^2} = K_0\delta \quad .....(6.8)$$

where h is the height, ζ and ω_0 are the damping ratio and undamped natural frequency of the motion, K_0 is a gain constant and δ is the elevator deflection. State variables are defined as follows:

$$x_1 = h$$
$$x_2 = dh/dt$$
$$x_3 = d^2h/dt^2 \qquad ....(6.9)$$
$$x_4 = d^3h/dt^3$$

and the state equations become,

$$\dot{x}_1 = x_2$$
$$\dot{x}_2 = x_3$$
$$\dot{x}_3 = x_4 \qquad ....(6.10)$$
$$\dot{x}_4 = -2\zeta\omega_0 x_4 - \omega_0^2 x_3 + K_0\delta$$

The technique can clearly be extended to systems defined by transfer functions.

6.3.3 Least-order and Model Reduction

Multivariable systems in which several transfer functions are present give rise to a problem in determining the minimal set of state equations which are necessary to describe the complete system (Rosenbrock (1)). For example, the system,

$$y = \frac{G_1(s)}{1+sT} \cdot u_1 + \frac{G_2(s)}{1+sT} \cdot u_2 \qquad ....(6.11)$$

where G_1 and G_2 are arbitrary but different transfer functions may be realised either as in Fig.6.5(a) or Fig. 6.5(b).

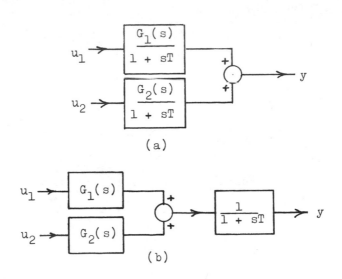

(a)

(b)

Fig. 6.5 Equivalent systems

The state equations of system Fig. 6.5(a) will contain one more state variable, and hence one more equation, than system Fig.6.5 (b) in which the sub-system $\frac{1}{1+sT}$ is included once only.

A further aspect is that of making a simplification by reducing the number of state equations below that required by strict representation of the physical system. The problem is to ensure that the reduced set of equations adequately represents the system for control purposes. This amounts to removing equations which correspond to modes with small time constants (Ellis and Roberts (9)).

6.3.4   Choice of State Variables

The set of state variables which may be used to represent a given system is not unique. Some combinations of variables are more convenient than others however. In particular with feedback control as an ultimate objective it is preferable to use variables which can be measured on the system. For example, in the aircraft altitude system described in Section 6.3.2 the state variables were defined as the derivatives of height. These variables are difficult to measure particularly in a noisy environment. An alternative is to use height, vertical velocity, pitch angle and pitch rate as the four state variables, the pitch angle and rate being readily measurable with gyroscopes (4).

6.4 CONCEPTS BASED ON STATE SPACE

The following concepts have emerged as distinctive features of the development of system theory based on state equations.

6.4.1 State Diagram

The linear state equations 6.2 can be represented as a set of interconnected integrators. For example,

$$
\begin{vmatrix} \dot{x}_1 \\ \dot{x}_2 \end{vmatrix} = \begin{vmatrix} a_{11} & a_{12} \\ a_{21} & a_{22} \end{vmatrix} \begin{vmatrix} x_1 \\ x_2 \end{vmatrix} + \begin{vmatrix} b_1 & 0 \\ 0 & b_2 \end{vmatrix} \begin{vmatrix} u_1 \\ u_2 \end{vmatrix}
$$

$$
y = \begin{vmatrix} c_1 & c_2 \end{vmatrix} \begin{vmatrix} x_1 \\ x_2 \end{vmatrix} \qquad \dots\dots\dots(6.12)
$$

gives rise to the diagram,

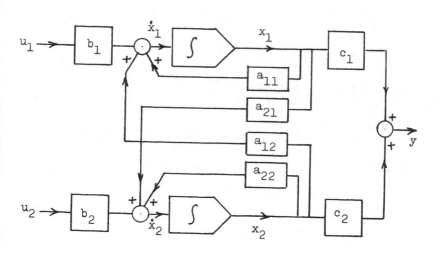

Fig. 6.6 State diagram

The role of the coefficients of matrix A in providing the feedback gains round the integrators is significant as these elements determine the dynamic modes of the system, i.e. the stability of the system depends on matrix A.

6.4.2. Eigen Values

The natural modes of the system represented in the state equations are determined by the matrix $|A|$ and identified in terms of its "eigen values". These are values of λ which satisfy det $[\lambda |I| - |A|] = 0$ for example, given the circuit

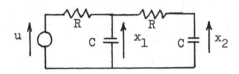

Fig. 6.7 Example

the state equations are,

$$\begin{vmatrix} \dot{x}_1 \\ \dot{x}_2 \end{vmatrix} = \begin{vmatrix} -2/RC & 1/RC \\ 1/RC & -1/RC \end{vmatrix} \begin{vmatrix} x_1 \\ x_2 \end{vmatrix} + \begin{vmatrix} 1/RC \\ 0 \end{vmatrix} u \quad \ldots(6.13)$$

forming $\lambda |I| - |A| = \begin{vmatrix} \lambda + 2/RC & -1/RC \\ -1/RC & \lambda + 1/RC \end{vmatrix}$

then det $[\lambda |I| - |A|] = (\lambda + 1/RC)(\lambda + 2/RC) - 1/(RC)^2$

the two values of λ which make this zero are,

$$\lambda_1 = \frac{1}{2RC} (-3 + \sqrt{5}) = -0.382/RC$$
$$\ldots\ldots(6.14)$$
$$\lambda_2 = \frac{1}{2RC} (-3 - \sqrt{5}) = -2.62/RC$$

These values determine the modes of the form $e^{\lambda t}$ which will be present in the responses of x_1 and x_2. In this case the modes are simple exponentials with time constants $-1/\lambda_1$ and $-1/\lambda_2$.

Alternatively, the transfer function x_2/u may be derived and we get,

$$\frac{x_2}{u} = \frac{1}{s^2(RC)^2 + 3sRC + 1} \quad \ldots\ldots\ldots\ldots(6.15)$$

which has poles at the values $s = \lambda_1$ and $s = \lambda_2$
indicating the same modal structure in the response of x_2
to changes of u. Thus eigen values correspond to pole
positions of the system transfer function in the s-plane.
When eigen values are complex the corresponding modal
component is oscillatory and when the real part of any
eigen value is positive the system response is unstable.

6.4.3   State Space

The state of the system at any instant in time is
completely defined once the values of all the state
variables are known, together with the system input values.
The rate of change of state given by the state equations
can then be evaluated. It is useful to imagine a graphical
representation in which a given combination of state
variable values is located as a point in a space where the
coordinates are the state variable values plotted along
orthogonal axes. Any change of state is then seen as a
trajectory of a curve in the state space.

6.4.4   Eigen Vectors

The motion of a system from any prescribed initial
state, with zero input $|U|$ is of general interest. We then
have $|\dot{x}| = |A| \, |X|$ and if only one mode $e^{\lambda t}$ is present the
components of $|X|$ must satisfy

$$\left[\lambda|I| - |A|\right]|X| = 0 \quad \ldots\ldots\ldots\ldots\ldots(6.16)$$

In the above example we then have,

$$\begin{vmatrix} \lambda + 2/RC & -1/RC \\ -1/RC & \lambda + 1/RC \end{vmatrix} \begin{vmatrix} x_1 \\ x_2 \end{vmatrix} = 0 \quad \ldots\ldots\ldots(6.17)$$

this requires $\dfrac{x_2}{x_1} = (\lambda RC + 2)$ from the first equation and

$\dfrac{x_2}{x_1} = \dfrac{1}{\lambda RC + 1}$ from the second. With λ equal to either

of the two eigen values these two ratios are the same.
This implies that a motion in state space involving only
one of the natural modes of the system is a straight line of

slope x_2/x_1. Substituting the values of λ_1 and λ_2 obtained in Section 6.4.2 we get $x_2/x_1 = 1.618$ for λ_1 and -0.618 for λ_2. This gives lines AB and CD as shown in the state space diagram Fig. 6.8

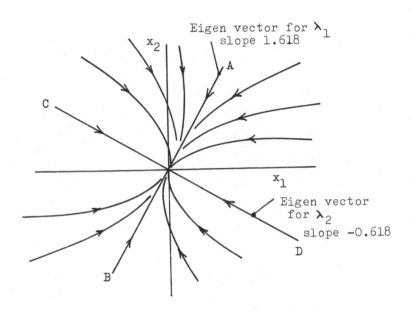

Fig. 6.8   State space diagram

The other lines are trajectories on which the state of the system changes in moving from any initial value to the origin. In terms of the circuit Fig. 6.7 the initial state represents the initial voltages of the capacitors and the trajectories show how the voltages vary together as the capacitors discharge. Significantly, if the initial values lie on either eigen vector the trajectory follows the vector line to the origin and only one of the two modes is present in x_1 and x_2 as this occurs. The mode λ_2 has a smaller time constant than that of λ_1 so that the trajectories move towards the eigen vector AB as time increases. The eigen vectors are lines which divide the state space into sectors with differing trajectory patterns and are therefore of primary interest in identifying the system behaviour in state space.

6.4.5 Canonical Variables

The partitioning of the state space by the eigen vectors in Fig. 6.8 suggest that it might be advantageous to measure the positions in state space using the eigen vectors AB, CD as coordinates as shown in Fig. 6.9

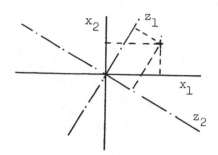

Fig. 6.9 State space

The conversion from variables x_1, x_2 to z_1, z_2 is obtained by a matrix transformation

$$|z| = |T| \ |x| \quad \dots\dots\dots\dots\dots(6.18)$$

where the rows of $|T|$ are eigen vectors. When this same transformation is applied to the state equations we get

$$|\dot{z}| = |T| \ |A| \ |T|^{-1} |z| + |T| \ |B| \ |U| \quad \dots\dots(6.19)$$

which simplifies to the canonical form,

$$|\dot{z}| = |\lambda||z| + |T| \ |B| \ |U| \quad \dots\dots\dots(6.20)$$

where $|\lambda|$ is a diagonal matrix of eigen values.

The state diagram Fig. 6.6 may be drawn as a multi-variable block diagram Fig. 6.10 (a)

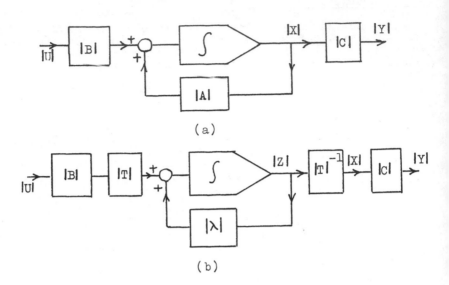

Fig. 6.10 Multivariable block diagrams

The canonical form of equations may be represented as in Fig. 6.10(b) where, because $|\lambda|$ is diagonal the integrators are no longer cross-connected as shown in Fig. 6.6. The feedback gains round the integrators are the eigen values λ, each of which determines a natural mode of the system.

6.4.6 Controllability

The general requirement for a system to be controllable is that the input variables $|U|$ must be capable of moving the state of the system from a given initial position to a required final position in state space. Within this general requirement there are a number of special definitions of controllability (1); for example, the effect of limitations on the control input magnitude may be included. In its simplest terms controllability requires that in the state diagrams Fig. 6.10 at least one element of $|U|$ must connect to the input of each integrator. This means that the transmission matrix $|T| \, |B|$ must have no rows with all elements zero.

6.4.7 Observability

Referring to the state diagrams Fig.6.10 the measured output variables $|Y|$ are combinations of the state variables $|X|$ which are in turn combinations of the canonical variables $|Z|$. Observability requires that all modes of the system should be present in the output variables. This means that the transmission matrix $|C||T|^{-1}$ must have none of its columns with all zero elements. If we further require that the state $|Z|$ should be observable at any instant from the output signals $|Y|$ it must be possible to construct a measurement matrix $|M|$ such that $|M||C||T|^{-1}$ is a unit matrix, i.e. $|M| = |T||C|^{-1}$. This requires that the output variables $|Y|$ are the same in number as the state variables. Normally this is not so, but it may still be possible to observe the state of the system using a dynamic state observer, for example a Kalman filter (Kuo (2), Tou (4)), in which a knowledge of the system dynamics is used to set up a model from which estimates of the state variables may be derived.

6.5 CONTROL IN STATE-SPACE

The state-space formulation of system dynamics provides the basis for three principal concepts related to control.

6.5.1 Stability

In state-space a stable system response is one in which the motion converges asymptotically towards a required position in state-space when initially displaced. This excludes "limit cycle" behaviour in which the motion traverses a closed trajectory repetitively in a periodic oscillation.

The motion can in principle be investigated by solving the state equations and examining the trajectories produced. A more elegant procedure is that due to Liapunov (Porter (10) In this a single scalar variable is defined as a quadratic measure of the displacement in state-space. For example a function $V = ax_1^2 + bx_2^2$ would be a possible Liapunov function for the circuit Fig. 6.6 and would relate directly to the total energy stored. The free motion in such a circuit will be accompanied by a continued loss of energy i.e. dV/dt must be negative. A divergent motion would be recognised by dV/dt going positive. Stability can be tested by observing the behaviour of the Liapunov function in state-space without the need to specifically compute the system trajectories.

6.5.2 Trajectory Optimisation

The time optimal control problem (4) resides in the concept of a "cost function" associated with the motion along trajectories in state-space. The cost is an accumulated value, expressed as an integral of the form,

$$J = \int_0^T f(|X|, |U|, t)dt \quad \ldots\ldots\ldots\ldots (6.21)$$

A control policy is required to determine how the system input $|U|$ should be varied during the control interval T to move the system from the initial state $|X|_o$ to the final state $|X|_T$ and at the same time to minimise J. In two-dimensional state-space this may be visualised as shown in Fig. 6.11.

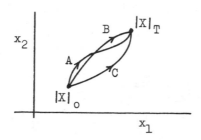

Fig. 6.11   Trajectories

The three trajectories A,B,C represent different excursions between the initial and final states which result from the application of different control signals $|U(t)|$. The cost function may typically be of a quadratic form,

$$J = \int_0^T (ax_1^2 + bx_2^2 + cu_1^2 + du_2^2)dt \ \ldots\ldots\ldots(6.22)$$

which constrains the choice of u_1 and u_2 so that on minimising J a single optimal trajectory is identified. This topic is discussed more fully in Chapter 8.

6.5.3 Modal Control

Feedback can be used to change the natural modes of a system; generally to reduce time constants or provide adequate damping of oscillatory modes. In terms of state-space concepts this amounts to changing the eigen values of the system (10) and is also referred to as 'pole assignment' (1),(2).

The principles on which this is based can be observed by reference to the canonical form of state diagram given in Fig. 6.10(b). Here a particular eigen value λ_n is represented as the gain in the feedback path round an integrator. To change λ_n to a value $(\lambda_n + g)$ we require to set up a parallel feedback path of gain g round this same integrator. This must be achieved, however, by a connection from the system output $|Y|$ to the input $|U|$. For this the structure of the control system is shown in Fig. 6.12

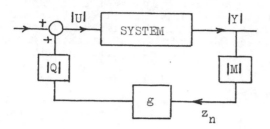

Fig. 6.12   Feedback system

The matrix $|M|$ represents a measurement vector designed to isolate the canonicle variable z_n. This variable is then multiplied by the required gain g and applied to the control vector $|Q|$ which connects through $|U|$ to the inputs of the integrators in the state diagram. If $|Q|$ is properly chosen we may be able to arrange that the signal gz_n is applied to the input of the integrator n alone. This then establishes the required feedback connection around integrator n. For this to be possible the system must be completely controllable and observable.

6.6   DISCRETE STATE EQUATIONS

In digital control systems the system output variables are measured, possibly using a multiplixed A/D converter to give discrete values. The sampling operation is carried out repeatedly with a time interval T between sampling times. The scan time required to convert the variables is presumed to be negligibly small. Input to the system is also of a discontinuous nature, being derived from a digital controller. The values of the inputs are held constant between sampling points which are presumed to coincide with the measurement times.

Given the continuous state equations of the system we require to relate output and input sample values. For example, with a single first order equation $\dot{x} = ax + bu$ we can approximate,

$$
\begin{aligned}
x_{n+1} &= x_n + T\dot{x}_n \\
&= x_n + T(ax_n + bu_n) \\
&= (1+aT)x_n + Tbu_n \\
&= \phi x_n + \Theta u_n
\end{aligned}
\qquad \ldots \ldots \ldots \ldots (6.23)
$$

This latter equation relates the next output sample to the current input and output samples. The approximation used above applies when the sample interval T is short compared with the system time constant, e.g. $aT<0.1$. A more precise formulation is necessary when the sample interval is large (2), (4), to take account of the variation in the state derivative during the sample interval. When the system is described by the matrix form of equations, $|\dot{X}| = |A| \; |X| + |B| \; |U|$ the discrete equations, in matrix form are derived using an extension of equation 6.23,

$$|X|_{n+1} = |X|_n + T \; |\dot{X}|_n \dots\dots\dots\dots(6.24)$$

and on substituting for $|\dot{X}|$ we get,

$$|X|_{n+1} = |\Phi||X|_n + |\Theta||U|_n$$

with
$$|\Phi| = (|I| + T|A|) \quad \dots\dots\dots\dots(6.25)$$
$$|\Theta| = T \; |B|$$

These equations provide what in many ways is a more simple formulation of the system dynamics than the continuous equations. For example, the response to a given set of input samples of $|U|$ can be computed numerically. From a given initial state $|X|_0$ the equations are applied recursively to advance the solution one step at a time.

For example, take the system of Fig. 6.7 for which the equations are given in equations 6.13. In formulating the discrete form of these equations it should be noted again that the input voltage u is presumed to be clamped between the sampling instants.

We take the time constant RC to be 1 second and the sample interval T as 0.25 second, then,

$$|A| = \begin{vmatrix} -2 & 1 \\ 1 & -1 \end{vmatrix} \qquad B = \begin{vmatrix} 1 \\ 0 \end{vmatrix} \quad \dots\dots(6.26)$$

$$|\Phi| = \begin{vmatrix} 0.5 & 0.25 \\ 0.25 & 0.75 \end{vmatrix} \qquad |\theta| = \begin{vmatrix} 0.25 \\ 0 \end{vmatrix}$$

and equation 6.25 becomes,

$$\begin{vmatrix} x_1 \\ x_2 \end{vmatrix}_{n+1} = \begin{vmatrix} 0.5 & 0.25 \\ 0.25 & 0.75 \end{vmatrix} \begin{vmatrix} x_1 \\ x_2 \end{vmatrix}_n + \begin{vmatrix} 0.25 \\ 0 \end{vmatrix} u_n \; ..(6.27)$$

The concept of eigen values can be introduced by taking the z-transform of equation 6.25,

$$z \, |X(z)| = |\Phi||X(z)| + |\theta||U(z)|$$

then

$$|X(z)| = |z \, |I| - |\Phi||^{-1} |\theta||U(z)|$$

The eigen values are given as solutions of $\det \, |z \, |I| - |\Phi|| = 0$ and the corresponding modes are identified by locating these values as points in the z-plane. With the above example,

$$\det \left|z \, |I| - |\Phi|\right| = \begin{vmatrix} (z-0.5) & -0.25 \\ -0.25 & (z-0.75) \end{vmatrix} = 0$$

which generates the polynomial,

$$z^2 - 1.25z + 0.3125 = 0$$

with solutions z=0.905 and z=0.345. These two modal values are located on the positive real axis of the z-plane indicating an exponentially decaying natural response, as would be expected from the circuit of Fig. 6.7. The mode corresponding to z=0.905 is a slow one as this point is close to z=1, while that due to z=0.345 decays more rapidly.

6.7 CLOSED LOOP SYSTEMS

When feedback is applied to a process using a digital controller the signal sampling operation, introduced by the analogue to digital converter, results in measurements of the system state at discrete sampling intervals. Typically a single variable control loop may be represented as shown in Fig. 6.13.

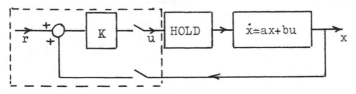

Fig. 6.13 Digital control system

Here the function of the controller K is that of a proportional control element. The sampling operation, represented by the switches, implies that the feedback equation takes the form,

$$u_n = K(r_n - x_n) \quad \ldots\ldots\ldots\ldots\ldots(6.28)$$

With the process state equation expressed in discrete form,

$$x_{n+1} = \phi \, x_n + \theta \, u_n \quad \ldots\ldots\ldots\ldots(6.29)$$

the closed-loop response is given by,

$$x_{n+1} = (\phi - K\theta)x_n + K\theta r_n \ldots\ldots\ldots(6.30)$$

This indicates a closed-loop eigen value given by
$z = \phi - K\theta$. Feedback therefore alters the eigen value of
the system and it is possible to select the controller gain
K to provide a specified value. It is important to note that
ϕ and θ are functions of the sample interval T so that the
final system design depends in the sampling rate of the
controller.

When a system with multiple state variables is involved
an equivalent form of closed-loop system is as shown in
Fig. 6.14.

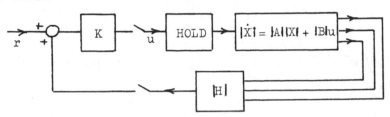

Fig. 6.14 Control based on state feedback

Here the matrix $|H| = |h_1 \; h_2 \; h_3 \; \ldots h_n|$ applies weighting
factors to the state variables. These weighting factors
represent a set of feedback gain values. The feedback
equation is now,

$$u_n = K(r_n - |H| \, |X|_n) \ldots\ldots\ldots\ldots\ldots(6.31)$$

and the closed-loop response is given by,

$$|X|_{n+1} = (|\Phi| - K|\theta||H|)\,|X|_n + K|\theta|r_n \ldots\ldots(6.32)$$

from which we get the eigen values as values of z which
satisfy,

$$\det \left| \; z|I| - (|\Phi| - K|\theta||H|) \right| = 0 \ldots\ldots\ldots(6.33)$$

The feedback matrix $|H|$ and the controller gain K may now
be chosen to give prescribed eigen values.

To design a system with specified eigen values alone
it is not necessary to have the controller gain K available
for adjustment; there are enough variable parameters in $|H|$.
However K may be used to set a required steady-state error
condition in the system.

Following on the example used in section 6.6 we can
examine the requirements of a digital feedback system to
control the system Fig. 6.7 to give a prescribed set of

eigen values. Here $|H| = |h_1 \; h_2|$ and if we set K=1 equation 6.33 becomes,

$$\det \begin{vmatrix} (z - (0.5-0.25h_1)) & (0.25-0.25h_2) \\ 0.5 & (z - 0.75) \end{vmatrix} = 0 \quad ..(6.34)$$

If we specify eigen values $z_1 = 0.5$ and $z_2 = 0.2$ then with the first of these inserted in equation 6.34 we obtain,

$$h_1 + h_2 - 1 = 0 \quad ....(6.35)$$

then on using the second,

$$1 - 3.44h_1 + 3.125h_2 = 0 \quad ....(6.36)$$

We may now solve these equations for h_1 and h_2 and obtain $h_1 = 0.629$, $h_2 = 0.371$ as the required feedback gain values.

The digital control function to be implemented in the controller of Fig. 6.14 is then,

$$u_n = r_n - (0.629x_{1n} + 0.371x_{2n}) \quad ....(6.37)$$

This simple example serves to illustrate the facility which the discrete state equations offer in system design. The design objectives have been limited to eigen value adjustment but the extension to impose further design constraints can be readily achieved (2).

6.8 REFERENCES

1. Rosenbrock, H.H., 1970, 'State space and multivariable theory', Nelson, London.

2. Kuo, B.C., 1980, 'Digital control systems', Holt, Rinehart and Wilson, New York.

3. Macfarlane, A.G.J., 1964, 'Engineering systems analysis', Harop, London.

4. Tou, J.T., 1964, 'Modern control theory', McGraw-Hill, New York.

5. Kuo, B.C., 1967, 'Linear networks and systems' McGraw-Hill, New York.

6. Dorf, R.C., 1980, 'Modern control systems', Addison Wesley, New York.

7. Gordon, G., 1978, 'System simulation', Prentice-Hall, Englewood Cliffs. N.J.

8. Jorgensen, V., 1974, Int.J.E.E.E., 11, 367-376

9. Ellis, J.E. and Roberts P.D., 1981, I.E.E. Conf. Pub. No. 194, 191-195.

10. Porter, B., 1969, 'Synthesis of dynamical systems', Nelson, London.

Chapter 7

System identification

Dr. K. Warwick

7.1 INTRODUCTION

In the majority of physical sciences of primary importance in the study of system behaviour, prediction and design is the formulation of a mathematical model which describes the system under consideration. The class and accuracy of a particular model is, however, dependant on its required application.

A model may be obtained by examining the internal structure of the system, although it is often the case that a complete picture cannot be achieved due to an unknown factor, an element which is not directly measurable or an extremely complicated process. It is likely, however, that by carrying out a set of detailed experiments on the overall plant, the gaps in the model can be filled and an exact mathematical description of the plant obtained which is representative under all conditions. This can, though, prove to be a time consuming and expensive procedure and may be rather overdoing what is actually required because the plant, when in normal operating mode, may well only encounter a small proportion of the conditions accounted for or certain factors included in the model might contribute only a negligible effect within the operating range. An example is given when modelling the climate of Britain, for which there would not be much point in ensuring that the model could cope with temperatures above 100°F. It is the case however, that most controller designs, found to be satisfactory, are based on a very simple model of the plant which is obtained as a representation over one particular set of operating conditions only. The most common approach taken to find such a model is to make use of data from the plant input(s) and output(s) when operating dynamically in its normal range. Unfortunately this data is invariably noisy because of random fluctuations or disturbances, some of which are due to the measuring devices employed. The process of making use of the data, despite the added noise, to obtain estimates of parameters within a model of the plant is called System Identification.

System Identification is effectively, therefore, the procedure of modelling systems mathematically with due consideration being given to the associated problems. Inclusive of processing noisy data in order to make it directly relevant, problems are found in the choice of model form, the statistical analysis of the processed data and testing of the completed model to assess how good a representation of the actual system it is.

In this chapter it is intended to look at the methods available for System Identification and how they can be linked in with information which may already have been obtained relating to the plant. Generally this involves applying statistical tests to the set of plant input-output data in order to estimate both the model order and the respective values of the parameters within a model of that order. Many software packages are however available, e.g. Denham and Abaza (1975), with which one can carry out the identification procedure on available data. The main cost in terms of time is therefore spent on carrying out experimentation on the plant in order to achieve sufficient data that will be of use to the package.

7.2   DYNAMIC SYSTEM MODELLING

The main aim of identification is to accurately model the system under consideration. Certain model structures must therefore be defined and estimations made of the parameters contained within each particular model. A good model structure should approximate the system to a good degree and contain all the known relevant information about system operation. It must also be flexible and lead to simple parameter estimation procedures.

Models may be divided into two main classes, parametric and non-parametric. With parametric models the system order must be specified such that errors are eliminated, in non-parametric models, however, order specification is unnecessary. But non-parametric models are infinite dimensional and hence it becomes virtually impossible to match the output from a model to that of the system. Similar difficulties arise with noise signals, which are not averaged out in the non-parametric case. Thus, in this chapter only parametric models are considered.

For the case of a linear system under discrete-time operation, possibly because of a computer, which has a single input and single output, the output may be expressed in terms of previous inputs and outputs as:

$$y(k) + \sum_{i=1}^{n} a_i\, y(k-i) = \sum_{i=1}^{n} b_i\, u(k-i)$$

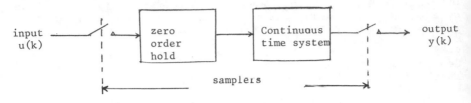

Fig.7.1: Sampled continuous time system

where y(k) is the system output at time t=k, y(k-1) is the
system output at time t=k-1, etc., in which the time period
t=k-1 to t=k signifies one sample period or length of time
between each output sample. Also u(k) is the system input
at time t=k, such that for the system model introduced an
assumption has been made that at least one sample period
will elapse before anything occurring at the system input
will affect the system output.

     The   most   useful class of models for parameter
estimation is that of prediction error models in which a
mathematical   representation   is   made   of   the system
description and the parameters a_1, a_2.., a_n and b_1, b_2,..,b_n
are estimated. If the same input, u(k), is then applied to
both the system and the mathematical model of the system,
the outputs of system and model can be compared giving rise
to an error which is dependent on how poor the parameter
estimates are, or rather the error gives an indication of
how close to the actual system the mathematical
representation is.   It must be remembered that of primary
importance is the choice of mathematical model structure, or
how many a_i, b_i parameters are included. If the structure
is incorrect, e.g. not enough parameters are used, no matter
how comprehensively the remaining parameters are estimated a
large error between system and model outputs will most
likely occur.

     Unfortunately, in practice the system output is
corrupted by noise due to disturbances, measurement devices
and interfacing.   This means that extremely complicated
models, e.g. those with many a_i and b_i parameters, are not
necessarily of great use because variations in some of the
parameters may well have less of an effect on the error
value   than   the   noise   signal.

     If $\epsilon(k)$ is the error value (the prediction error) at
time t=k, an attempt to obtain a better mathematical model,
i.e. better estimates of a_i and b_i, by trying to minimise

the error would be of little direct use as such a procedure would result in a large 'negative' error. What is really of interest is the magnitude of the error, rather than its sign, such that as the error tends to zero the model becomes a good approximation of the system. The square of the error $\epsilon(k)^2$, rather than the error itself is therefore of far more use and this is in fact used as the basis for the vast majority of identification procedures. The general idea is to sum a number of samples, N, of the error signal squared and to use this sum of squares as a cost function to be minimised by selection of a_i and b_i estimates. The effect of noise on the model can by this means be reduced considerably.

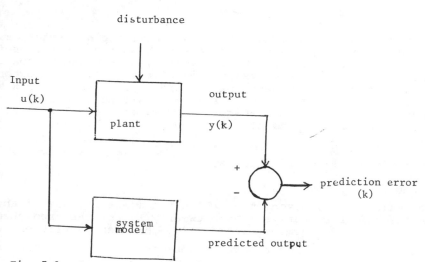

Fig. 7.2: Propagation of prediction errors in estimation

Numerous procedures exist for the minimization of the cost function described in the previous paragraph. Often these are referred to as 'hill climbing' techniques, despite the fact that this particular name more directly describes maximizing, rather than minimizing, a function. The first, and perhaps simplest method - Linear Least Squares, is the

subject of the next section. Unfortunately in some cases
the cost function may well have several minima and it is
dependant on the starting point of the procedure used as to
which minima is finally found. When this sort of cost
function is encountered a more complicated minimizing
procedure must be used such that a global minimum, i.e. the
minimum minimum, is found. For the purpose of the
explanation of Least Squares techniques, it is assumed that
no such problems of many minimum values exist.

7.3   LINEAR LEAST SQUARES ESTIMATION

Essentially the sum of the squares of the differences
between the model output and actual system output data is
minimized such that the parameter estimates in the model are
as near as possible to their actual values.

Consider the linear model:

$$Y = X\theta + E$$

in which it is assumed that the N dimensional vector, E,
includes the effects of measurement noise and disturbances
affecting the system. Y is a vector, also N dimensional,
containing N output measurements and θ is a vector
containing the M system parameters. X is then an N x M
matrix which consists of terms which are functions of the
input signals, and therefore are known values. The single
output measurement taken at time instant k, can thus be
written as:-

$$y(k) = \sum_{i=1}^{M} \theta_i x_i (k) + e(k)$$

The value of θ, given as $\hat{\theta}$, which minimises the sum of the
squares of the derivations between y(k) and the predicted
output; $\theta_i x_i(k)$ summed from i=1,..,M; can be written in
vector form, such that the cost function

$$S = (Y - X\theta)^T (Y - X\theta)$$

must be minimized by a selection of system parameters, $\hat{\theta}$ for
a given set of input/output information.

By differentiating S with respect to θ and setting the
result equal to zero for a minimum, one obtains:

$$\hat{\theta} = (X^T X)^{-1} X^T Y$$

on condition that $X^T X$ is invertible, in order to find a
unique solution.

Notes.

1.   The individual elements contained within the matrix X
     are known values dependant on the .input signals

applied to the system. For the identification procedure the selection of input signals is part of the design procedure.

2. Although X is of dimension N x M, the matrix $X^T X$ is of dimension M x M and as the number of system parameters, M, is generally much smaller than the number of data values, N, the necessary matrix inversion is not as much of a computational burden as it might at first appear.

3. The results given here for parameter estimation of linear systems are also applicable to non-linear systems although in this case a linearisation at each step is necessary and several iterations of the algorithm are usually required.

If E is a vector of random variables, such that $\{e(k)\}$ is a zero-mean white sequence; i.e. $e(k)$ is independent of $e(k-1)$, $e(k-2)$,...; and is assumed to be caused by data fluctuations alone, then the mean sequence value is

$$\xi\{E\} = 0$$

where $\xi\{\cdot\}$ signifies the expected value.

Also the vector variance is given by:

$$\xi\{EE^T\} = \sigma^2 I$$

in which σ is a real number and I is the identity matrix.

Subject to the assumption that the elements of X are uncorrelated with those of E, the least squares esimate, $\hat{\theta}$, then has the following properties.

(1) It is a linear function of the data.

(2) It is unbiased.

i.e. $\xi\{\hat{\theta}\} = \xi\{(X^T X)^{-1} X^T (X\theta+E)\} = \xi\{\theta + (X^T X)^{-1} X^T E\}$

or $\xi\{\hat{\theta}\} = \theta + (X^T X)^{-1} X^T \xi\{E\} = \theta$

(3) The Covariance matrix, $Cov(\hat{\theta}) = \sigma^2 (X^T X)^{-1}$;

as $Cov(\hat{\theta}) = \xi\{(\theta-\hat{\theta})(\theta-\hat{\theta})^T\}$.

(4) It is the BEST linear unbiased estimator (BLUE), and this means that in the class of all linear unbiased estimators it is the method which produces the

smallest variation in $\hat{\theta}$ when different data sets, Y and X are used.

(5)   If E is normally distributed then $\hat{\theta}$ is the minimum variance unbiased estimator (MVUE) of θ, since $\hat{\theta}$ is unbiased and achieves the Cramer-Rao lower bound.

(6)   If E is not a white sequence, then the least squares estimation $\hat{\theta}$ is biased, whereas if E is a white sequence and is correlated with X then the estimation $\hat{\theta}$ is 'strongly consistent' according to regularity conditions such as the input being 'suitably exciting'.

A better guide, in terms of how good a particular algorithm is, can be found by consideration of statistical confidence regions corresponding to each method. For this purpose the 'residuals' vector $\hat{E}$ must be obtained from:

$$\hat{E} = Y - X\hat{\theta}$$

such that the sum of squares of the residuals is calculated by means of the equation

$$\hat{S} = \sum_{k=1}^{N} \epsilon(k)^2 / \sigma^2$$

and this is distributed in the form of $\chi^2(N-M)$, where M is the number of parameters and N is the number if data points. It is, however, fairly straightforward to apply a t distribution if this is preferred to the Chi-square test.

7.3.1 Application of Least Squares Estimation

The method of Least Squares parameter estimation has been described. Consideration has however not yet been fully given to the types of systems which may be encountered and the practical problems to be overcome.

It is generally the case that the purpose of carrying out system identification on a plant is to obtain a plant model which can be used as the basis for regulator and/or controller design. This particular discussion is concerned with obtaining single input, single output discrete-time representations of continuous-time or discrete-time systems which can be subject to disturbances of random and invariably unknown magnitudes. The objective of a regulator based controller is then to reduce or even minimize the effect of these disturbances on the measured output signal. Further controller design parameters can however be incorporated to deal with a change of desired output value (set-point), to ensure a reasonable rate of change in the output response, possibly such that no overshoot occurs. Application areas are found in paper and steel production, level control in fluid storage areas and mixture percentages in combustion engines.

Although identification theory has now been applied to many industrial problems, giving rise to much available practical expertise, it is found in many cases that straightforward identification of the parameters in an input/output system transfer function model does not by itself lead to good controller design. In these instances it is in fact found beneficial to also obtain as much information as possible with regard to the nature of the disturbances. A second transfer function, relating disturbances, considered random, to the measured output must then be included such that improved controller performance is achieved.

The input/output system transfer function was considered earlier for an nth order plant. This is now generalised slightly to allow for a pure time delay (transport delay), such that r sampling periods occur before any plant input fluctuation has an effect which is observed at the output port.

Let $A(z^{-1})y(k) = z^{-r}B(z^{-1})u(k)$

where $y(k)$ and $u(k)$ are the plant output and input signals respectively at time instant k. Also, the polynomials A and B are defined as:

$$A(z^{-1}) = 1 + a_1 z^{-1} \ldots .+a_n z^{-n}$$

and $$B(z^{-1}) = b_0 + b_1 z^{-1} + \ldots .. + b_n z^{-n}$$

in which the operator z^{-i} has effect such that $z^{-i}y(k-i)$, and so here $z^{-r}u(k) = u(k-r)$.

The disturbance affecting the plant output, $w(k)$, is considered to be obtained by passing a white noise sequence, $e(k)$, through a second transfer function,

$$D(z^{-1})w(k) = C(z^{-1})e(k)$$

where $C(z^{-1})$ and $D(z^{-1})$ are monic mth order polynomials, of similar form to $A(z^{-1})$

The overall, actual, measured output variable is then regarded as being the sum of the deterministic transfer function output and the disturbance transfer function output, such that

$$y(k) = z^{-r} \frac{B(z^{-1})}{A(z^{-1})} u(k) + \frac{C(z^{-1})}{D(z^{-1})} e(k)$$

Assuming the total plant needs to be identified, this will involve model order selection in order to specifiy both n and m, calculation of the transport delay r and finally estimates of the parameters contained within the A,B, C and D polynomials. Also if an offset if present, such as that caused by a non-zero mean value in the disturbance term, i.e. a d.c noise level, this can be taken into account by the inclusion of one extra parameter in the system model, which is estimated by extending the parameter vector, θ, by one term, as well as the data vector, x(k)

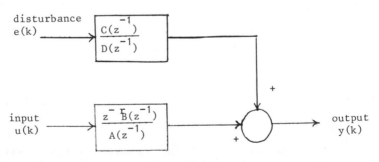

Fig.7.3: General form of system description

If the denominator polynomial of a transfer function is stable, i.e. all its roots lie within the unit circle of the z- plane, the parameters contained within its inverse power series decay to zero at a rate dependant on how near the roots are to the unit circle. The deterministic part of the system transfer function $A^{-1}B$ then, for a stable A polynomial, i.e. an open loop stable plant, is usually of a form such that its parameters decrease at a rapid enough rate such that cross-correlation requires only a finite number of parameters in order to obtain a representative model.

Consider the system model, written in terms of its power series:

$$y(k) = \sum_{i=0}^{p} q_i \, u(k-i-r) + w(k)$$

in which the parameters q_i tend to zero as the index i increases, in such a way that $q_i \approx 0$ for $i > p$. Using cross correlation on the vectors θ and $x(k)$, where $\theta = [q_0, q_1, \ldots, q_p]$ and $x^T(k) = [u(k-r), u(k-r-1), \ldots, u(k-p-r)]$, the linear in the parameters model will result in optimal estimates when the term $X^T X$, obtained after collecting N data points, is diagonal.

Unfortunately cross correlation usually requires a large number of parameters in order to obtain a reasonable model of the system, and the general form of power series plant representation is not particularly suitable for control law calculations. Further, a model of the disturbance is not obtained and a signal such as PRBS, which has an impulse-like autocorrelation function, must be used.

An alternative approach is to consider the application of least squares to the general system equation when multiplied throughout by the open-loop denominator $A(z^{-1})$, such that

$$A(z^{-1})y(k) = z^{-1}B(z^{-r})u(k) + v(k)$$

in which $v(k) = AD^{-1}C\ e(k)$

The system can also be written as:

$$y(k) = -a_1 y(k-1),\ldots-a_n y(k-n) + b_0 u(k-r) + \ldots$$
$$+ b_n u(k-n-r) + v(k)$$

where $x^T(k) = [-y(k-1),\ldots, -y(k-n);\ u(k-r)\ldots,u(k-n-r)]$

Although the model appears to be linear in the parameters the data vector $x(k)$ is unfortunately not known exactly due to the fact that $y(k-1)$, $y(k-2)$, etc., are the measured output values and hence are subject to measurement errors. Because of these errors the least squares method can lead to a considerable bias in the estimated parameters. However if the sequence $v(k)$ is a white noise signal, the bias will not be present, this is though not a regular occurence as it necessitates that $v(k) = e(k)$ and hence $AD^{-1}C$ must be equal to unity.

The problem of bias in the least squares parameter estimates has led to the development of other identification algorithms in which no bias is present. The most widely encountered of these methods are discussed in the following section.

7.4 ALTERNATIVE ESTIMATION SCHEMES

7.4.1 Generalised Least Squares

Biased parameter estimates are obtained when using Linear Least Squares estimation if the disturbance term is correlated. Consider then the system equation

$$A(z^{-1})y(k) = B(z^{-1})u(k) + w(k)$$

where $w(k)$ is correlated, but the autocorrelation function is known such that

$$w(k) = C(z^{-1})e(k)$$

in which $\{e(k)\}$ is an uncorrelated sequence and $C(z^{-1})$ is invertible. The system equation can therefore be rewritten as:

$$A(z^{-1})y^*(k) = B(z^{-1})u^*(k) + e(k)$$

where $y^*(k) = C^{-1}(z^{-1})y(k)$

and $u^*(k) = C^{-1}(z^{-1})u(k)$

Then the Least Squares can be applied, once the values $y^*(k)$, $u^*(k)$ are calculated, in order to obtain estimates of the parameters within the polynominals $A(z^{-1})$ and $B(z^{-1})$. Unfortunately, though, the colouring polynomial $C(z^{-1})$ is usually not known and must also be estimated. On the assumption that $R(z^{-1}) = C^{-1}(z^{-1})$, the overall algorithm can be described as:

1. Fit a model of the form $[A(z^{-1})y(k) = B(z^{-1})u(k)+w(k)]$ to the data by using Linear Least Squares parameter estimation.

2. Test the residuals obtained for whiteness. If they are white then stop.

3. Using Linear Least Squares estimate $R(z^{-1})$ for the autoregressive model $[R(z^{-1})w(k) = e(k)]$.

4. Filter the input and output data through $R(z^{-1})$, i.e. $u^*(k) = R(z^{-1})u(k)$.

5. GOTO 1.

Note:

This method can take several iterations before convergence (whiteness in step 2) is achieved, however, more complicated versions are possible in which the rate of convergence is increased, i.e. fewer iterations are required.

7.4.2 Instrumental Variables

The method of instrumental variables is another way of dealing with correlated disturbances. Consider a matrix W, whose elements are functions of the data, and which satisfies

$$\det[\xi\{W^T X\}] \neq 0$$

and $\xi\{W^T E\} = 0$

Then   if   the standard linear model is multiplied throughout by W transposed,

$$W^T Y = W^T X\theta + W^T E$$

The   parameter   estimate   vector, $\hat\theta$, can be found by solving the equation

$$W^T Y = W^T X\hat\theta$$

and this leads to an unbiased estimate.

Any matrix W having the properties described is called an   instrumental   matrix and the estimator $\hat\theta$ is in this case termed an instrumental variable estimator.

7.4.3 Maximum Likelihood Estimation

The   Linear   Least Squares parameter estimation method deals   with   models   that   are   linear   in   the   parameters. Maximum Likelihood Estimation however, covers a more general class of problems.

Consider a single input, single output model with correlated noise and zero initial conditions

$$A(z^{-1})y(k) = B(z^{-1})u(k) + C(z^{-1})e(k)$$

where   e(k)   is   a   sequence of zero-mean independent random variables with variance σ^2.

Let the unknown parameter vector be θ, such that:

$$\theta^T = (a_1, a_2, \ldots, a_n; b_0, \ldots, b_n; c_1, \ldots, c_n)$$

Then,   if e(k) and y(k) are normally distributed for fixed θ and σ^2:

$$\xi\{y(k)\} = A^{-1}B \ u(k)$$

in   which   the z^{-1}qualifier   has   been   left   out for ease of explanation, i.e. $A = A(z^{-1})$.   Also,

$$\text{Cov}\{y(k)\} = \sigma^2(A^{-1}C)(A^{-1}C)^T$$

Thus the probability density of $y(k)$ can be described in logarithm form as:

$$L = -\frac{N}{2}\log(2\pi\sigma^2) - \frac{1}{2\sigma^2}\ \epsilon^T(CC^T)^{-1}\epsilon$$

where

$$\epsilon(k) = A\ y(k) - B\ u(k)$$

If $\quad V(\Theta) = \frac{1}{2}\ \epsilon^T(CC^T)^{-1}\epsilon$

then the maximum likelihood estimates of Θ and σ are obtained separately from

$$V(\hat{\Theta}) = \min V(\Theta)$$

and

$$\hat{\sigma} = \left[\ \frac{2V(\hat{\Theta})}{N}\ \right]^{1/2}$$

Notes:

1. $\hat{\Theta}$ is the least squares estimate when $c_i = 0$; $i=1,\ldots,n$, and Θ is a weighted least squares estimate when the c_i values are known. The estimates are however non-linear when the c_i values are not known.

2. The Newton-Raphson iteration procedure is perhaps the easiest to apply in calculating the Maximum Likelihood Estimator, although many methods are suitable.

3. Maximum Likelihood Estimators have good asymptotic properties and if there exists an unbiased estimator which achieves the Cramer-Rao lower bound, then it is also the Maximum Liklihood Estimator.

7.5   EXPERIMENT DESIGN

Selection of experiments to be performed and their mode of operation plays a major role in any identification process. Information about the structure of the plant must be maximised to provide data for statistical analysis.

The design procedure must take into account several variables, which may well be interactive, e.g. test signals and sampling instants. But, what often seems like, a plethora of constraints on the conditions of the experiment, such as the amplitude and power of inputs and outputs must also be complied with. Some of these constraints will however appear more relevant than others in a particular application, and it is therefore more usual to respect a specified number of them, and to check at a later time as to whether or not other constraints have been violated.

A measure of the 'goodness' of an experiment should be defined so that different experiments may be compared. As the parameter accuracy is dependant on both the estimator form and the experimental conditions a sensible choice is a measure related to both of these factors. Nevertheless, it must be remembered that the exact choice of measure is unimportant, since if numerous criteria are applicable a good experiment will be judged to be good by all of them.

The choice of input signal in terms of the time domain system representation is largely dependant on the disturbance affecting the plant. For a straightforward white noise disturbance it is usually sufficient to apply a PRBS input signal where the number of data points is finite, however, as the number of data points increases so a white noise input sequence may well be required. For a coloured disturbance though the control problem becomes non-linear and optimization procedures must be used.

In terms of the frequency domain system representation, optimal inputs can be found which consist of a limited number of sinusoids with various amplitudes. The number of sinusoids depends on the order of the model and the number of parameters. When available input power is a constraint this must be taken into account if the input is generated exogenously. More generally however, the input may be partly of wholly generated by feedback, and this can be advantageous when output power constraints exist.

In some cases sampling instants and presampling filters can be adjusted as well as the input signals. Various aspects may then be considered, such as (a) uniform sampling interval design, (b) non uniform sampling and (c) suboptimal design of non-uniform sampling.

Structure discrimination is a term used to describe the choice between rival models as opposed to accurate parameter estimation within a model of specified structure.

This topic is dealt with, to an extent, in the following section under the heading of Model Order Testing.

A final point with regard to experiment design must be made as far as data preparation is concerned. For the identification procedures considered the input/output data collected is assumed to be unbiased, thus its expectation must be zero and no d.c. levels must be present. However, in practice it is usually the case that a certain amount of bias exists and can be placed in one of two categories. Either the disturbance affecting the plant has an unknown non-zero mean value or a d.c. level is present in the output signal which cannot be accounted for in terms of any disturbance. A justifiable procedure for the removal of bias involves finding the mean of the output and subtracting this from the actual output for each data point. On subtraction the new data will approximately satisfy the same model but no bias will be apparent.

7.5.1 Model Order Testing

Invariably it is possible to fit models of different order to the data obtained for the identification procedure. If the wrong model order is employed serious errors can occur, or a redundancy of terms will be apparent. Thus a test is required to find the 'correct' or 'best' model order.

As the model order increases the sum of squares of the residuals will decrease due to a better fit being obtained. If the decrease is very slight between any two models, the order of the second model being an increase on that of the first, then the use of a model of higher order does not significantly reduce the sum of squares.

Defining the prediction error term as that given by:

$$\epsilon(k) = \hat{A}(z^{-1})y(k) - \hat{B}(z^{-1})u(k)$$

The sum of the squares is then:

$$S_n = \sum_{k=1}^{N} |\epsilon(k)|^2$$

where N is the number of data points and n is the model order.

For a certain set of data, i.e. a set of input and output values, u(k) and y(k) respectively, the coefficients within the polynomials $\hat{A}(z^{-1})$ and $\hat{B}(z^{-1})$ can be found, for a particular model order, by means of one of the methods described earlier, e.g. generalised least squares. The same procedure can then be carried out with the data by employing models of different orders. By plotting S_n against n, a model order may be chosen, say $\bar{n}$, where the slope of S_n/n is steep for $n < \bar{n}$ and shallow for $n > \bar{n}$, see fig. 7.4.

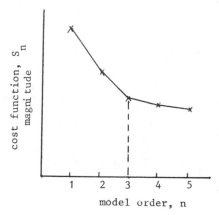

Fig.7.4: Test for 'correct' model order

7.6   PRACTICAL ASPECTS

As mentioned earlier, the identification procedures described are only intended to be used with data which has no bias or offset. Limitations also occur insofar as non-stationarity of data is concerned, i.e. parameter drift, and this is apparent in many transfer processes in which a flow of energy is involved, such as heating systems. The disturbance acting on the plant can though be modelled by an integrating function operating on white noise, and this has been found to account for many offset problems.

The offset, affecting the plant, can be removed in several ways. The most obvious of these is perhaps to apply an integrator to the control input signal, that is the input signal becomes $(1-z^{-1})u(k)$, an example of this type of action being found in a PID controller. This particular method can however be extended to ensure stationarity. Another, and not completely unrelated, approach though is to consider the offset terms as a low frequency element, which if it remains approximately constant with respect to time, it certainly is. The data can then be fed through a high-pass filter to remove the undesirable low frequency elements.

Once the plant response data has been suitably prepared, the identification procedure proper can be set in progress by selecting a model of a particular structure and calculating the parameters within that model such that it produces a response which is as near as possible to that of the plant. The various identification methods; LLS, GLS,

Maximum Likelihood; then produce a set of parameters which provide the 'best' fit within the confines of that one model structure. Once one parameter set has been obtained a model of an alternative structure, i.e. with a different number of parameters, is selected and its respective parameter set found by means of the same identification procedure. The properties of the differently structured, or ordered, models can then be compared to find the most appropriate. For this purpose a series of residuals is formed, such that at time instant k the residual is defined to be the difference between the plant output at that time and the output, at the same time, of a model of the plant constructed from the parameter set obtained for the structure under consideration. The larger the residuals in magnitude, the worse the model.

If the residuals point to a high error value or if a pattern appears in what should be an uncorrelated error sequence, then the model structure is not adequate in terms of allowing for a close fit with the response data. The residuals can in fact be used in the calculation of a residual sequence auto-correlation function, and this can in turn be used to test for whiteness conditions, i.e. if the residuals are not white the model structure is not suitable.

The overall identification procedure may then be summarized as follows:

(a) Collect data and consider possible model order limitations.

(b) Start with the lowest possible, realistic, model order.

(c) Estimate the parameters, using one of the techniques discussed, for the specified model order (n). Note, where one of a number of pure time delays is possible, these must be fitted in turn and the best delay selected for a model of the order under consideration.

(d) Test residual sequence for whiteness. If not white filter data and go to (c).

(e) Calculate the sum of squares (S_n) of the prediction errors. Plot latest point on graph of S_n against n.

If necessary, increase model order and go to (c).

(f) List results of parameter estimates. Check sensitivity of the type of control to be employed.

In the section which follows the results of an identification exercise are discussed. The system to be identified is simulated within the package and hence when a 'good' model of the system is obtained, the parameters can be compared directly with their 'true' values.

One problem encountered when attempting to identify systems is the need for some systems to either (a) operate continually on-line, where shut down is expensive, or (b) to operate only within the confines of a feedback loop, i.e.

closed-loop control, in order to retain a steady output signal. To directly employ plant input and output information from on-line operation, such that everyday plant life is not affected, will not give a true representation of the plant alone since the controller transfer function will also be accounted for. Although not a lot can be done in these situations, it is sometimes possible to provide an external input signal which can be summed with the control input signal provided by the feedback loop. The plant response to the external input can then be monitored.

7.7   AN IDENTIFICATION EXAMPLE

In order to clarify the theory given, this section includes a worked example carried out using the system identification package - SYSID (see references, Denham and Abaza). This package includes facilities for data generation as well as a choice between identification procedures.

As the example given here is more for explanation purposes rather than for a deep technical insight, a relatively simple second order model, with no pure time delays, was simulated such that

$$\bar{y}(k) = \frac{0.61 + 0.48z^{-1}}{1 - 1.14z^{-1} + 0.29z^{-2}} \, u(k)$$

where $\bar{y}(k)$ is the deterministic system output, i.e. the system transfer function assuming no disturbances.

In a physical system though, the observed output is affected by noise such that

$$y(k) = \bar{y}(k) + w(k)$$

in which $w(k)$ is a disturbance affecting the plant and $y(k)$ is the actual, measured plant output.

For this example the disturbance $w(k)$ was generated as a correlated noise signal by passing a white noise input, $e(k)$, through the second order noise model transfer function

$$w(k) = \frac{e(k)}{1 - z^{-1} + 0.25z^{-2}}$$

It is therefore assumed that the plant transfer function and noise model described above, are in fact the actual plant, i.e. the plant is simulated by them. The purpose of this identification exercise is then to start with no knowledge regarding the plant model given, other than any data which can be obtained from it. For this purpose {u(k)} was chosen to be a white noise sequence with variance equal to 2.05, whereas the disturbance itself, w(k), also white noise, was specified as having a variance equal to 0.256. Both sequences were applied to the plant for 500 data points, leading to 500 values of plant output, y(k). Three models were obtained in the identification procedure and these are detailed as follows.

(a) First Order Model:

Least Squares Estimation was applied to the first order model of the form

$$y(k) = \frac{b_o}{a_o + a_1 z^{-1}} u(k)$$

The three parameter estimates were obtained as

$b_o = 0.6336$; $a_o = 1.0000$; $a_1 -0.9263$

On testing for whiteness however the residuals were found to be coloured, thus, to avoid biased estimates, the Generalised Least Squares technique had to be applied. To this end a tenth order autoregressive model was applied to the error sequence resulting in:

GLS iteration	b_o	a_o	a_1	Whiteness test
1	0.4994	1.0000	-0.9289	not white
2	0.3996	1.0000	-0.9315	not white
3	0.3180	1.0000	-0.9341	white

(b) Second Order Model:

Least Squares Estimation was applied to the second order model

$$y(k) = \frac{b_o + b_1 z^{-1}}{a_o + a_1 z^{-1} + a_2 z^{-2}} u(k)$$

and this resulted in the parameter estimates

b_0 = 0.6169; b_1 = 0.4871 and a_0 = 1.0000; a_1 = 1.1320; a_2 = 0.2827

The residuals were found to be coloured on applying the whiteness test and hence the GLS technique was once again called upon, via a tenth order autoregressive model, to give the following results.

GLS iteration	b_0	b_1	Whiteness test
1	0.6155	0.4836	white
2	0.6147	0.4823	white
3	0.6142	0.4817	white

GLS iteration	a_0	a_1	a_2
1	1.0000	−1.1371	0.2873
2	1.0000	−1.1375	0.2876
3	1.0000	−1.1377	0.2878

Although the residuals were white after only one GLS iteration, further iterations were carried out in order to obtain better parameter estimates.

(c)   Third Order Model:

Least Squares Estimation was applied to the third order model

$$y(k) = \frac{b_o + b_1 z^{-1} + b_2 z^{-2}}{a_o + a_1 z^{-1} + a_2 z^{-2} + a_3 z^{-3}} u(k)$$

resulting in the parameter estimates:

$b_o = 0.6170$; $b_1 = 0.5912$; $b_2 = 0.0874$ and $a_o = 1.0000$;
$a_1 = -0.9630$; $a_2 = 0.0926$; $a_3 = 0.0471$

In this case the residuals were found to be white at this point, thus removing the necessity for GLS parameter estimation. However to obtain 'better' estimates by means of a tenth order autoregressive model, the following were achieved.

GLS iteration	b_o	b_1	b_2
1	0.6167	0.5878	0.0814
2	0.6164	0.5832	0.0783
3	0.6161	0.5797	0.0755

GLS iteration	a_0	a_1	a_2	a_3
1	1.0000	-0.9691	0.0964	0.0464
2	1.0000	-0.9754	0.1035	0.0455
3	1.0000	-0.9813	0.1103	0.0446

At each iteration the residuals remained white.

(d) Model Order Testing:

First, second and third order models have been found which approximate to the 'unknown' system. Within the confines of a particular model order, several different models have been obtained by the use of Least Squares Estimation or by a certain number of iterations of the GLS

procedure.   It is now desirable to choose one of the models
obtained as being the 'best', in terms of simplicity as well
as how closely it approximates the plant. For this reason
the sum of squares cost function described previously is
reconsidered as:

$$S_n = NV_n + NM_n$$

in which S_n is the sum of squares of the error terms for an
nth order model, N is the number of data points considered
(500), V_n is the variance and M_n the mean of the error
sequence. Note that this is not a different value of S_n, it
is in fact just another way of writing S_n as described.

    A comparison will be made in the first instance of the
three models obtained by the simple application of Linear
Least Squares.   From the error data obtained the following
table can be drawn up:

Model Order (n)	Mean (M_n)	Variance (V_n)	S_n (LLS)
1	0.00585	1.99	995
2	-0.000956	0.0677	33.85
3	-0.000962	0.0655	32.75

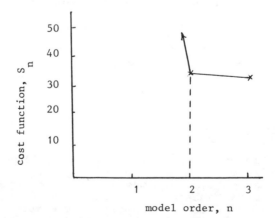

Fig. 7.5: LLS model order test

Figure 7.5 shows S_n plotted against model order (n) for the Linear Least Squares case. It is obvious that the slope changes greatly where n = 2, indicating that a model of order 2 is the best choice, as was expected.

A comparison can also be made in terms of the error sequence obtained from the final (third) iteration of the GLS technique, which gives the following results:

Model Order (n)	Mean (M_n)	Variance (V_n)	S_n (GLS)
1	-0.0117	1.51	755
2	-0.000918	0.0646	32.3
3	-0.000896	0.0643	32.15

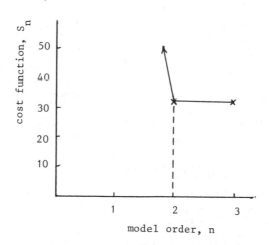

Fig.7.6: GLS model order test

Figure 7.6 shows S_n plotted against model order (n) for the GLS case. Once again it is obvious that the slope changes greatly where n = 2, confirming the choice of a second order model, as before. The values for S_n obtained for the GLS procedure are however consistently lower than the corresponding values found for the LLS case. Hence the model chosen should be of order 2, with parameters equal to those calculated in the final GLS iteration, thus minimizing error.

(e)   Remarks on the experiment:

In consideration of the second order model, although the residuals were eventually white, the parameter estimates altered drastically at each GLS iteration. Thus one would expect the error between the estimated parameter values and their 'true' values to be large. In fact the sum of squares model order test gave figures for S_n which were greater than those obtained for the higher order models.

With the second order model, although the Least Squares Estimate gave biased parameters, the residuals were white after only one GLS iteration. After three iterations the parameter estimates were nicely converging to their 'actual' values and in fact by taking each parameter to two decimal places, identical figures are achieved.

For the third order model the Least Squares Estimation procedure resulted directly in white residuals. Despite this, convergence to actual values was rather slow and may probably have led to biased estimates.

Finally, with the models and data values employed, the mean of the residuals in each case had negligible effect on the sum of squares S_n, and this therefore became almost entirely dependant on the variance of the residuals.

7.8   RECURSIVE PARAMETER ESTIMATION

Assuming that the structure of the plant to be controlled is either known or can be found, the objective of the identification exercise is then to find values for the parameters within a model of similar structure to the plant, such that it behaves as nearly as possible like the real plant for a set of input signals.

In many practical situations though, variations of the plant parameters occur. Where parameter variations are small in magnitude or take place over a long period of time, the controller operating on the plant need only be retuned occasionally to ensure reasonable performance and this is the standard procedure carried out on many industrial PID controllers. However, the variations can be large in magnitude and occur quite rapidly, and almost always they are unpredictable. In these cases it is often necessary and/or desirable to modify the controller in line with the changes that have occurred in the plant, thus providing an adaptive control scheme. The controller modifications need an updated version of the system model and hence new estimated parameters are required at any time instant, k.

In this section a form of the Least Squares parameter estimation procedure is considered such that the parameters in a model of the plant are continually (recursively) updated. A basic requirement of the recursive algorithm is that the information stored be only of a finite amount. Hence, once the data is of a certain age, i.e. once a number of time periods have elapsed, it is automatically given the boot. At any one time instant therefore, a window of past and present input/output data is taken into account. Indeed this is true for recursive forms of all the identification techniques previously considered.

In order to further consider the Recursive form of Least Squares (RLS) parameter estimation it must be remembered that the system output at time instant k can be written as:

$$y(k) = \theta^T x(k) + e(k)$$

in which the parameter vector $\theta^T = (\theta_1, \theta_2, \ldots, \theta_M)$ and the data vector $x^T(k) = (x_1(k), x_2(k), \ldots, x_M(k))$. Also, let the vector of parameter estimates obtained at time instant k be denoted by $\tilde{\theta}$, such that at that instant the model prediction error is:

$$\epsilon(k) = y(k) - \tilde{\theta}^T(k-1) \, x(k)$$

The idea behind this is that if the estimated parameter vector $\tilde{\theta}$ is very close to its true value θ, the error $\epsilon(k)$ should be very small. A better vector $\tilde{\theta}$ can however be obtained by taking into account the magnitude of difference, i.e. just how large $\epsilon(k)$ is, such that a new version of $\tilde{\theta}$ can be found from:

$$\tilde{\theta}(k) = \tilde{\theta}(k-1) + K(k)\epsilon(k)$$

where $K(k)$ can be chosen to improve the estimated values, hopefully by means of an optimal path. The error between the estimated and true parameter values is then defined as:

$$\lambda(k) = \Theta - \tilde{\Theta}(k)$$

such that

$$\lambda(k) = \{I - K(k) \ x^T(k)\} \ \lambda(k-1) - K(k)e(k)$$

in which $\lambda(k)$ is a column vector. If $K(k)$ is chosen to contain elements of high magnitude, although this results in rapid updating it also leads to the disturbance term $e(k)$ causing large pertubations. A choice of $K(k)$, the Kalman gain, to contain elements of small magnitude, results in better noise rejection but a slow response in terms of parameter updating. In practice $K(k)$ is selected, in terms of the covariance matrix of the error, as a time varying gain.

Let the covariance matrix formed by the parameter estimates at time instant k, be $P(k)$ where:

$$P(k) = \xi\{\lambda(k), \ \lambda^T(k)\}$$

The elements of the covariance matrix are then minimized by choosing the Kalman gain vector as:

$$K(k) = \frac{P(k-1)x(k)}{\beta + x^T(k)P(k-1)x(k)}$$

which means that we also have:

$$P(k) = [I - K(k) \ x^T(k)] \ P(k-1)$$

where β is a scalar gain factor which can be made equal to σ^2 the variance of the disturbance, if this is known. The RLS parameter estimator then consists of three equations to be carried out every time instant k, these are the equations for $P(k)$ and $K(k)$ given directly above, along with the equation for $\tilde{\Theta}$.

It is common practice with many adaptive controllers to use β as an exponential forgetting factor such that new data is relatively important and its importance declines exponentially. For this purpose the value for $P(k)$ must also be divided by β. However where the data is not persistently exciting enough the algorithm may 'blow up', i.e. the parameter estimates tend to infinity, therefore the factor β can be made variable and dependant on the prediction error in order to 'wake up' the estimator when necessary.

A final point to note is that the equations given for the update of $K(k)$ and $P(k)$ may result in divergent values because of numerical problems. Hence, it is best to use a numerically stable updating method such as square root extraction or UD factorization in order to achieve a working algorithm.

7.8.1 A Program for Recursive Least Squares Estimation

In this section a FORTRAN program is described which can be employed as a subroutine to carry out Recursive Least Squares Parameter Estimation with UD factorisation. The program follows that given in Clarke and Gawthrop (1979) and requires the following information to be input:

Y -	New system output value
X(1) to X(N) -	Past data vector: previous system input and output signals
FF -	Value of forgetting factor: can be constant e.g. 0.98,or time varying:- see Warwick (1986) for a further discussion.

The routine gives as its output:

THETA(1) to THETA(N) -	The set of estimated parameters

Initial values are also required for the following:

N -	The number of parameters to be estimated
THETA(1) to THETA(N) -	Initial parameter estimates (anything) but zero
U -	Unit upper triangular matrix (zero for all elements will do)
D -	Diagonal matrix (If uncertain try 1000)

Note also that PE is the error between the actual and the predicted outputs. The program is then as follows:

```
          PE = Y
          DO 100 I = 1,N
100       PE = PE - THETA(I)*X(I)
          R = X(1)
          V = D(1)*R
          K(1) = V
          S = 1.0 + V*R
          D(1) = D(1)/S
          IF (N.EQ.1) GO TO 500
          T = 0
          W = 0
          DO 400 J = 2, N
          R = X(J)
          DO 200    I = 1, J - 1
          T = T + 1
200       R = R + X(I)*U(T)
          V = R*D(J)
          K(J) = V
          L = S
          S = L + V*R
          D(J) = D(J)*L/S/FF
          P = - R/L
          DO 300 I=1, J-1
          W = W + 1
          Z = U(W) + K(I)*P
          K(I) = K(I) + U(W)*V
300       U(W) = Z
400       CONTINUE
500       CONTINUE
          DO 600 I = 1,N
600       THETA(I) = THETA(I) + PE*K(I)/S
          END
```

It must be remembered that:

a) if being used for a subroutine, vectors and matrices must be dimensioned within the routine; also a RETURN command is required before END.
b) The past data vector X(I) must be updated somewhere else in the program i.e. X(1) → X(2) etc.

7.8.2 A Simplified Recursive Estimator

Computation time necessary to complete a recursion of a parameter estimation procedure can, in many cases, provide the vast majority of the total necessary time, the remainder being primarily for controller parameter calculation. This is especially true when the system is of high order, resulting in a large number of parameters to be estimated. Problems can therefore occur if computing power is limited, a low cost implementation is desired or, perhaps most importantly, a high sampling frequency is required. This latter point is particularly significant in terms of robot manipulator control, where it would be rash to not consider applying certain control strategies because of the time taken to carry out parameter estimation. It seems sensible because of these factors to attempt to reduce the computing time for parameter estimation , where possible, if no loss of estimator accuracy or convergence rate results. In this section a much simplified form of the RLS estimator is briefly introduced.

It is normally the case with recursive estimators that on start up the elements of the covariance matrix are given values such that all terms on the diagonal are positive and fairly large (~100) whereas all off-diagonal terms are set to zero. Under normal conditions the diagonal terms reduce considerably as the parameters converge, and the off-diagonal terms become of very small magnitude. However, if the assumption is made that the diagonal retains enough information, the covariance matrix $P(k)$ may be approximated by a simple diagonal matrix, hence resulting in a considerable reduction in computational effort. The rest of the simplified algorithm is then as for that described in the previous section.

7.9   CONCLUSIONS

An attempt has been made to bring together the main topics and methods involved in linear system identification. In general good, and often simple techniques exist, the final choice of method being dependant on the problem for which the model is to be employed.

The exposition was based largely on the theory behind Linear Least Squares parameter estimation. This was because of the fewer number of steps involved in this particular mathod and its essentially straightforward relationship with the response data. However, because of the possibility of bias in the Least Squares parameter estimates, other methods which result in unbiased estimates were also considered, i.e. GLS, Recursive Maximum Likelihood. The identification schemes discussed here are by no means the only ones available, many more functional identification schemes exist, e.g., prediction error methods, however they are less commonly encountered in practice and hence were not considered.

Although a very simple transfer function model was considered in the identification example, application of the theory given was shown. Firstly if correlated noise is present, straightforward Linear Least Squares parameter estimation gives biased results and secondly if the correct model order has been chosen, the Generalised Least Squares technique results in parameter estimates which converge rapidly to their true values. However if an incorrect model is chosen large errors can be apparent, hence the need for model order testing in order to select the preferred number of parameters.

Recursive estimation schemes were considered, via the example of Recursive Least Squares, and these are extremely useful for dealing with systems which contain parameters which vary slowly with respect to time. However, without modification such algorithms are inappropriate for direct use due to the swamping of the new, more important, data by old data. The old data can though be discounted by the use of an exponential window in which the effect of new data is recognised exponentially. It is also possible for adaptive control algorithms to be made more robust in order to cope with plant non-linearities or sudden parameter variations. In these cases it is imperative that the parameter estimator responds rapidly to the change in plant conditions, this can often be achieved or aided by the inclusion of a lively variable forgetting factor.

REFERENCES

1.   Goodwin, G.C. and Payne, R.L.: 'Dynamic system identification', Academic Press, 1977.

2.   Eykhoff, P.: 'System identification', John Wiley and Sons, 1974.

3.   Astrom, K.J., Borisson, U., Ljung, L. and Wittenmark, B.: 'Theory and application of self-tuning regulators', Automatica, Vol. 13, pp. 457-476, 1977.

4.   Box, G.E.P. and Jenkins, G.M.: 'Time series analysis, forecasting and control', Holden-Day, 1970.

5.   Denham, M.J. and Abaza, B.A.: (revised by Bloomer, D.C.J. and Rigby, L.) 'System identification package - SYSID', Department of Computing and Control, Imperial College, London, Publication No. 75/28, 1975.

6.   Sage, A.P. and Melsa, J.L.: 'System identification', Academic Press, 1971.

7.   Clarke, D.W.: 'Generalised least squares estimation of the parameters of a dynamic model', Proc. IFIP Symposium on Identification in Automatic Control Systems, Prague, Paper 3.17, 1967.

8. Barker, H.A.: 'Choice of pseudo random binary signals for system identification', Electronics Letters, Vol. 3, pp. 524-526, 1967.

9. Farsi, M., Karam, K.Z. and Warwick, K.: 'A simplified recursive identifier for ARMA processes', Electronics Letters, Vol. 20, pp. 913- 915, 1984.

10. Papoulis, A.: 'Probability, random variables and stochastic processes', McGraw-Hill.

11. Astrom, K.J. and Eykhoff, P.: 'System identification - a survey', Automatica, Vol. 7, pp. 123-162, 1971.

12. Young, P.C.: 'Recursive estimation and time-series analysis', Springer-Verlag, Berlin, 1984.

13. Ljung, L. and Soderstrom, T.: 'Theory and practice of recursive identification': MIT Press, 1983.

14. Moroney, M.J.: 'Facts from figures', Penguin, 1977.

15. Warwick, K., Farsi, M. and Karam, K.Z.: 'A simplified pole placement self-tuner', Proc. Control 85, Cambridge Univ., 1985.

16. Warwick, K: 'Recursive methods in identification', in Signal processing for control, eds. Jones R.P. and Godfrey K., Springer-Verlag, 1986.

17. Clarke, D.W. and Gawthrop, P.J.: 'Implementation and application of microprocessor based self-tuners', Proc. 5th IFAC Symposium on Identification and System Parameter Estimation, Darmstadt, 1979.

Chapter 8

Introduction to discrete optimal control

R.A. Wilson

8.1 Introduction

This chapter introduces the principles of optimal
control for discrete systems. The optimisation task is
quantified, and the general optimal control problem stated
algebraically. The problem is likened to a route planning
exercise, and the Principle of Optimality proved; the
Dynamic Programming method is applied to find the best route,
so suggesting a solution method for the optimal control
problem.

The method is then applied to the particular case of a
linear system with a quadratic performance cost function;
the solution is shown to necessitate state feedback and
require the solution of a matrix Riccati equation (MRE). A
method of solving the MRE in the regulator case is presented.

8.2 The General Problem

The general design problem encountered in classical
control theory is to select a controller that endows the
complete (closed loop) system with certain attributes, as
listed in a performance specification.

Open loop margins, closed loop bandwidth and damping
factor, overshoot, rise time and settling time are typical
elements of such a specification (some of which are closely
interrelated). In many cases, however, these elements are
derived using our experience (and intuition?) from more
general ideas such as 'the final temperature is to be
achieved quickly' or 'the radar is to acquire the new target
within 1 second' or, in a more commonplace case 'the dart is
to land in the double top'.

In these examples the time response is being addressed
directly, and there is an implied cost of failing to achieve
the goal; a low temperature may cause loss of product, the
inability to acquire the target may cost lives, and a missed
double top may cost a whole round of drinks! The cost
function and its reduction to a minimum value is thus fundamental to the system design; the optimal control problem is:
given a system, and a cost function, determine a controller
for the system that minimises the cost. Note that equally
well we might occasionally wish to maximise the 'profit' from
some systems; in this case we can minimise '-profit'.

8.3   The System and Cost Functions

The systems we are interested in are discrete and, since the system response in the time domain is of importance, we naturally use the state description. In the general case we propose to use:

$$\text{system } \Sigma : \quad \underline{x}_{k+1} = \underline{f}_k(\underline{x}_k, \underline{u}_k, k)$$

and where necessary note that the output $\underline{y}_k = \underline{g}_k(\underline{x}_k, \underline{u}_k, k)$.

Here $\underline{x}_k$ is the state at (sampling) instant k given by $t = kT$, where T is the sampling period.

$\underline{u}_k$ is the control signal at 'time' k.

and $\underline{f}_k$ is the functional relationship between future and present state.

The cost function is also a function of state (temperature, target motion, miss distance) and in very many cases we will be concerned at the cost of the fuel, energy etc used in achieving the goal; the cost may also be 'time' dependent and so we generalise the cost function to

$$\text{cost } \underline{\mathfrak{c}}: \quad J_N = \sum_{k=0}^{N} j_k(\underline{x}_k, \underline{u}_k, k)$$

Here the summation extends over the full range of k values of interest - often $N \to \infty$ is the case, especially in regulating systems. The variation of the function j with $_k$ reflects the need to perhaps change the cost function structure; for example we might wish to restrict fuel usage once the state is close to the target value, and this might be accomplished by changing the weighting against $\underline{u}_k$ in j_k.

The general optimal Control Problem is then:

$\mathbb{P}$:   find $\underline{u}_k^o$, k = 0, 1, --N-1 that minimises J_N

subject to the system constraints Σ and to any given initial (and/or final) conditions on $\underline{x}$.

8.4   The Route Planning Problem

In approaching the solution of $\mathbb{P}$ we remind ourselves that, if we were given $\underline{u}_k^o$, then the state equations Σ allow the determination of the optimal path (or trajectory) of the state, ie. $\underline{x}_k^o$, say. Inversely, then we might ask that the path minimises J_N - we now have a 'route-planning' problem, which is perhaps a rather more familiar idea to us.

Stated algebraically, we are to find the route $\underline{x}_k^o$, k = 0, 1 -- N such the cost of the route, J_N is minimised; we identify j_k as the cost of that part of the route from $\underline{x}_k^o$ to $\underline{x}_{k+1}^o$, and call it the kth stage cost. Note that J_{N-1} is the cost from $\underline{x}_{N-1}^o$ to $\underline{x}_N^o$ and that we often then incur a different final cost $j_N(\underline{x}_N^o)$, there being no further action to take (ie. $\underline{u}_N$ is irrelevant, and not needed).

Using these ideas, we visualise P as a journey along a map such as Figure 8.1; the objective is to travel from A to B, incurring the least cost. Sometimes too, there may be several B's, any one of which will be acceptable (they all serve the same beer!).

We can now find the optimal ("cheapest") route; we travel all possible paths from A to B, working out their costs and finally choose the cheapest route as the one with least cost. Such is the <u>enumeration method</u>; if there are N stages, n states and m control actions possible at each state, then there are nm^N paths to test! There is surely a better way.

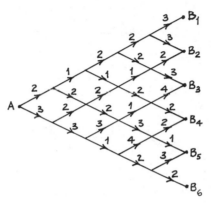

How many possible routes are there? Which is the shortest? The longest? Which is the shortest route to B_4?

Figure 8.1 An optimal control problem
         arranged as a route finding
         problem.

8.5   The Principle of Optimality

In Figure 8.2 consider a point C on a possible path and suppose we already know the <u>best</u> route from C to B, and that this route passes through D. Then, the best route from A to B cannot be A to C to E to B because such a result would contradict the optimality of the section C-D-B.

This apparently obvious statement is an example of the Principle of Optimality: <u>any subsequent part of an optimal path is in itself optimal (under the same conditions</u>).

Figure 8.2 Illustrating the Principle of
         Optimality.

8.6   Dynamic Programming

The Principle of Optimality now allows the solving of the route planning problem - the method is known as Dynamic Programming. Essentially we travel the route backwards! In Figure 8.3, if at stage k we are at C_{ik} then we already know the optimal routes from the C_{ik} to the end B (we just worked them out) and now we want to decide what to do at stage k-1 from point C_{k-1}. This is a simple problem - just evaluate the cost of each route from C_{k-1} and then add the cost from C_{ik} to the end B. The best route from stage k-1 is the one which has lowest total cost to the end.

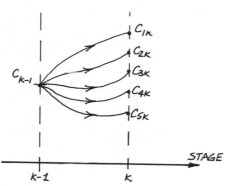

Figure 8.3 The basic idea of Dynamic Programming.

We can 'mechanise' this approach by marking each map node with (a) the cost to the end from that node and (b) the optimal thing to do at that node (up, down, left, right, etc, etc). The node values are evaluated from end to beginning, and a few simple examples show us that the number of calculations to be made is very much smaller than in the enumeration method. (Figure 8.4)

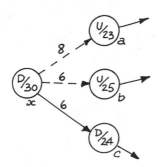

The routes from a, b, and c are already marked. We can thus choose the move from x.

Figure 8.4 Annotating the nodes of a route.

8.7   Embedding and Constraints

It is apparent that in these route planning problems, the solution is the decision (control) at each point (state or stage)-the optimal route. However, we have also worked out the best route to the end from any point in the map i.e. the solution embeds all the sub-problems.

Note also that a constraint such as 'we must go via M' (the darts team captain's house!) can actually simplify the problem, Figure 8.5, since other states at that stage can be immediately ignored in our calculations.

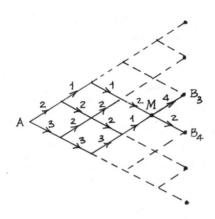

Figure 8.5 The problem of Figure 8.1 when the
route is constrained to pass through
a point such as M.

The effect of constraints immediately arises in the optimal control problem; the state equations constrain the possible succeeding state.

Similarly, the control u may be constrained within given bounds, though this is not often tackled directly, but rather the possible violation of such bounds is examined after the unconstrained problem has been solved.

We are now in a position to solve the optimal control problem by the method of Dynamic Programming.

3.8   The Control Problem and DP

Above we proposed the use of Dynamic Programming to solve the optimal control problem:

given Σ : the discrete system $\underline{x}_{k+1} = \underline{f}_k(\underline{x}, \underline{u}_k, k)$

find $\underline{u}_k$, $k = 0, 1 \ldots N-1$ so that the cost function,

$$q: \quad J = \sum_{k=0}^{N} j_k(\underline{x}_k, \underline{u}_k, k)$$

is minimised.

We attack this problem by starting from the end; the optimal) minimum cost at stage N must be

$$J^o(\underline{x}_N, N) = j_N(\underline{x}^o_N, N)$$

since a control $\underline{u}_N$ is not available to us, and arriving at $\underline{x}^o_N$ determines $J^o(\underline{x}_N, N)$.

t the previous stage

$$J_{N-1} = J(\underline{x}_{N-1}, N-1) = j_{N-1} + J(\underline{x}^o_N, N)$$

$$= \underbrace{j_{N-1}(\underline{x}_{N-1}, \underline{u}_{N-1}, N-1)}_{\text{stage cost}}$$

$$+ \underbrace{j_N(\underline{f}_{N-1}(\underline{x}_{N-1}, u_{N-1}, N-1), N)}_{\text{cost to end}}$$

nd this cost is minimised by the choice u_{N-1} given by ifferentiating (if this is possible). Thus

$$\frac{\partial J_{N-1}}{\partial \underline{u}_{N-1}} = 0 \quad \text{gives } \underline{u}^o_{N-1} \quad \text{and hence } J^o_{N-1}.$$

imilarly, at the N-2 stage

$$J_{N-2} = j_{N-2} + J_{N-1} \quad \text{and} \quad \frac{\partial J^o_{N-2}}{\partial \underline{u}_{N-2}} = 0 \text{ gives}$$

$\underline{u}^o_{N-2}$ which gives J^o_{N-2} and so on.

lus we can find $\underline{u}_k^o$ stage by stage by solving

$$J^o_k = \underset{\underline{u}_k}{\text{Min}} \left[j_k + J^o_{k+1} \right]$$

8.9   The Linear Quadratic Performance Problem

Obviously we won't get very far analytically with such a general problem; in fact very few problems yield a completely analytic solution.

The linear system quadratic performance index (LQP) problem does yield somewhat. The problem is by no means unrealistic; the linear system may be a small signal model, and the quadratic PI is a useful practical form:

$$\Sigma: \quad \underline{x}_{k+1} = F\underline{x}_k + G\underline{u}_k$$

$$\mathcal{G}: \quad J_N = \tfrac{1}{2} \sum_{k=0}^{N-1} (\underline{x}_k^T Q\underline{x}_k + \underline{u}_k^T R\underline{u}_k) + \tfrac{1}{2}\underline{x}_N^T Q_N \underline{x}_N$$

Here we note the terminal cost (Q_N), that both Q and R are symmetric (and may be indexed, as Q_k) and finally that the the time varying case (F_k, G_k) is also covered by the following analysis. The $\tfrac{1}{2}$'s reduce some arithmetic.

8.9.1 Scalar version. Let us firstly deal with a scalar problem;

$$\sigma: \quad x_{k+1} = fx_k + gu_k$$

$$\mathcal{L}: \quad j_N = \tfrac{1}{2} \sum_{k=1}^{N-1} (qx^2{}_k + ru^2{}_k) + \tfrac{1}{2}q_n x^2{}_n$$

Using Dynamic Programming, we have:

At k = N we are forced to accept the cost of the state we arrive at

$$J_N^o = \tfrac{1}{2} q_N x^2{}_N$$

(there is no choice of u_N available).

Then, at N-1: $J_{N-1} = \tfrac{1}{2}(qx^2{}_{N-1} + ru^2{}_{N-1}) + J_N$

Substituting from J_N, above, and using the state equation to expand x_N:

$$J_{N-1} = \tfrac{1}{2}(qx^2{}_{N-1} + ru^2{}_{N-1}) + \tfrac{1}{2}q_N(fx_{N-1} + gu_{N-1})^2$$

Differentiation is possible here and thus

$$\frac{\partial J_{N-1}}{\partial u_{N-1}} = ru_{N-1} + q_N(fx_{N-1} + gu_{N-1}) = 0 \text{ for a minimum}$$

(maxima do not occur because J is quadratic!).

Thus the optimal control $u_{N-1}^o = \dfrac{-q_n f\, x_{N-1}}{r + q_N g} \triangleq -d_{N-1}\, x_{N-1}$ say

Thus, the optimal control is state feedback - an important result.

Substituting u^o_{N-1} back in J_{N-1} gives

$$J^o_{N-1} = \tfrac{1}{2}(qx^2_{N-1} + r\, d^2_{N-1}\, x^2_{N-1}) + \tfrac{1}{2}q_N(f - gd_{N-1})\, x^2_{N-1}$$

$$= \tfrac{1}{2}\, p_{N-1}\, x^2_{N-1}, \text{ quadratic in } x_{N-1}, \text{ just as } J_N$$

was in x_N.

Similarly, at the previous stage, $N-2$:

$$J_{N-2} = \tfrac{1}{2}(qx^2_{N-2} + ru^2_{N-2}) + J_{N-1}$$

$$= \tfrac{1}{2}(qx^2_{N-2} + ru^2_{N-2}) + \tfrac{1}{2}p_{N-1}(fx_{N-2} + gu_{N-2})^2$$

from the state equations and from the stage $N-1$ result.

Differentiating:

$$\frac{\partial J_{N-2}}{\partial u_{N-2}} = 0 \text{ gives } u_{N-2} = \frac{p_{N-1}\, f}{r + p_{N-1}g}\, x_{N-2} \overset{\Delta}{=} -d_{N-2}\, x_{N-2}$$

(again state feedback)

and eventually $J^o_{N-2} = \tfrac{1}{2}p_{N-2}\, x^2_{N-2}$, again quadratic.

We can continue in this fashion; at each stage we find that $u^o_k = -d_k x_k$ and $J^o_k = \tfrac{1}{2}p_k x^2_k$, which suggests that the same will be true in the multivariable case.

8.9.2 The Multivariable Case. Let us therefore propose that the MV case has solution:

$$\underline{u}^o_k = -D_k \underline{x}_k$$

with $J^o_{k+1} = \tfrac{1}{2}\underline{x}^T_{k+1} \cdot P_{k+1} \cdot \underline{x}_{k+1}$.

Since J^o_{k+1} is assumed quadratic then P_{k+1} is symmetric.

Consider the k th stage:

$$J_k = \tfrac{1}{2}(\underline{x}^T_k Q \underline{x}_k + \underline{u}^T_k R \underline{u}_k) + J_{k+1}$$

$$= \tfrac{1}{2}(\underline{x}^T_k Q \underline{x}_k + \underline{u}^T_k R \underline{u}_k) + \tfrac{1}{2}\underline{x}^T_{k+1} P_{k+1} \underline{x}_{k+1}, \text{ from}$$

the quadratic assumption

$$= \tfrac{1}{2}(\underline{x}^T_k Q \underline{x}_k + \underline{u}^T_k R \underline{u}_k)$$

$$+ \tfrac{1}{2}(F\underline{x}_k + G\underline{u}_k)^T P_{k+1}(F\underline{x}_k + G\underline{u}_k),$$

from the state equations at k.

Minimising:

$$\frac{\partial J_{k-1}}{\partial \underline{u}_{k-1}} = R\underline{u}_k + G^T P_{k+1} G\underline{u}_k + G^T P_{k+1} F\underline{x}_k = 0$$

gives the optimal control as:

$$\underline{u}_k^o = -\left[R + G^T P_{k+1}\right]^{-1} G^T P_{k+1} F \ \underline{x}_k \quad \triangleq \ -D_k \underline{x}_k$$

and $\quad J_k^o = \frac{1}{2} \underline{x}_k^T (Q + F^T P_{k+1} G \left[R + G^T P_{k+1} G\right]^{-1} G^T P_{k+1} F)\underline{x}_k$

after some manipulation!

Since, by definition $J_k^o = \frac{1}{2}\underline{x}_k^T P_k \underline{x}_k$ then P_k satisfies

$$P_k = Q + F^T P_{k+1} F - F^T P_{k+1} G \left[R + G^T R_{k+1}\right]^{-1} G^T P_{k+1} F,$$

the discrete matrix Riccati Equation.

By construction, we can see that $P_N - Q_N$ and that P_k is symmetric.

8.9.3 The Closed loop.

The MRE can alternatively be written:-

$$P_k = Q + D_k^T R D_k + (F - GD_k)^T P_{k+1} (F - GD_k) \tag{8.1}$$

which is useful on the computer, and also highlights the fact that the closed loop is:

$$\underline{x}_{k+1} = F\underline{x}_k + G\underline{u}_k = F\underline{x}_k + G(-D_k\underline{x}_k)$$

ie. $\quad \underline{x}_{k+1} = (F - GD_k)\underline{x}_k$

which generally stabilises the system. Note that all the above is still true in the time varying case

$$(F = F_k, \ G = G_k, \ Q = Q_k, \ R = R_k)$$

8.9.4 The Regulator Problem.

An interesting sub-problem is the infinite horizon case, when it can be shown that

$$P_k \succ \overline{P} \text{ as } N-k \to \infty .$$

Then $\overline{P}$ satisfies:

$$\overline{P} = Q + F^T \overline{P} F - F^T \overline{P} G (R + G^T \overline{P} G)^{-1} G^T \overline{P} F \tag{8.2}$$

and

$$D_k \to \overline{D} = (R + G^T \overline{P} G)^{-1} G^T \overline{P} F \tag{8.3}$$

Note that since J must be finite (to be minimisable) then $\overline{P}$ is finite (strictly $\overline{P} > 0$ and is UNIQUE!), $\underline{x}_k \to 0$ as $k \to \infty$ and the closed loop $F - G\overline{D}$ is asymptotically stable.

8.10   Solving the Matrix Riccati Equation

We can obviously find $\bar{P}$ by solving the original MRE (backwards) from (any) P_N and waiting for P_k to become constant at $\bar{P}$; better is to use Kleinmann's Algorithm:

1) Choose any D_1 such that $F_1^+ = F - G D_1$ is stable

   ($D_1 = 0$ may suffice if F is stable) and set i = 1 (This feedback minimises some unknown LQP!)

2) Calculate $F_i^+ = F - GD_i$

$$Q_i^+ = Q + R_i^T RD_i$$

   and solve $S_i = Q_i^+ + (F_i^+)^T S_i F_i^+$ for S_i,   (8.4)

   which is the MRE of (8.1) as though $P_i = S_i$.

3) Treat S_i as optimal (like P_i) and set

$$D_{i+1} = (R + G^T S_i G)^{-1} G^T S_i F$$

4) If D_{i+1} is not constant, set i = i+1, go to (1).

Under reasonable conditions this algorithm produces a sequence of S_i's that decrease towards $\bar{P}$ (since the D_i's are not optimal, the S_i's must be $> \bar{P}$), and hence $D_i \rightarrow \bar{D}$ as i increases. [Note that (8.4), the Lyapunov equation, is linear in S_i and fairly easy to solve].

REFERENCES

.   FRANKLIN, G.F. and POWELL, J.D.: "Digital control of dynamic systems", Addison Wesley, 1980

.   KUO, B.C.: "Digital control systems", Holt, Rinehart and Winston, 1980.

.   BRYSON, A.E. and HO, Y.C.: "Applied optimal control", Blaisdell, 1969 (2nd Edition, 1975)

.   DENN, M.M.: "Optimization by variational methods", McGraw-Hill, 1969.

.   McCAUSLAND, I.:   "Introduction to optimal control" Wiley, 1969.

Chapter 9

Multivariable control

Dr. I. Postlethwaite

9.1 INTRODUCTION

These notes have been written to supplement two
lectures on multivariable control given at the IEE Vacation
School on Industrial Digital Control Systems held annually
at Oxford.   Accordingly the notes have been divided into two
Parts.

In Part 1, an introduction to multivariable control is
given.   It begins with an example – a helicopter model – to
illustrate what is meant by a multivariable system and how
we model such systems.   Next we discuss the primary
objectives of control and present a feedback configuration
commonly used to achieve them.   The bulk of Part 1 is then
taken up with the presentation of analysis techniques for
assessing the significant properties of a multivariable
control system design.   Bode diagrams of singular values are
seen to have a similar role in characterising feedback
properties of multivaribable systems as the Bode magnitude
diagram does for scalar systems.

In Part 2, we look directly at the design problem
and outline some of the more popular design methods
available for multivariable systems.   Some industrial design
examples will be described in the lecture but there is
insufficient space to cover them properly in this chapter.
Case-studies are included elsewhere in the book.

The digital implementation of controllers is not
covered in this chapter and therefore the presentation will
be for continuous systems only, which is the usual framework
for presenting analysis and design of multivariable systems

PART 1: AN INTRODUCTION TO MULTIVARIABLE CONTROL

9.2 A MULTIVARIABLE EXAMPLE: THE HELICOPTER

We may consider a multivariable plant to be a system in
which there is more than one output to be controlled and

more than one control input to manipulate. Furthermore, a signal on any one of the inputs produces a response on more than one of the outputs; that is, the inputs and outputs are interactive.

A helicopter is an example of a multivariable sytem. Typically it has the following control inputs which are used by the pilot to produce the required forces and moments:

1. Collective pitch of main rotor   (rads)
2. Longitudinal cyclic pitch   (rads)
3. Lateral cyclic pitch   (rads)
4. Tail rotor collective pitch   (rads)
5. Throttle/fuel control for engine

If we assume the rotor is working at a fixed speed and ignore the throttle as an input we have four inputs to control four outputs which might be:

1. Total speed   (ft/sec)
2. Descent angle   (rads)
3. Sideslip angle   (rads)
4. Bank angle   (rads)

A mathematical model relating the helicopter's inputs and outputs will generally be nonlinear but for the purpose of controller design lineariztions are often obtained about operating points. Several linear controllers can then be designed for these equilibria and gain scheduling used to switch from one controller to the next. Sometimes it is possible to design a fixed gain controller which is robust enough to perform satisfactorily over many operating points. If significant nonlinearities are present, as in the helicopter, an important part of the design process involves simulation of the overall nonlinear control scheme. These lectures will be concerned with the analysis and design of <u>linear</u> control systems.

On linearizing the nonlinear equations of motion for the helicopter the linear model can be put into state-space form

$$\dot{x} = Ax + Bu \qquad\qquad (9.1)$$

$$y = Cx \qquad\qquad (9.2)$$

where u and y are the vectors of inputs and outputs respectively, and x is the state vector describing the internal dynamics of the system. The A, B, C matrices for a lynx helicopter operating at 100 knots, straight and level are shown in Appendix 1 courtesy of the Royal Aircraft Establishment, Bedford. By taking Laplace transforms of (9.1) and (9.2.) we obtain the following input-output description of the plant.

$$y(s) = G(s) u(s) \qquad\qquad (9.3)$$
where

$$G(s) = C(sI - A)^{-1}B \qquad (9.4)$$

To illustrate the severe interaction properties of the helicopter the responses of the linear model of appendix 1 to a 0.1 radian step input in the collective pitch of the main rotor are shown in Figs. 9.1 and 9.2. The responses indicate that the uncontrolled helicopter is an unstable system.

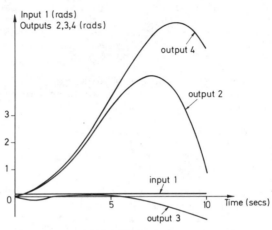

Fig 9.1

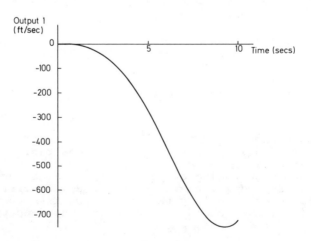

Fig 9.2

9.3 CONTROL OBJECTIVES

Control systems are designed to "perform well" but it is not always possible to state precisely what the control objectives are. Nevertheless, in any reliable design most (if not all) of the following objectives will have been influential in its development:

* Stability
* Low sensitivity
* Disturbance rejection
* Noise mitigation
* Command following/low interaction
* Integrity
* Robustness of the above in the presence of uncertainty

Qualitatively, these design objectives can be described as follows:

Stability. Generally, a system is said to be stable if bounded inputs produce bounded outputs.

Low sensitivity. Sensitivity refers to the effects that small perturbations have on system response.

Disturbance rejection. This is the ability of a control system to reject unwanted signals such as load disturbances from the outputs.

Noise mitigation. This is the ability of a control system to reduce the effect on the outputs of sensor noise (typically high frequency signals).

Command following/low interation. This refers to the accuracy with which the output signals follow the reference inputs.

Integrity. A control system is said to have integrity if it can sustain component failures, for instance in the actuators or sensors, without becoming unstable.

Robustness. Because of unavoidable uncertanties in a nominal plant model used for design purposes it is important that the above objectives are maintained for a representative set of plant models in the neighbourhood of the nominal model. A control system which possesses this property is called robust.

All of these objectives except command following /low interaction require the use of feedback control and are therefore commonly referred to as feedback properties.

9.3.1   A Typical Feedback Configuration

The major difficulties in control system design can be attributed to the presence of uncertainty: uncertaity in th mathematical models used to represent the process and uncertainty in disturbance signals which impinge on the system.   It is well known that feedback can result in a satisfactory   control system despite these uncertainties but that its design is often a demanding taste.   For this reason feedback design is of fundamental importance to control engineers.   In this chapter we will consider the feedback configuration shown in Fig. 9.3.

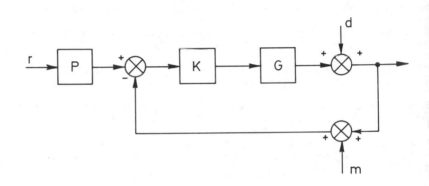

Fig 9.3

We will assume that G, K and P are linear lumped systems in which case their Laplace transforms are rationa transfer functions.   For practical reasons we assume that G(s) is strictly proper (approaches 0 as s approaches ∞), and that K(s) and P(s) are proper (stay finite as s approaches ∞).   We also assume as we must, that there are hidden (unobservable and uncontrollable) unstable modes in G.

From Fig. 9.3 it is straight-forward to show that

$$y = TPr - Tm + Sd \qquad (9.5)$$

where

$$S(s) = (I + G(s)K(s))^{-1} \qquad (9.6)$$

and

$$T(s) = (I + G(s)K(s))^{-1}G(s)K(s) = I - S(s) \qquad (9.7)$$

S(s) and T(s) are of considerable importance in control
system analysis and design; they are called the sensitivity
matrix and complementary sensitivity matrix respectively.
It is clear from (9.5) how disturbance rejection is
influenced by S and noise mitigation by T and how both are
functions of the feedback controller K. Given a feedback
controller K and hence T we also see that P can be used to
shape the command response.

We now look in detail at some methods for analysing a
feedback design.

9.4 ANALYSIS METHODS (1) - (16)

In this section we describe some stabiltiy tests and
some tests for assessing the robustness of the feedback
properties (especially stability) of a control system
design. The latter are dependent on a knowledge of singular
values which are also introduced and discussed in some
detail. Bode diagrams of singular values are seen to have a
similar role in multivariable feedback design as the Bode
magnitude diagram does for scalar systems.

9.4.1 Stability tests

We begin with some fundamental relationships for the
feedback system of Fig. 9.3 Note that because P(s) (which
must be stable) does not affect the feedback properties of
the control system we will for convenience let it be equal
to the identity operator I. Also , for simplicity, and
without loss of generality we assume that the plant is
square. We then have

$$\det [I + G(s)K(s)] = \frac{x_c(s)}{x_o(s)} \qquad (9.8)$$

and

$$\det [I + (G(s)K(s))^{-1}] = \frac{x_c(s)}{Z_G(s)Z_K(s)} \qquad (9.9)$$

where $X_o(s)$ and $X_c(s)$ are the open and closed-loop
characteristic polynomials respectively and $Z_G(s)$ (respec.
$Z_K(s)$) is the zero polynomial of G(s) (respec. K(s)) defined
by

$$Z_G(s) = \det [sI-A_G] \det G(s) \qquad (9.10)$$

where A_G is the state-space A-matrix of G(s).

A straightforward application of the Argument
Principle of Complex Analysis to each of (9.8) and (9.9)
leads directly to the following stability criteria;

Theorem 1. Multivariable Nyquist Criterion (Rosenbrock (3))

The feedback system is stable if, and only if, the
image of the standard Nyquist D-contur D_D under det $[1 +$
$G(s)K(s)]$ encircles the origin P times anticlockwise where P
is the number of open-loop poles in D_D.

Theorem 2. Multivariable Inverse Nyquist Criterion
(Postlethwaite and Foo (2))

The feedback system is stable if, and only if, the
image of the standard inverse Nyquist D-contour D_I under

det $[I + (G(s)K(s)K(s))^{-1}]$ encircles the origin Z times
anticlockwise, where Z is the total number of zeros of $G(s)$
and $K(s)$ in D_I.

If we denote the eignevalues of $G(s)K(s)$ as s goes
around D_D the characteristic gain loci and the eigenvalues of

$(G(s)K(s))^{-1}$ as s goes around D_I the inverse characteristic

gain loci, then theorems 1 and 2 can be restated as follows

Theorem 3. (Postlethwaite and MacFarlane(4))

The feedback system is stable if, and only if, the
characterisistic gain loci encircle the critical point, -1,
P times anticlockwise where P is the number of open-loop
unstable poles in D_D.

Theorem 4. (4)

The feedback system is stable if, and only if, the
inverse characteristic gain loci encircle the critical
point, - 1, Z times anticlockwise, where Z is the total
number of zeros of $G(s)$ and $K(s)$ in D_I.

Theorems 3 and 4 are more useful than 1 and 2 for
assessing the robustness of the stability properly since
reliable gain and phase margins can be defined with respect
to a scalar gain common to each of the control channels.
However, even these are far from satisfactory when
considering general matrix perturbations of the system   to
take accound of uncertainty in the nominal models.   In
general therefore, although theorems 1 to 4 accurately
predict stability (and instability) they are unsuitable fo
assessing the robustness of the stability property to plant
perturbations.   To overcome this problem we use singular
values to characterize uncertainty, and then derive tests i

terms of these quantities. First let us define singular values.

9.4.2 Singular values, maximum energy gain and uncertainty representation

The singular values of a complex nxm matrix M, denoted $\sigma_i(M)$, are the k largest non-negative square roots of the eigenvalues of $M^H M$ where $k = \min(n,m)$ and superscript H denotes complex conjugate transpose; we order the σ_i such that $\sigma_i \geq \sigma_{i+1}$.

The maximum and minimum singular values may alternatively be defined by

$$\bar{\sigma}(M) = \sigma_1(M) = \max_{x \neq 0} \frac{||Mx||_E}{||x||_E} = ||M||_E \qquad (9.11)$$

$$\underline{\sigma}(M) = \sigma_k(M) = \min_{x \neq 0} \frac{||Mx||_E}{||x||_E} = (||M^{-1}||_E)^{-1} \qquad (9.12)$$

where $||x||_E$ denotes the usual Euclidean length of the vector x and $||M||_E$ denotes the matrix norm induced by the Euclidean vector norm.

Let $F(j\omega)$ be a frequecy response matrix, then from (9.11) $\bar{\sigma}(F(j\omega))$ is the maximum vector gain from input to output at frequency ω. Similarly $\underline{\sigma}(F(j\omega))$ is the minimum vector gain at frequency ω.

A physical interpretation of the maximum singular value of $F(j\omega)$ can also be given in terms of maximum energy gain. Let u(t) be the input to F(s) and y(t) the corresponding output and assume that u(t) is bounded in energy so that its L_2-norm is finite i.e.

$$||u(t)||_2 \triangleq (\int_0^\infty u^T(t)u(t) \, dt)^{1/2}$$

$$= (\text{ Total input energy})^{1/2}$$

$$< \infty \qquad (9.13)$$

Then the maximum energy gain from input to output is

$$\sup_{||u(t)||_2 < \infty} \frac{||y(t)||_2}{||u(t)||_2} = \sup_\omega \bar{\sigma}(F(j\omega)) \qquad (9.14)$$

i.e. it is the maximum singular vaue of $F(j\omega)$ over all frequencies. The right-hand side of (9.14) is called the H^∞-norm of the stable transfer function $F(s)$. Later we will describe a design approach based on the minimization of H^∞-norms.

As a matrix norm, the maximum singular value is useful in characterizing uncertainty. For instance, one might define a set of possible plant models $\left\{ G_p(s) \right\}$ about a nominal model $G(s)$ by

$$\left\{ G_p(s) \right\} = \left\{ G(s) + E(s) : \bar{\sigma}\,[E(j\omega)] \leqslant e(\omega) \right\} \qquad (9.15)$$

Representations like this are used in robustness tests for assessing feedback stability as is now shown.

Robust stability

Several representations of uncertainty have been used in the literature and correspondingly several robustness tests are available. In these notes we consider just two:

$$\left\{ G_p(s) \right\}_\Delta = \left\{ (I+\Delta(s))G(s) : \bar{\sigma}[\Delta(s)] \leqslant \delta(s), \; \forall s \in D_D \right\} \qquad (9.16)$$

and

$$\left\{ G_p(s) \right\}_{\tilde{\Delta}} = \left\{ (I+\tilde{\Delta}(s))^{-1}G(s): \bar{\sigma}[\tilde{\Delta}(s)] \leqslant \tilde{\delta}(s), \; \forall s \in D_I \right\} \qquad (9.17)$$

From theorem 1, we see that if $G(s)$ is replaced by $G_p(s)$, a perturbed plant model taken from the set $\left\{ G_p(s) \right\}_\Delta$, and if $G_p(s)$ has the same number of poles in D_D as $G(s)$, then the perturbed feedback system will remain stable if the image of D_D under $\det [I + G_p(s) K(s)]$ encircles the origin the same number of times as the image of D_D under $\det [I + G(s)K(s)]$. This argument leads to the following robustness test.

Theorem 5. Direct Nyquist Robustness Test (Doyle and Stein (5))

The feedback system will remain stable when the nominal plant model is replaced by any perturbed plant model taken from the set $\left\{ G_p(s) \right\}_\Delta$ if:

(i) $G_p(s)$ and $G(s)$ share the same number of poles in D_D, and

(ii) $\bar{\sigma}[T(s)] > \dfrac{1}{\delta(s)}$, $\forall\ s\epsilon D_D$.

Similar arguements based on theorem 2 and the perturbed model set $\left\{G_p(s)\right\}_{\tilde{\Delta}}$ lead to the folowing robustness test:

Theorem 6. Inverse Nyquist Robustness test (2)

The feedback system will remain stable when the nominal plant model is replaced by any perturbed plant model taken from the set $\left\{G_p(s)\right\}_{\tilde{\Delta}}$ if;

(i) $G_p(o)$ and $G(s)$ share the same number of zeros in D_I, and

(ii) $\bar{\sigma}\ [S(s)] < \dfrac{1}{\tilde{\delta}(s)}$, $\forall s\epsilon D_I$

We see that theorem 5 allows the plant and perturbed plant model to have different numbers of non-minimum phase zeros, while theorem 6 allows different numbers of unstable poles. But what if the plant and perturbed plant models have different numbers of unstable poles and different numbers of non-minimum phase zeros. This situation can be handled by a test which combines the two inequalities of theorems 5 and 6 at the expense of a more refined (yet no less practical) representation of uncertainty $\left\{G_p(s)\right\}$.

The set $\left\{G_p(s)\right\}$ is assumed to be arcwise connected (2) in a topology (the graph topology) for unstable plants defined by Vidyasagar, Schneider and Francis (1). Very simply this means that if we perturb $G(s)$ to $G_p(s)$ along a path there cannot be any sudden appearance or disappearance of unstable poles or non minimum phase zeros except for those which move continuously across the $j\omega$-axis.

Theorem 7. Combined Nyquist and Inverse Nyquist Robustness Test (2)

Let $\left\{G_p(s)\right\}$ be an arcwise connected set of transfer function matrices in the graph topology of unstable plants. Assume that for every member $G_p(s)$ in the set there exists $\Delta(jw)$ and $\tilde{\Delta}(j\omega)$ such that

$$G_p(j\omega) = (I+\Delta(j\omega))\, G(j\omega) = (I+\tilde{\Delta}(j\omega))^{-1}G(j\omega)$$

where $G(s)$ is a member of $\left\{G_p(s)\right\}$, known as the nominal model, such that the nominal feedback system is stable. Define:

$$\delta(\omega) = \sup_{G_p(s)\,\epsilon\,\left\{G_p(s)\right\}} \left\{\bar{\sigma}[\Delta(j\omega)]\right\}$$

$$\tilde{\delta}(\omega) = \sup_{G_p(s)\,\epsilon\,\left\{G_p(s)\right\}} \left\{\bar{\sigma}[\tilde{\Delta}(j\omega)]\right\}$$

Then under these condidtions, the perturbed feedback systemis stable for any member $G_p(s)$ from $\left\{G_p(s)\right\}$ if, at each $\omega\epsilon$ $[0,\infty]$, either

(i) $\bar{\sigma}[T(j\omega)] < \dfrac{1}{\delta(\omega)}$, or

(ii) $\bar{\sigma}[S(j\omega)] < \dfrac{1}{\tilde{\delta}(\omega)}$.

This robustness test allows us to switch arbitarily between the Nyquist based inequality (i) and the inverseNyquist based inequality (ii). This is particularly useful at frequencies for whih poles cross the imaginary axis because $\bar{\sigma}$ $[\Delta(j\omega)]$ then approaches infinity and hence inequality (i) cannot be satisfied. Similarly, at frequencies for which zeros cross the imaginary axis $\bar{\sigma}$ $[\Delta(j\omega)]$ approaches infinity and inequality (ii) cannot be satisfied.

The robustness tests of theorems 5,6 and 7 emphasize the importance of the size (maximum singular value) of the sensitivity and complementary senstitvity matrices, S and T The same functions are also of importance in assessing the robustness of other feedback properties such as disturbance rejection and noise mitigation.

9.4.3 Robust Performance

Consider (for simplicity) the case of a scalar system in which the plant output is required to stay close to a se point despite the presence of an uncertain disturbance signal. Suppose that the disturbance signal is known to be one of a set of narrow band signals whose spectra are restricted to a common frequency band. Let $S(s)$ be the transfer function from the disturbance to the regulation error (set point minus output) and let $W(s)$ be a weighting function which is large over the common frequency band and small off if. Then minimising the H^∞ -norm $\|WS\|_\infty$ is equivalent to minimising (the square root of) the maximum possible energy in regulation error over the set of all possible disturbances. In this regulation problem therefore the H^∞ -norm provides a good measure of robust performance.

Also from equation (9.5) and the interpretation of the maximum singular value of a frequency response matrix as maximum vector gain, it is clear that disturbances will be rejected over frequencies for which $\bar{\sigma}\ [S(j\omega)]$ is made small.

The above discussion highlights the important role that Bode diagrams of singular values can play in the analysis of multivariable feedback systems.

PART 2: MULTIVARIABLE CONTROL SYSTEM DESIGN

9.5 DESIGN OR SYNTHESIS

A distinction is often made between design and synthesis (MacFarlane(17)). A synthesis technique, it is argued, strives to formulate a sharply defined problem with a rigorous mathematical solution. On the other hand, in a design technique one is given a set of manipulative and interpretive tools for building up, modifying and assessing a design put together with the help of physical reasoning and engineering experience. Optimal linear quadratic control (e.g. Anderson and Moore(18)) is a good example of a synthesis technique while the Characteristic Locus Design Method (MacFarlane and Kouvaritakis (19)). is a good example of a design technique.

I prefer not to make this distinction since there appears to be no real reason why a synthesis technique should not be considered a good design technique, or why it should not, at least, form part of a good design procedure. In what follows therefore we will consider some design methods which others might prefer to describe as synthesis methods.

9.6 DESIGN METHODS (3), (4), (17-32)

9.6.1. Classical SISO Extensions

Much effort has been devoted to extending the classical frequency response techniques of Nyquist, Bode and Evans to multivariable systems, e.g. (3), (4) (19). We will briefly describe two of these.

In the Inverse Nyquist Array (INA) Design Method pioneered by Rosenbock (3) the pre-compensator K(s) and

usually a post-compensator L(s) are used to make the loop
transfer function L(s)G(s)K(s) approximately diagonal. By
doing this the multivariable system looks approximately like
a set of single-input single-output (SISO) systems for which
controllers can be designed independently using classical
techniques. Mathematical rigour was given to the approach
by the derivation of Nyquist-type stability tests which are
applicable if L(jω)G(jω)K(jω) is made diagonally dominant at
all frequencies.

The method was found attractive in its simplicity by
practising engineers, although for some problems it was
found difficult, if not impossible, to achieve diagonal
dominance at all frequencies. Another disadvantage of the
method is that it does not explicity include robustness as a
design objective. The process of making the INA design
method robust is under study (Arkun et al(20), Kidd (21)).

In the Characteristic Locus Design Method (19) the
characteristic gain loci (introduced in part I) are treated
as the natural generalisation of the SISO Nyquist diagram.
Multivariable design then consists of manipulating the shape
of the characteristic gain loci as one would do to a Nyquist
diagram in classical control. To shape the individual
branches of the characteristic gain loci is generally a
difficult problem, but by making K(s) approximately commute
with the plant in a frequency range it is often possible to
manipulate satisfactorily the loci over the same
frequencies. Again,   the simplicity of the approach
was found attractive by engineers. However, like the INA
method it does not give reliable information about the
robustness of feedback properties. This is largely due to
the sensitivity of the characteristic gain loci which in
certain circumstances can be very bad, (Postlethwaite (23).
Atempts have been made (Postlethwaite et al (6)) and
research is currently in progress (Daniel and Kouvaritakis
(24)) to develop a truely robust design method based on the
characteristicgain loci. The idea is to replace the
characteristic gain loci by regions which include the
characteristic gain loci and also reflect the degree of
uncertainty in the plant model. That is, the larger the
uncertainty, the larger the regions. Design then proceeds
on the basis of manipulating the regions rather than the
loci.

9.6.2 LQG Loop Shaping

Introduced by Doyle and Stein (5) the aim here is to
shape the singular values of the frequency response of the
loop transfer function G(s)K(s) to obtain desirable feedbac
properties. For good sensitivity $\underline{\sigma}$ [G(jω)K(jω)] is require
to be large, whilst $\bar{\sigma}$ [G(jω)K(jω)] is required to be small
to reduce the effects of measurement noise on the outputs
and to maintain stability in the presence of moelling
errors. Assuming the former to be of importance at low

frequencies and the latter at high frequencies the necessary trade-off is illustrated in Fig. 9.4.

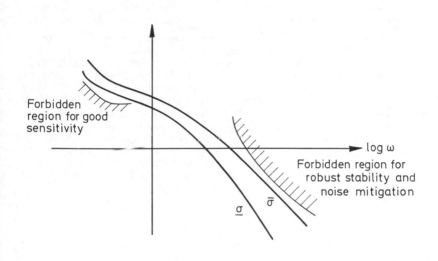

Fig 9.4

An LQG controller (eg Kwakernaak(25)) is used to shape the singular value plots. This has the advantage of nominal feedback stability but the "shaping" is done iteratively by trial and error selections of the associated weighting matrices. Guidelines for the choice of the weighting functions are given in a "robustness recovery procedure" (5) which asymptotically recovers the guarenteed gain and phase margins whcih would be present in the input control channels of a full state feedback regulator (18). A "sensitivity recovery procedure" (25) also exists which recovers the gain and phase margins at the output of the plant. Note that both these recovery procedures are based on asymptotic results and are therefore of a limited use; also they are only applicable to minimum phase systems.

9.6.3 H^{∞} Design

The latest developments in feedback design are centred on H^{∞} optimization and originate from the influential work of Zames (26). He argued that the poor robustness properties of LQG could be blamed on the integral criterion and criticised the use of white noises to represent uncertain disturbances as unrealistic. For a single-input single-output system, he considered the apparently straightforward problem of designing a feedback controller to minimise the effect of a disturbance on the plant output, subject to the constraint of closed loop stability. The disturbance was assumed to belong to a prespecified set of signals and the energy of the output was to be minimised for the worst disturbance in the set. The solution involved the minimisation of the H^∞ - norm (maximum modulus) of the sensitivity function i.e. the transfer function from the disturbance to the output, over the set of all stabilizing controllers. This minimax approach to feedback design has been extended to multivariable systems and has been rapidly developed to cope with more general control problems than sensitivity minimization. For example, in Part 1 we saw how robust stability and robust performance could be characterised by $\bar{\sigma}[S(j\omega)]$ and $\bar{\sigma}[T(j\omega)]$. To shape these maximum singular values the following optimization problem (Foo and Postlethwaite (27)) has been posed: find a stabilizing controller K(s) which minimises the cost function

$$\left|\left| \begin{array}{c} W_1 S \\ W_2^{-1} T \end{array} \right|\right|_\infty \qquad\qquad (9.18)$$

where $W_1(s)$ and $W_2(s)$ are weighting functions chosen by the engineer to meet the design specifications.

Loosely speaking W_1 should be "large" at frequencies where performance (accuracy, disturbance rejection, etc) is important and where there are no uncertain zeros close to the imaginary axis. W_2 should be "large" at frequencies where measurement noise is significant and where there are no uncertain poles close to the imaginary axis.

For an excellent expository article on the solution of the above and many other H^∞ design problems see Francis and Doyle (28).

9.6.4 Semi-infinite Optimization

In this approach it is assumed that a controller structure has been selected (perhaps designed using one of the previous methods) and that "semi-infinite" optimization is to be used to compute compensator parameter values to satisfy design specifications and to minimise a cost function. "Semi-infinite" refers to an infinitely constrained optimisation problem in the finite dimensional parameter, x say, which represents the free compensator coefficients. Such problems arise for example if the design specifications are in terms of frequency dependent bounds on a maximum singular value; time domain examples would be constraints on rise time, settling time and overshoot. Mathematically these design problems may be expressed in the form

$$\min\left\{f(x):g_j(x) \leqslant 0, \ j=1,..,m; \phi_k(x,y_k) \leqslant 0, \ y_k \epsilon \ Y_k, k=1,,..,\ell\right\}$$

$$(9.19)$$

where f is a cost function, $g_j(x)$ and ϕ_k are constraints and the Y_k are finite time or frequency intervals.

The development of algorithms to solve semi-infinite optimization problems is an active area of research (Wardi and Polak (29), Polak and Mayne (30)). Polak, Mayne and Stimler [3] conclude that the implementation of these techniques for control system design is best carried out in a highly interactive computing environment. In response to this a software system called DELIGHT. MIMO is currently being developed by reasearch teams at the University of California, Berkeley and Imperial College, London. DELIGHT.MIMO is a member of a family of optimisation based packages for engineering design being implemented in a new and highly portable operating system called DELIGHT (Nye (32)) which stands for DEsign Laboratory with Interaction and Graphics for a Happier Tomorrow.

9.7 CONCLUDING REMARKS

In this chapter, I have given a brief description of many of the important analysis and design techniques for multivariable control systems. In attempting to cover a wide range of topics I have overlooked much of the necessary detail for a complete understanding. However, it is hoped that the reader will be sufficiently interested to want to learn more of the fine detail and for this purpose an extensive list of references is included.

Appendix 1:   State-space model for a Lynx helicopter operating at 100 knots, straight and level

$$\dot{x} = Ax + Bu$$
$$y = Cx$$

Inputs (nominal control angles):
1. Main rotor collective pitch   0.2221 rads
2. Logitudinal cyclic pitch   -0.0603 rads
3. Lateral cyclic pitch   0.0271 rads
4. Tail rotor collective pitch   0.0422 rads

Trim conditions (nominal state variables):
1. x-component (forward) of total velocity 169.9 ft/sec
2. z-component (down) of total velocity   -0.696 ft/sec
3. Pitch angular velocity   0 rads/sec
4. Pitch angle   -0.004 rads
5. y-component (right) of total velocity   0 ft/sec
6. Roll angular velocity   0 rads/sec
7. Bank angular velocity   0 rads/sec
8. Yaw angular velocity   0 rads/sec

Outputs (nominal flight conditions):
1. Total speed   168.8 ft/sec
2. Decent angle   0 rads
3. Sideslip angle   0 rads
4. Bank angle   -0.038 rads

System A matrix

-0.039	0.040	2.908	-32.186	-0.001	-0.327	0.0	0.0
0.010	-0.857	168.411	0.132	-0.021	-1.923	1.229	0.0
0.008	0.010	-2.469	0.0	0.002	0.415	0.0	0.0
0.0	0.0	0.100	0.0	0.0	0.0	0.0	0.038
0.005	0.013	-0.421	-0.005	-0.200	-2.964	32.16	-167.044
-0.010	0.069	-1.933	0.0	0.051	-10.409	0.0	-0.349
0.0	0.00.	0.000	0.0	0.0	1.0	0.0	-0.004
-0.007	0.002	-0.005	0.0	0.036	-1.757	0.0	-1.563

Control B matrix

13.29669	-23.88917	6.87120	0.00000
-416.64496	-130.76123	0.00066	0.00000
17.48406	29.63527	-5.91284	0.00000
0.00000	0.00000	0.00000	0.00000
4.96504	-4.86813	-30.884312	24.86813
32.96783	-22.42182	-153.24533	-1.51585
0.00000	0.00000	0.00000	0.00000
14.74352	-5.60603	-26.58559	-20.41646

Output C matrix

1.0	-0.00412	0.0	0.0	0.0	0.0	0.0	0.0
-0.00002	-0.00591	0.0	1.0	0.00023	0.0	0.00016	0.0
0.0	0.0	0.0	0.0	0.00592	0.0	0.0	0.0
0.0	0.0	0.0	0.0	0.0	0.0	1.0	0.0

REFERENCES

1. Vidyasagar, M., Schneider. H,and Francis B.A., 1982
    "Algebraic and topological aspects of feedback
    stabilization", IEEE Trans. Autom Control, AC-27, 880-
    894.

2. Postlethwaite I. and Foo, Y.K., 1985, "Robustness with
    simultaneous pole and zero movement across the jω -
    axis", Automatica, vol 21, 433-444.

3. Rosenbrock, H.H 1974.Computer-Aided Control System
    Design, Academic Press.

4. Postlethwaite I. and MacFarlane,A.G.J.,1979, A Complex
    Variable Approach to the Analysis of Linear
    Multivariable Feedback Systems, Springer.

5. Doyle, J.C. and Stein, G., 1981 "Multivariable Feedback
    Design: Concepts for a Classical/Modern Synthesis",
    IEE Trans. Autom. Control, AC-26, 4-16.

6. Postlethwaite, I., Edmunds, J.M. and MacFarlane, A.G.J,
    1981,"Principal Gains and Principal Phases in the
    Analysis of Linear Multivariable Feedback Systemsx,":
    IEE Trans. Autom Control, AC-26, 32-46.

7. Safonov, M.G., Laub, A.J. and Hartmann, G.L, 1981.
    "Feedback Properties of Multivariable Systems: The role
    and Use of the Return Difference Matrix", IEEE Trans.
    Autom. Control, AC-26, 47-65.

8. Cruz, J.B. Jr., Freudenberg, J.S. and Looze, D.P., 1981
    A Relationship between Sensitivity and Stability of
    Multivariable Feedback Systems", IEEE Trans. Autom.
    Control, AC-26, 66-74.

9. Lehtomaki, N.A., Sandell, N.R. Jr. and Athans, M, 1981,
    "Robustness Results in Linear Quadratic Gaussian based
    Multivariable Control Design", IEEE Trans. Autom.
    Control, AC-26, 75-92.

10. Lehtomaki, N.A. 1981, Practical Robustness Measures in
    Multivariable Control System Analysis, Ph.D. Thesis,
    Massachusetts Institute of Technology.

11. Barrett, M.F.,1980, Conservatism with Robustness Tests
    for Linear Feedback Control Systems, Ph.D. Thesis,
    University of Minnesota.

12. Safonov, M.G., 1980, Stability and Robustness of
    Multivariable Feedback Systems, MIT Press.

13. Doyle, J.C., 1982, "Analysis of feedback systems with structured uncertainties", IEE Proceedings D, Vol 129, 242-250.

14. Horowitz, I., 1982, "Quantitative feedback theory", IEE Proceedings D, Vol 129, 215-226.

15. Foo, Y.K., 1985, Robustness of multivariable feedback systems: analysis and optimal design, Ph.D. Thesis, University of Oxford.

16. Kouvaritakis, B. and Latchman, H. "Singular value and eigenvalue techniques in the analysis of systems with structured perturbations", Int. J. Control, to appear.

17. MacFarlane, A.G.J., 1980, editor Complex variable Methods for Linear Multivariable Feedback Systems, Taylor and Francis Ltd.

18. Anderson, B.D.O. and Moore, J.B., 1971, Linear Optimal Control, Prentice-Hall.

19. MacFarlane, A.G.J. and Kouvaritakis, B., 1977, "A design technique for linear multivariable feedback systems", Int. J. Control, Vol 25, 837-874.

20. Arkun, Y., Manousiothakis, B. and Putz, P.,1984, "Robust Nyquist array methodology:a new theoretical framework for analysis and design of robust multivariable feedback systems, Int. J. Control, Vol 40, 603-630.

21. Kidd, P.T., 1984, "Extensions of the direct Nyquist array design technique to uncertain multivariable systems subject to external disturbances", Int. J. Control, Vol 40, 875-902.

22. Limebeer, D.J.N.,1984, "The application of generalized diagonal dominance to linear system stability theory", Int. J. Control, Vol 36, 185-212.

23. Postlethwaite, I.,1982, "Sensitivity of the Characteristic Gain Loci", Automatica, Vol 18, 709-712.

24. Daniel, R. and Kouvaritakis, B., 1983, "Analysis and design of linear multivariable feedback systems in the presence of additive perturbations", Int. J. Control, Vol 59, 551-580.

25. Kwakernaak, H. and Sivan, R. 1972, Linear Optimal Control Systems, Wiley.

26. Zames, G., 1981, "Feedback and optimal sensitivity: model reference transformations, multiplicative seminorms and approximate inverses", IEEE Trans. Autom. Control, AC-26, 301-320.

27. Foo, Y.K. and Postlethwaite, I., 1984, "An H^∞ minimax approach to the design of robust control systems", Systems and Control Letters, Vol 5, 81-88.

28. Francis, B.A. and Doyle, J.C. 1985, "Linear Control Theory with an H Optimality Criterion", University of Toronto Systems Control Group Report No. 8501. To appear in SIAM J. of Control and optimization.

29. Wardi, Y.Y. and Polak, E., 1982, "Computer-Aided Design of Structures Subject to Eigenvalue Inequality Constraints", University of California, Berkeley, Memo. No. UCB/ERL M82/91.

30. Polak, E. and Mayne, D.Q., 1984, "Theoretical and Softward Aspects of Optimization Based Control System Design", Memo. No. UCB/ERL M84/23.

31. Polak, E., Mayne, D.Q. and Stimler, D.M., 1984, "Control System Design Via Semi-Infinite Optimization", Memo. No. UCB/ERL M 84.

32. Nye, W.T., 1983, DELIGHT: An Interactive System for Optimization Based Engineering Design, Ph.D. Thesis, University of California, Berkeley.

Chapter 10

Adaptive control

Dr. D.W. Clarke

10.1 INTRODUCTION

Servomechanism design underlies much current control theory but its success depends on having reasonably accurate system models. In process applications these models are expensive to derive mathematically or by identification experiments, so PID (three-term) regulators are used which can give satisfactory performance if the corresponding coefficients have been properly tuned. This can often be difficult because:

o The plant may have complex dynamics due to excessive phase lag or dead-time.

o Nonlinearities may make the effective plant gain vary with the set-point. Actuators such as valves generally exhibit dead-bands, hysteresis and saturation which add to the complexity of the loop.

o The dynamics may vary with time, such as with the decay of a catalyst or the fouling of a heat-exchanger.

o Interactions between loops are often such that they cannot be tightly tuned independently.

o The material inputs into a process may vary in quality and there may be environmental disturbances such as changes in cooling water temperature. These variations cause 'load-disturbances' on the outputs which generally are non-stationary. This feature of disturbances acting on physical processes is the principal reason for use of integral action in process control. The output measurement can also be corrupted by noise and quantization errors which restrict the maximum allowable phase advance of the controller.

Some of these problems can be overcome using PID regulators, and it is probably true to say that the three-term algorithm is more robust than most simple alternatives. However, although systematic manual tuning rules such as those proposed by Ziegler and Nichols (56) are available, the problem of finding the 'best' set of PID coefficients can be difficult and time consuming. Currently on the market are PID self-tuners based on mechanizing the Ziegler-Nichols' approach, on using an experimental 'reaction-curve' or step-response (e.g. Honeywell), on looking for patterns in the error response to a step demand (e.g. Foxboro), or on estimating a second-order plant model from input/output data (e.g. Turbull).

It can be shown that under quite general requirements, such minimising quadratic costs or with prespecified closed-loop dynamics, the PID form is optimal for second-order plant subjected to load-disturbances of a Brownian motion type. On the other hand with more complex plant, such as those with dead-time, the PID algorithm is too simple and has to be detuned (e.g. reduced gain). The self-tuning approach described below, which is based on more advanced control laws, offers a successful alternative.

A self-tuning controller consists of a recursive parameter estimator (real-time plant identifier) coupled with an analytical control design procedure, such that the currently estimated parameters (if adjudged to be valid) are used to provide feedback controller coefficients, as shown in Fig.1. The user's input is no longer the values of the PID 'knobs' K, Ti and Td, but rather the <u>closed-loop performance objective</u> which is required by the analytical design procedure. The important feature of self-tuning is that even if the plant model is not known the adaptation mechanism attains the desired closed-loop performance by automatically adjusting the controller parameters. For complex plant the number of parameters that might be involved would be greater than the three of the PID form, and so could not practically be manually tuned. On the other hand proofs of convergence (Ljung (39),(40)) of a self-tuner to the 'true' controller depend on assumptions about the process (linearity, stationarity, model-order,...) which in general cannot be fully justified. As with other 'intelligent' algorithms, the viability of a self-tuner in practice depends as much on the <u>robustness</u> of the basic design procedure and on appropriate 'jacketting' software as on theoretical convergence considerations.

Although a self-tuner can be derived by combining any reasonable controller design procedure with a recursive parameter estimator, this tutorial will concentrate on two particular algorithms. The first is based on predictive control theory (Clarke and Gawthrop (14,15), Clarke (11)) as it can provide a wide class of useful performance objectives. The second is the pole-placement method (Wellstead et al (51), Clarke (12)) which can be effective for plant with variable dead-time.

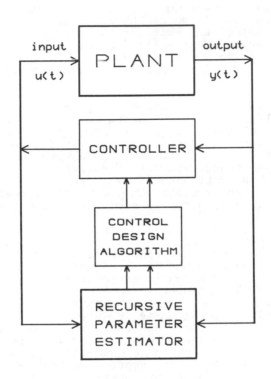

Fig.1 <u>Self-tuning controller structure</u>

The tutorial is organised as follows. Discretization of continuous-time plant is briefly described, particularly with respect to the discrete-time zeros of the transfer-function and to possible models of typical load-disturbances. The general theory of predictive control is then shown to have interpretations as Smith Predictive, 'detunable' model-reference and generalised minimum-variance laws. Extensions to include feedforward compensation for known disturbances are provided. The pole-placement method is derived to cover both the servo and the regulator problem, and in the latter case it is shown that a simple estimation algorithm (RLS) is sufficient. General recursive estimation methods based on minimising the prediction error are discussed with particular emphasis on numerically stable versions of the Recursive Least Squares (RLS) method. A simple self-tuner and its programming is provided as a combination of an estimator and a basic control law. Applications of self-tuning are summarized together with a discussion of the important 'jacketting' software which is used to make the theory work reliably.

10.2 THE ASSUMED PLANT MODEL

To apply a self-tuner the structure (though not the parameters) of the process model is first assumed with a practical proviso that it should be capable of at least approximating the behaviour of a broad class of processes. As in the main the set-point of a process control loop is constant over long periods, a locally–linearized model is often adequate:

$$y(t) = \frac{B_1(s)}{A_1(s)} u(t-T) + x(t) \qquad \dots (1)$$

where $u(t)$, $y(t)$ are the plant input and output and T is the dead-time or transport-delay. A_1 and B_1 are polynomials in the differential operator $s = d./dt$ for which:

$$\text{degree } A_1(s) = n_1 > \text{degree } B_1(s) = m_1 \qquad \dots (2)$$

In practice values of n_1 and m_1 in the range 1 to 3 are generally acceptable.

The signal $x(t)$ is a general disturbance term (though it would also include the effect of modelling errors) with components:

1. a constant x_1 as processes do not generally have signals with zero mean, and the incremental gain $\partial y/\partial u$ does not equal the static gain y/u

2. a load-disturbance $x_2(t)$ which might slowly vary or consist of random steps at random times

3. a measured auxiliary signal $x_3(t)$ suitable for feedforward compensation

4. a stationary stochastic process $x_4(t)$ given by:

$$x_4(t) = \frac{C_2(s)}{A_2(s)} \xi(t) \qquad \dots (3)$$

where A_2 and C_2 are further polynomials in s of degree n_2 and m_2, and $\xi(t)$ is a white-noise process. Most self-tuning theory concentrates on $x_4(t)$ whereas in practice $x_2(t)$ is more common.

In computer control the plant is preceded by a zero-order-hold which provides a 'staircase' input (Franklin and Powell (24), Isermann (33)). If $G(s)$ is the transfer-function relating $y(s)$ to $u(s)$, the discrete-time model is given by:

$$G(z^{-1}) = (1-z^{-1}) Z\{G(s)/s\}$$

and the Z-transform of $G(s)/s$ is obtained by using a partial-fraction expansion of $G(s)/s$ and then transforming term by term. If now t is defined to be the discrete-time sample instant and z to be the forward shift operator, the discrete-time model corresponding to eqn.1 where $x(t)$ is given by eqn.3 becomes the CARMA (Controlled AutoRegressive Moving-Average) representation:

$$A(z^{-1})y(t) = B(z^{-1})u(t-k) + C(z^{-1})e(t) \qquad \ldots (4)$$

Here the dead-time is reflected in the k samples between a control action and the corresponding effect on the output; if h is the sample interval then $k = INT(T_d/h) + 1$. The polynomials A, B and C are all of degree $n = n_1 + n_2$ in the backward shift operator z^{-1}, so the corresponding difference equation is:

$$y(t) + a_1 y(t-1) + \ldots + a_n y(t-n) = b_0 u(t-k) + \ldots + b_n u(t-k-n)$$

$$+ e(t) + c_1 e(t-1) + \ldots + c_n e(t-n)$$

The sequence $\{e(t)\}$ is serially uncorrelated with a common variance σ^2. As such model 4 is the one generally assumed is the self-tuning literature in which a linear dynamic process with dead-time is subjected to stationary disturbances with rational spectral density.

Many related models can be deduced depending on the assumed nature of $x(t)$. Early work (Clarke and Gawthrop (14)) added a constant d to model $x_1(t)$ giving:

$$A(z^{-1})y(t) = B(z^{-1})u(t-k) + d + C(z^{-1})e(t) \qquad \ldots (5)$$

The value of d can be estimated along with the other plant parameters. However a Brownian motion model is more practical (as well as leading naturally to integral action in the control law), leading to the CARIMA (Integrated) plant representation:

$$A(z^{-1})y(t) = B(z^{-1})u(t-k) + C(z^{-1})e(t)/\Delta \qquad \ldots (6)$$

where $\Delta = 1-z^{-1}$ is a differencing operator (Clarke et al (17)). In the following the general model:

$$A(z^{-1})y(t) = B(z^{-1})u(t-k) + x(t) \qquad \ldots (7)$$

is assumed, where the general disturbance $x(t)$ is interpreted according to context.

The CARMA/CARIMA model is not the only one usable in deriving adaptive controllers. For example, eqn.4 can be divided by the polynomial A to give:

$$y(t) = A^{-1}B\, u(t-k) + A^{-1}C\, e(t)$$

in which long-division yields the __weighting-sequence__ model:

$$y(t) = H_1(z^{-1})\, u(t) + H_2(z^{-1})\, e(t) \qquad \ldots (8)$$

where the polynomials H are in principle of infinite degree, though if A is stable and therefore has roots within the unit circle the expansion of 1/A converges and the polynomials H can be truncated after a finite number of terms. This is used in the IDCOM approach of Richalet et al (46) and is also the basis of the DMC algorithm; note that many more parameters may be required to give an accurate plant representation.

10.2.1 Discrete-time Zeros.

Some self-tuners (e.g. model-reference type) are based on cancellation controller designs, in which the plant poles and zeros are first cancelled by terms in the controller and further terms provide the desired closed-loop performance. If the plant has zeros (roots of B) outside the unit circle the corresponding controller poles induce instability as cancellation is inevitably inexact. Hence for these designs it is important to know when such zeros exist - in the literature they are somewhat loosely termed as nonminimum-phase. Now whereas there is a one-to-one relationship between continuous-time and discrete-time poles (given by the mapping z = exp(sh)), there is no such correspondence between the zeros. Indeed a continuous-time plant with no zero is discretized to give a B polynomial with n zeros in general (n-1 if the dead-time is an integral multiple of h). The following simple rules in which n is the order of the continuous-time plant cover some of the cases:

1. If the continous-time plant has n-1 or n zeros, the mapping z = exp(sh) applies to zeros also. Hence a minimum-phase plant has a minimum-phase discrete-time model. This result is used in Gawthrop's 'hybrid' method (26).

2. If G(s) has j more poles than zeros, then for small h the discrete-time model is proportional to that of a j-integrator plant. In particular if j > 2 at least one zero lies outside the unit circle.

3. If G(s) is strictly proper (less than n zeros) then for large h the discrete-time model tends to G(0)/z, which has no zero. Hence slow sampling 'overcomes' the problem; recall that a plant is effectively in open loop between samples.

4. Let δ be the fractional dead-time T-(k-1)h. Then if G(s) is proper and δ tends to h at least one zero of the discrete model becomes nonminimum-phase. A well-known example is the plant exp(-sδ)/s whose discrete-time model has a zero outside the unit disc if h/2 < δ < h. Fractional dead-times often arise because of computational delays in the controller algorithm.

In practice plant engineers wish to sample as rapidly as possible (as they are used to PID regulators). The above rules imply that there is a strong possibility of a nonminimum-phase plant model; slower sampling to 'cure' the problem is at the cost of inferior control performance. What is required is a controller design which is relatively insensitive to the positions of the zeros. This can be achieved, for example, by developing algorithms based on the operator s' = (z-1)/h which approximates the differential operator s for small h.

10.3   PREDICTIVE MODELS

Predictive control theory has its roots in the work of Smith (49) in which a model with no dead-time is used to predict the an advanced version of the output of a plant with dead-time. This allows a PID controller in the forward path to be more tightly tuned as the term exp(-sT) is eliminated from the characteristic equation. A Smith predictor, however, depends on a good plant model and does not predict the effects of the disturbance on the future output. More general predictors which form the basis of many self-tuning methods overcome these deficiencies. Let T be a __design__ polynomial:

$$T(z^{-1}) = 1 + t_1 z^{-1} + \ldots + t_n z^{-n}$$

and consider the __identity__ (or Diophantine equation):

$$T(z^{-1}) = E(z^{-1})A(z^{-1}) + z^{-k}F(z^{-1}) \qquad \ldots (9)$$

This equation can be used to obtain coefficients of the E and F polynomials given T, A and the dead-time k. Multiplying eqn.7 by E, replacing EA using eqn.9, and considering time t+k gives:

$$[T - z^{-k}F] \, y(t+k) = EB \, u(t) + E \, x(t+k)$$

or: $$T \, y(t+k) = F \, y(t) + G \, u(t) + E \, x(t+k) \qquad \ldots (10)$$

where the polynomial $G(z^{-1})$ = EB.   Define __filtered__ signals u'(t) and y'(t) to be the known data u(t) and y(t) passed through the (all-pole) filter 1/T, giving u'(t) = u(t)/T and y'(t) = y(t)/T.   Then a model which gives a prediction p(t) of the future output y(t+k) from currently available data u' and y' is:

$$p(t) \quad = F(z^{-1}) \, y'(t) + G(z^{-1}) \, u'(t) \qquad \ldots (11a)$$

and: $$y(t+k) = p(t) + E(z^{-1}) \, x(t+k)/T(z^{-1}) \qquad \ldots (11b)$$

In the purely deterministic case x(t+k) = 0 and thus the prediction is exact if the polynomials F and G are known. In the general case the quality of the predictor depends on the choice of the polynomial T. Often the 'best' predictor is required, and it is seen that if $x = Ce$ and $T = C$ the prediction error in eqn.11b becomes simply $Ee(t+k)$. Now eqn.9 shows that the degree of E is $k-1$, so $Ee(t+k)$ is a moving-average of order k whose last term contains $e(t+1)$ and is therefore independent of $u'(t)$ and $y'(t)$. This is the predictor which has minimal variance, and for this case $p(t)$ is often written as $y^*(t+k|t)$, i.e. the optimal prediction of $y(t+k)$ given data up to and including time t. However, $x(t)$ is not always describable by the simple form $Ce(t)$, and eqn.11 is merely considered here to be a 'good' predictor.

10.3.1 An Example.

Consider the plant model:

$$(1 - 0.8z^{-1}) \, y(t) = 0.2 \, u(t-2) + (1 + 0.7z^{-1}) \, e(t)$$

where $k=2$, with optimal prediction for $T = C$ in eqn.9:

$$(1 + 0.7z^{-1}) = (1 + e_1 z^{-1})(1 - 0.8z^{-1}) + z^{-2} f_0$$

Comparing powers of z^{-1} gives $e_1 = 1.5$, $f_0 = 1.2$ and so:

$$(1 + 0.7z^{-1}) \, y^*(t+2|t) = 1.2 \, y(t) + (0.2 + 0.3z^{-1}) \, u(t)$$

$$\tilde{y}(t+2|t) = e(t+2) + 1.5 \, e(t+1)$$

The prediction variance is that of the remnant $\tilde{y}$, i.e. $\sigma^2(1+1.5^2) = 3.25\sigma^2$. If $k = 1$ instead of 2 the prediction variance is simply σ^2; this confirms our intuition that the more distant the prediction horizon the less the expected accuracy. This reduction of accuracy leads to a loss of performance in a closed-loop control system in which output fluctuations can become large if the dead-time is significant; only feedforward can then improve matters.

10.3.2 General Predictive Models

In general prediction (Yaglom (54), Clarke and Gawthrop (15)) we are concerned with an auxiliary plant output $\psi(t)$ produced by passing $y(t)$ through a chosen transfer-function P:

$$\psi(t) = P(z^{-1}) \, y(t) = Pn(z^{-1})/Pd(z^{-1}) \, y(t) \qquad \dots (12)$$

The identity 9 then becomes:

$$C(z^{-1})Pn(z^{-1}) = E(z^{-1})A(z^{-1}) + z^{-k}F(z^{-1}) \qquad \dots (13)$$

Proceeding with the same development as before gives:

$$\psi(t+k) = F(z^{-1})y''(t) + G(z^{-1})u'(t) + Ex(t+k)/T \qquad \dots (14)$$

where $y''(t) = y(t)/PdT$ and $u'(t) = u(t)/T$ are filtered signals. Hence:

$$p(t) = F(z^{-1})\ y''(t) + G(z^{-1})\ u'(t) \qquad \ldots (15a)$$

and: $$\bar{\psi}(t+k|t) = E(z^{-1})\ x(t+k)\ /T(z^{-1}) \qquad \ldots (15b)$$

The argument for predicting an auxiliary output ψ is that the __augmented__ plant $G*P$ is 'easier' to control than the original plant.

Multi-step or 'long-range' predictors can be derived for which the prediction horizon exceeds k, and hence some assumption must be made about future control actions (De Keyser and van Cauwenberghe (19), Clarke et al (17)). Such predictors can be used with quadratic cost functions defined over a 'receding horizon', such as in LQG or the powerful 'Generalized Predictive Control' approach of (17). An important feature of these methods is that the adaptive controller does __not__ need to have prior knowledge of the dead-time k. Alternatively the predicted plant response could simply be displayed to the process operator, who could adjust the control actions accordingly (De Keyser and van Cauwenberghe (18)).

10.3.3   Incremental Predictors

Consider the plant model 6 with C = 1:

$$A(z^{-1})\ y(t) = B(z^{-1})\ u(t-k) + e(t)/\Delta, \text{ and:}$$

$$1 = E(z^{-1})A(z^{-1})\Delta + z^{-k}\ (1 + \Delta F(z^{-1})).$$

Multiplying the above model by $z^k EA\Delta$ we obtain:

$$y(t+k) = y(t) + F(z^{-1})\ \Delta y(t) + G(z^{-1})\ \Delta u(t) + E(z^{-1})\ x(t+k)$$

This model predicts __changes__ in the output y in terms of known changes in the current data Δy and Δu. This is important in applications as dc levels on the data do not affect the prediction performance. Moreover, when used in self-tuning the data-vectors have zero mean and hence give significantly better estimation properties. See Tuffs and Clarke (51) for a discussion of the use of incremental predictors in adaptive control.

10.4   PREDICTIVE CONTROL

Having derived predictors for both y(t+k) and $\psi(t+k)$ we are now in a position to develop a range of control laws with useful properties depending on the choice of P and T. Practical features of the controlled plant indicate which choices are appropriate. In the following the set-point is denoted as w(t); in many cases w(t) remains constant but it may undergo step changes at infrequent intervals. Note that the prediction of eqn.15a involves the <u>current</u> control u(t) which is to be chosen on the basis of its effect on the predicted variable.

A simple policy is to choose u(t) such that the prediction p(t) becomes equal to the set-point w(t):

i.e. $p(t) = w(t)$ ... (16)

So that, using eqn.11a, the algorithm leads to:

$$F(z^{-1}) \, y'(t) + G(z^{-1}) \, u'(t) = w(t).$$

Hence, recalling that u'(t) = u(t)/T, this implies a control:

$$u(t) = \frac{T}{G} \{ \, w(t) - \frac{F}{T} y(t) \, \}$$

Solving to get the closed-loop equations, we have:

$$y(t) = w(t-k) + E \, x(t)/T \qquad \text{... (17a)}$$

$$u(t) = \frac{A}{B} w(t) - \frac{F}{BT} x(t) \qquad \text{... (17b)}$$

Eqn.17 shows that this control law produces a dead-beat response to set-point changes (y(t+k) = w(t)). The effect of the disturbances x(t) is 'tailored' by the user-chosen polynomial T, so that the control signal amplitude following a step change in x can be damped by choosing a 'sluggish' characteristic in T. Note that the effect of the disturbance on the controlled loop is simply that of the error of the prediction model of eqn.11b, so good prediction implies good closed-loop performance. For example, if T = C and the process admits the stochastic model of eqn.4, the prediction and hence the control is minimum-variance, and output fluctuations are as small as possible. This is useful in quality control, such as when applied to a paper-making plant in which the thickness of the sheet is to be maintained within close limits.

Although the control policy is 'ideal', it achieves its objective by cancelling the plant dynamics, which as seen before can cause instability when there are nonminimum-phase zeros. Even if there are no zeros outside the unit circle, their positions might lead to badly damped control signals.

10.4.1 Model-following Control.

In practice an engineer knows that dead-beat performance leads to excessive control signals, and he has an intuitive idea of the ultimate achievable speed of the closed-loop (e.g. by knowing the open-loop response). Hence a more appropriate objective is to make the closed-loop respond at a rate determined by a prescribed model M, such that if x(t) = 0:

$$y(t) = M(z^{-1}) \, w(t-k) \qquad \qquad \ldots (18)$$

If M = 1 dead-beat control is specified, but if M is equipped with dynamics the closed-loop would respond to steps (and to load-disturbances) with the corresponding speed. One obvious requirement is that the closed-loop gain is unity, so that M(1) = 1. One way to arrive at eqn.18 is to prefilter the set-point to give a new signal w'(t) = Mw(t), and then to use dead-beat control with a set-point of w'(t), though this would not affect the response to loads. It is therefore better in practice to embed the model within the closed-loop, as it can be shown (Lim (38)) that the resulting controller is more robust. This is done by using the auxiliary output described above.

If in the generation of ψ = Py the transfer-function P is chosen to be the inverse model P = 1/M, and eqn.16 uses the prediction of ψ(t) instead of y(t), then if x = 0 the closed-loop will satisfy ψ(t+k) = w(t). Hence y(t+k) = ψ(t+k)/P = w(t)/P = Mw(t), as required. In the general case where x(t) is non-zero, the closed-loop equations become:

$$y(t) = M(z^{-1}) \, \{ w(t-k) + \frac{E}{T} x(t) \} \qquad \ldots (19a)$$

$$u(t) = M(z^{-1}) \, \{ \frac{A}{B} w(t) - \frac{F/Pd}{BT} x(t) \} \qquad \ldots (19b)$$

Note that the model M affects both the set-point and the disturbance responses. Hence P and T can be chosen in accordance with the expected variations in w(t) and x(t) and the loop then behaves as desired. However, eqn.19b shows that there are still cancellations in the forward path and the algorithm is therefore unsuitable for nonminimum-phase plant; this is a defect shared by 'model-reference adaptive controller' (MRAC) designs.

10.4.2   Control Weighting.

The Smith predictive approach inserts a controller (generally PID) into the forward path which is fed by the prediction error rather than the conventional error w(t) - y(t), and hence eliminates the dead-time from the characteristic equation. In the same way, the general predictive control is given by:

$$u(t) = \{ w(t) - p(t) \} / Q(z^{-1}) \qquad \ldots (20)$$

where p(t) is given by eqn.11a for a 'generalised Smith

controller or eqn.15a for a 'detunable model-following' controller. The user-chosen transfer-function Q is typically such that 1/Q is of PI form, and in the non-adaptive case its parameters could be selected as with the Smith method (e.g. hand tuned). The original Smith algorithm is seen to be a subset of the above for the case where $P = T = 1$ and $x(t)$ is assumed to be constant.

Substituting for $p(t)$ from eqn.15a gives:

$$Q(z^{-1})\, u(t) = w(t) - \{\ F(z^{-1})\ y''(t) + G(z^{-1})\ u'(t)\ \}$$

and solving for the current control $u(t)$ gives:

$$u(t) = [Q + G/T]^{-1}\{\ w(t) - F\ y(t)/TPd\ \}$$

The closed-loop equations are now:

$$y(t) = \frac{B}{PB + QA}\ w(t-k) + \frac{EB + QT}{PB + QA}\ \frac{x(t)}{T} \qquad \ldots (21a)$$

$$u(t) = \frac{A}{PB + QA}\ w(t) - \frac{F/Pd}{PB + QA}\ \frac{x(t)}{T} \qquad \ldots (21b)$$

The crucial point here is that this is no longer a cancellation law; the characteristic equation is instead $PB + QA = 0$. Often the transfer-function Q is written in the form λQ_1, and this gives the closed-loop poles to be the roots of $PB + \lambda Q_1 A = 0$. Hence λ is a root-locus parameter which moves the poles from B (small λ), which could be unstable, toward A (large λ). If the plant is open-loop stable this means that the plant will be closed-loop stable for large enough λ, even if it is nonminimum-phase. It is seen, though, that increasing λ 'detunes' the performance, as the closed-loop model is no longer specified by 1/P alone. In practice this has not been found to be significant. Note that the dead-time k has been eliminated from the loop, as seen in Fig.2 which shows the effect of the predictor and the P/Q design polynomials.

The predictive control method was developed following the seminal work of Astrom and Wittenmark (6), who invoked a minimum variance objective and thus produced a cancellation controller. The objective of Clarke and Gawthrop (14) was a generalisation of this, as it was shown that for $P = 1$, $T = C$ and $Q = \lambda$ the controller minimises a cost-function:

$$J = E\ \{\ (y(t+k) - w(t))^2 + \lambda g_0 u^2(t)\ |\ t\ \}$$

where the expectation is conditioned on data acquired up to time t. In the general case, the controller of eqn.20 can be shown to minimise:

$$J_1 = E\ \{\ (\psi(t+k) - w(t))^2 + g_0/q_0\ (Qu(t))^2\ |\ t\ \}$$

The effect of 'detuning' is seen both in the cost-functions and in the fact that the control $u(t)$ can be written as:

$$u(t) = u^*(t) / (1 + Q/g_0)$$

where $u^*(t)$ is the control required to attain exact model-following performance. For example, if Q is simply λg_0 with λ chosen to be 1, exactly half the model-following control is exerted. In general λ is used to 'trade' control variance against output variance, so that large control signals are not produced.

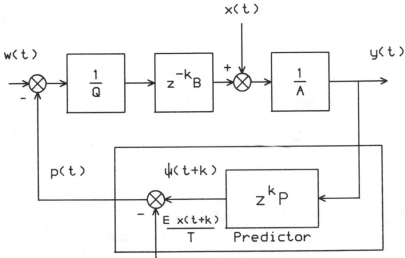

Fig.2 **Feedback loop of predictive control**

10.4.3 **Feedforward**.

One nice feature of predictive control is that measurable load-disturbances can easily be included in the algorithm. Suppose one component of $x(t)$ is a known signal $v(t)$, giving a system model (c.f. eqn.4):

$$A(z^{-1}) y(t+k) = B(z^{-1}) u(t) + D(z^{-1}) v(t+k) + x_1(t+k)$$

where the variance of $x_1(t)$ is clearly less than that of $x(t)$. This equation shows that the current control can be chosen to eliminate components of $v(t)$ which are known at time t, but not future components. Hence complete elimination of $v(t)$ is only possible if the delay in v is the same as or greater than the delay in u; one example is in metal rolling where the incoming guage can be measured. The control signal might still be large (see Allidina and Hughes (1)) and constraints might preclude complete removal of the effect of $v(t)$.

Following the development of eqns.9 to 15, and defining polynomials E_1 and F_1 from the identity:

$$ED = E_1 + z^{-k}F_1,$$

Then:

$$\psi(t+k) = Fy''(t) + Gu'(t) + F_1v'(t) + \{E_1v(t+k) + Ex_1(t+k)\}/T$$

This gives a prediction signal:

$$p(t) = F\ y''(t) + G\ u'(t) + F_1v'(t).$$

which can be used in the control law of eqn.20. If $Q = 0 = E_1$ then no component of v remains in the output under control: 'ideal' feedforward compensation.

Clearly as many feedforward signals as required can be included in the predictor. One interesting use is to reduce interactions between coupled loops by considering the input and output of loop A, for example, to be feedforward signals for other loops (Morris et al (42)). A single-loop self-tuner can therefore be easily extended to the multivariable case.

The predictive control method is very versatile: it can cope with dead-time, provide a model-following loop, and eliminate measurable disturbances. A parameter λ allows the user to 'tune' the performance on-line to balance control effort against output regulation. Moreover it is readily included in a self-tuning controller. However for adaptive control it suffers from the disadvantage that the dead-time k needs to be known reasonably well; current work (LQG, GPC) in self-tuning attempts to retain the design features whilst covering the varying dead-time case.

10.5   POLE-PLACEMENT CONTROL

In a series of papers (e.g. (51)) Wellstead and his coworkers developed the idea of using pole-placement as the underlying design algorithm in a self-tuner. The relationship between this approach and pure model following was explored by Clarke (12). The idea is that the offending zeros of the plant should not be cancelled by the controller, which instead should place only the poles of the closed-loop in desired locations (such as in 'minimum ripple' designs of classical discrete-time theory).

Let $T(z^{-1})$ be the filter (observer) polynomial defined before, and $u'(t)$ and $y'(t)$ be the corresponding filtered signals u/T and y/T. Model 7 then becomes:

$$A(z^{-1})\ y'(t) = B(z^{-1})\ u'(t-k) + x(t)\ /\ T(z^{-1})$$

Let w(t) be the set-point and consider the control in Fig.3:

$$G(z^{-1})\ u(t) = H(z^{-1})\ w(t) - F(z^{-1})\ y(t)$$

Eliminating u(t) from these equations gives:

$$(AG + z^{-k}BF) \; y'(t) = BH \; w'(t-k) + x(t)/T$$

The left-hand side term $(AG + z^{-k}BF)$ is the closed-loop characteristic polynomial, whose roots are the poles. Suppose $Am(z^{-1})$ is a user-chosen polynomial which has roots in the desired pole locations. Then let polynomials F and G be computed from the Diophantine equation:

$$A(z^{-1})G(z^{-1}) + z^{-k}B(z^{-1})F(z^{-1}) = T(z^{-1})Am(z^{-1})$$

The closed-loop equation then becomes:

$$Am(z^{-1}) \; y(t) = [B(z^{-1})H(z^{-1}) \; w(t-k) + x(t)] \; / \; T(z^{-1})$$

Now let the design polynomial H be given by $T(z^{-1})H_1(z^{-1})$,

so that: $$y(t) = [BH_1 \; / \; Am] \; w(t-k) + x(t) \; /[AmT].$$

In this way Am determines the modes of the set-point response and AmT the modes of the disturbance response. To get zero steady-state error the gain between w and y should be unity; this is achieved by choosing the gain of H to give $B(1)H_1(1) = Am(1)$. It would appear that the roles of T and Am are interchangable, and that T is not really necessary. The discussion below shows that T is indeed useful in the self-tuning application of the pole-placement approach.

Suppose F_0 and G_0 are polynomials given by a related Diophantine equation:

$$A(z^{-1})G_0(z^{-1}) + z^{-k}B(z^{-1})F_0(z^{-1}) = C(z^{-1})Am(z^{-1})$$

where the disturbance x(t) is given by the stochastic model $C(z^{-1})e(t)$. When the polynomials F_0 and G_0 are used in a _regulator_ – that is when w(t)=0, the output and the control signals are given by:

$$y(t) = G_0(z^{-1})/Am(z^{-1})e(t); \qquad u(t) = -F_0(z^{-1})/Am(z^{-1})e(t)$$

Now let $A_1(z^{-1})$ and $B_1(z^{-1})$ be polynomials which satisfy:

$$A_1(z^{-1})G_0 + z^{-k}B_1(z^{-1})F_0 = T(z^{-1})Am(z^{-1}) \qquad \ldots (22)$$

where Am and T are the design polynomials described above. Consider now the sequence $A_1y'(t)-B_1u'(t-k)$:

$$A_1y'(t) - B_1u'(t-k) = [A_1G_0+z^{-k}B_1F_0]/TAm \; e(t),$$

using the properties of u(t) and y(t) given above. But the term within the brackets [] is simply TAm, by the second Diophantine equation (22). Hence in closed-loop the input-output model is:

$$A_1(z^{-1}) \; y'(t) = B_1(z^{-1}) \; u'(t-k) + e(t) \qquad \ldots (23)$$

The crucial point is that for arbitrary stable T and C the error term on the RHS of eqn.23 is simply e(t), and the unknown polynomial C need not be accounted for in the design. Hence a self-tuner would consist of the following:

1. At each sample instant use filtered data u'(t-k) and y'(t) to estimate the parameters of the polynomials A_1 and B_1, using RLS as described below.

2. Use $\hat{A}_1$, $\hat{B}_1$ and the design polynomials T and Am in eqn.22 to obtain controller parameters $\hat{F}_0$ and $\hat{G}_0$ (this can be quite time-consuming, involving the solution of a set of linear equations).

3. Use $\hat{F}_0$ and $\hat{G}_0$ in the controller:

   $$u(t) = -\hat{F}_0/\hat{G}_0 y(t).$$

4. Return to 1.

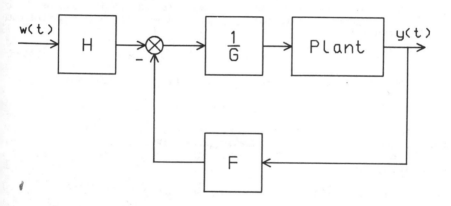

Fig.3 <u>Pole-placement feedback structure</u>

T does not appear in the closed-loop performance, but its role now is to improve the behaviour of the parameter estimator, as it can be shown that T = C gives the fastest rate of convergence. Alternatively one could consider T to act as a low-pass filter which removes high frequency noise from the data used in the estimation, so providing greater reliability. Note that the above discussion relates to the use of pole-placement as a pure regulator; in the more general case with varying w(t) a full estimation of A,B and C is required. Even so, T still has a role to play in robust estimation.

10.6   RECURSIVE PARAMETER ESTIMATION

The closed-loop model of eqn.23 and the predictor models of eqn.11a and eqn.15a contain parameters which depend on the unknown plant. The job of a parameter estimator is to provide these parameters for the control design algorithm based on available plant input/output data. Define $\theta = \{\theta_1, \theta_2, \ldots, \theta_n\}$ to be a vector of n unknown plant parameters and $\hat{\theta}(t)$ to be a vector of corresponding estimates available at time t. Let $s(t) = \{s_1, s_2, \ldots, s_n\}$ be a vector of available data, which is assumed to be __exactly__ known at time t. A __predictor model__ generates a prediction $\hat{y}(t)$ (or $\hat{y}$) depending on $\hat{\theta}(t-1)$ and $s(t)$:

$$\hat{y}(t) = F \{ \hat{\theta}(t-1); s(t) \}$$

and the scalar __prediction error__ is defined to be:

$$\epsilon(t) = y(t) - \hat{y}(t) = y(t) - f \{ \hat{\theta}(t-1); s(t) \}$$

where $y(t)$ is the new output measurement provided at time t. In what follows we restrict analysis to predictive models which are __linear in the parameters__, in which the prediction $\hat{y}$ is given by:

$$\hat{y}(t) = \Sigma_1^n \hat{\theta}_i(t-1)s_i(t) \qquad \ldots (24)$$

The plant is assumed to satisfy the linear equation:

$$y(t) = \Sigma_1^n \theta_i s_i(t) + e(t),$$

in which $e(t)$ is a disturbance term. (This need not correspond to a linear dynamic plant model as, for example, s_i could be $u^2(t-i)$.) The above equations can be written concisely as:

$$\hat{y}(t) = \hat{\theta}'(t-1)s(t), \quad \text{and} \quad y(t) = \theta's(t) + e(t),$$

so that if the parameter error $\bar{\theta}(t)$ is defined to be $\theta - \hat{\theta}(t)$ then:

$$\epsilon(t) = \bar{\theta}'(t-1)s(t) + e(t) = s'(t)\bar{\theta}(t-1) + e(t)$$

This useful result shows that the prediction error of the model depends on both the modelling error θ and the noise $e(t)$. Recall that the prediction is directly required in predictive control, so that this equation shows how a bad model can affect adaptive control performance.

A __recursive prediction-error algorithm__ provides a new estimate $\hat{\theta}(t)$ using an equation of the form:

$$\hat{\theta}(t) = \hat{\theta}(t-1) + a(t)M(t)s(t)\epsilon(t) \qquad \ldots (25)$$

where $a(t)$ is a scalar 'gain-factor' giving the step length and $M(t)$ is a matrix which modifies the parameter update direction. In general we want the algorithm to have the following properties:

1. When little is known about the plant, for example when the self-tuner starts up, the parameter estimates are allowed to move rapidly, hopefully to the 'true' values θ. Here $\epsilon(t)$ is interpreted as a modelling error and we are considering the tuning phase.

2. During steady-state operation the movements of the estimates should be small as then $\epsilon(t)$ is mostly noise.

3. The allowable rate of movement in the steady-state should depend on the expected speed at which the true parameters might change. This is the adaptive phase.

4. The computational burden per sample instant should not change with time, nor should the data vectors be allowed to expand indefinitely.

5. If n is large a simplified approach might be required for computational reasons; otherwise the 'best' method should be used.

Note that eqn.25 can be written as:

$$\bar{\theta}(t) = [I - a(t)M(t)s(t)s'(t)]\bar{\theta}(t-1) - a(t)M(t)s(t)e(t)$$

so that a value of $a(t)$, say, which rapidly reduces the error also increases the effect of $e(t)$ on the parameters.

10.6.1 Simple Algorithms.

Consider the 'cost-function' $J = \epsilon^2 = (\bar{\theta}'s+e)^2$. The gradient ∇J of J has components $\partial J/\partial\theta_i$ which equal $-2\epsilon(t)s_i u(t)$. To minimise J using the method of steepest descent, we choose to change the parameter estimates in the direction of $-\nabla J$. The simplest such method, known as the LMS algorithm, puts $a(t) = \mu$ and $M(t) = I$, giving:

$$\hat{\theta}(t) = \hat{\theta}(t-1) + \mu\, s(t)\, \epsilon(t)$$

This involves very little computation (one multiplication and addition per parameter), and is used when n is large, such as with weighting-sequence models. Typical applications are to adaptive equalizers in communications. However, the adaptive gain μ has to be chosen with care: divergence is obtained if μ is too large, and the critical value of μ depends on the autocorrelation matrix of the input data, which is not generally known. For adaptive control better methods are required.

A 'recursive learning' method is used by Richalet (46) and was developed from purely deterministic considerations in which the noise e(t) is ignored and the objective is to adapt as rapidly as possible:

$$\hat{\theta}(t) = \hat{\theta}(t-1) + \lambda/[s'(t)s(t)] \; s(t) \; \epsilon(t)$$

This can be shown to converge if $0 < \lambda < 2$ and 'optimally' for $\lambda = 1$. It scales the adaptive gain using known data, and can be thought of as an optimal LMS design. The computational requirements are approximately doubled but it produces much better performance. Table 1 gives an outline Fortran routine for implementing the method. SCAPRO is a function which computes the scalar product of two arrays. The vectors S and THETA are of dimension N, the number of parameters. SMALL is a constant which prevents overflow if the data vector s(t) is null (quiescent conditions).

TABLE 1 The recursive learning method

```
      SUBROUTINE LEARN(Y,S,THETA,N,RLAMDA)
C
      PERR = Y - SCAPRO(THETA,S,N)    !prediction error ε
      UNDER = SCAPRO (S,S,N)          !denominator
      IF (UNDER .LT. SMALL) GOTO 999  !little data?
      GAIN = RLAMDA*PERR / UNDER      !adaptive gain
C
      DO 1 I = 1,N                    !N parameters
1     THETA(I) = THETA(I) + GAIN*S(I)
```

Under special circumstances where the components of s(t) are uncorrelated with variance s^2, and the variance of e(t) is σ^2, it can be shown that the variance of parameter estimate i converges for large t to:

$$\text{Var } \{\tilde{\theta}_i\} \rightarrow \lambda \; \sigma^2 / [n(2-\lambda)s^2]$$

This result shows that a large signal/noise ratio s/σ and a small value of λ is required to get an accurate model. Hence to get good adaptive speed initially and a good model eventually a <u>variable</u> λ should be used.

10.6.2 <u>Recursive Least Squares (RLS)</u>.

If the identification problem is treated in a classical statistical framework of finding the <u>best</u> estimates of θ given data $\{s(t),y(t); t=1..N\}$, the method of least-squares is optimal. A function

$$S(\hat{\theta}) = \Sigma_1^N \; \epsilon^2(t,\hat{\theta})$$

is defined and minimised with respect to $\hat{\theta}$. The formulation however leads to a set of n equations in the n unknown parameters in which the vectors and matrices involved expand with N, the experiment's duration; this is unsuitable for on-line use. Moreover, the estimate $\hat{\theta}(N)$ tends to a constant vector as N increases (i.e. the variance of $\hat{\theta}$ tends to 0). This is acceptable if it is known that the plant is constant, but in general we want to track (slowly) varying parameters. To solve these problems we use a recursive method and modify S to weight 'old' data using a FADING MEMORY:

$$S_1(\hat{\theta}) = \Sigma_1^N \beta^{N-t}\epsilon^2(t,\hat{\theta})$$

Here β (which is a number slightly less than 1) is called the FORGETTING FACTOR, as the contribution of data j samples in the past is weighted by β^j. A useful measure here is the 'asymptotic sample length' or 'data window': $\alpha = 1/(1-\beta)$, being the approximate number of samples which are significant for the current estimate $\hat{\theta}(t)$.

It can be shown (see Clarke (11) that the recursive least squares (RLS) method which is an __exact__ equivalent of the nonrecursive version and which minimises S_1 is:

$$\hat{\theta}(t) = \hat{\theta}(t-1) + K(t) \epsilon(t)$$

... (26a)

$$K(t) = P(t-1) s(t) / [\beta + s'(t)P(t-1)s(t)]$$

... (26b)

$$P(t) = [I - K(t)s'(t)] P(t-1) / \beta$$

... (26c)

where K(t) is the 'Kalman-gain' vector and P(t) is an $n \times n$ symmetric matrix.

Algorithm 26 operates as follows. Initially P(0) is chosen, typically as γI where γ is 0.1-100, say. At each time t, $\epsilon(t)$ is computed using the the new data and eqn.24; then eqn.26b is used to update K(t), then eqn.26c to update P(t) and finally eqn.26a to obtain the new parameter estimates $\hat{\theta}(t)$. Note that the data storage ($\hat{\theta}$, s, P and K) and computational requirements stay constant with time.

One advantage of the RLS method is that the matrix P(t) is proportional to the covariance of the parameter estimates, so that a near-singular P indicates that there are badly estimated directions in parameter space. This could indicate, for example, that the number of parameters is too large or that the data is bad.

In practice there can be numerical problems with the updating, particularly with eqn.26c. P(t) should be symmetrical and positive definite, but round-off errors may induce P to lose symmetry or definiteness. In this case the algorithm will go unstable and the estimates 'explode'. This does not usually happen until about 5000 iterations, so that it would not be detected in simulations but would inevitably occur in a real plant controller. The cure, which is essential is practice, is to update a __factor__ of (t). As P is positive, it can always be written as UDU', where U is an upper-triangular matrix with units down the diagonal and where D is a diagonal matrix. The updating is now that of U and D, and P reconstructed (if desired) from UDU'. General ideas are in Bierman (8), and Fortran coding is in Clarke (10),(11); Peterka (44) describes a related 'square-root' method.

10.6.2.1 <u>Properties Of RLS</u>. - If $P(0) = \gamma I_n$, where γ is large <u>and</u> the plant is noise-free <u>and</u> if the s(t)-vector is 'persistently exciting' (e.g. with an injected test signal) <u>and</u> if the plant is accurately modelled by a constant parameter vector θ, then:

$$\hat{\theta}(t) \rightarrow \theta \text{ \underline{exactly} in n steps.}$$

This is the fastest possible rate of convergence. For example, if the self-tuner has 5 parameters then 5 samples (after the data vectors have been filled) are sufficient to get good estimates. With this quality of estimation there is plenty of leeway to deal with drifting plant and with noise.

10.7   <u>EXPECTED ACCURACY OF THE ESTIMATES</u>

The accuracy of the estimated model is a function of several aspects of the design and of the practical aspects of the plant:

1. The signal/noise ratio: if this is small then either slow convergence (small adaptive gain) or a poor model (large adaptive gain) is obtained. If the method is to be used as part of a self-tuner (which requires suitable precautions) a small value of λ or a value of β near to 1 is needed.

2. The particular method used. Simple methods can be 'tuned' to give either reasonably fast convergence or good noise-rejection. The RLS/UDU is better as its gain is chosen optimally from the data. For example if P(t) is large, which implies poor estimates, then K(t) would be large and thus allow the estimates to move quickly.

3. The number of parameters estimated. In general the fewer the number of parameters the better the model is as a predictor, provided that it is adequate. If too many parameters are involved there will be poorly estimated directions; this is more relevent to methods such as pole-placement than to predictive controllers.

4. The inadequacy of the model structure. e.g.:

o choosing n too small
o rapid fluctuations in θ
o nonlinearities in the plant and/or actuators
o stiction in the actuator which stops the control signal affecting the plant
o quantization noise in the output transducer
o unmeasured load-disturbances not modellable as 'white noise'.

10.8 SELF-TUNING

A self-tuner connects the estimator to a control design algorithm. To show how this is done the minimum variance self-tuner of Astrom and Wittenmark (6) will be described in which model 4 is assumed and where the predictor model 11 with T = C can be written:

$$C(z^{-1})p(t) = F(z^{-1})y(t) + G(z^{-1})u(t) \qquad \ldots (27a)$$

$$y(t+k) = p(t) + E(z^{-1})e(t+k) \qquad \ldots (27b)$$

The corresponding control law for w(t) = 0 is from eqn.16:

$$F(z^{-1})y(t) + G(z^{-1})u(t) = 0 \qquad \ldots (27c)$$

Now if $H(z^{-1}) = z[1-C(z^{-1})]$ the prediction at t-k becomes:

$$p(t-k) = F(z^{-1})u(t-k) + G(z^{-1})u(t-k) + H(z^{-1})p(t-k-1)$$

But control 27c has set p(t-k-1), p(t-k-2) .. to 0, so:

$$y(t) = F(z^{-1})u(t-k) + G(z^{-1})u(t-k) + E(z^{-1})e(t)$$

i.e. $$y(t) = f_0 y(t-k) + ..+ g_0 u(t-k) + ..+ \epsilon(t) \qquad \ldots (28)$$

We can consider this model to predict the present outcome y(t) based on previous data y(t-k), u(t-k). As $\epsilon(t)$ is simply a moving average of order k, it contains no term which is correlated with the data. Hence RLS (or any other estimation algorithm) can be used to obtain estimates of the unknown parameters $\{f_i, g_i\}$. The certainty-equivalent self-tuner of Fig.1 simply accepts these estimates as if they were exact and exerts a control:

$$\hat{f}_0 y(t) + \hat{f}_1 y(t-1) + \ldots + \hat{g}_0 u(t) + \hat{g}_1 u(t-1) + \ldots = 0$$

Note that the control u(t) is unaffected if an arbitrary factor α multiplies the above equation, implying that the estimates will not be unique. To avoid 'wandering' of the parameters, a value of any parameter can be assumed (this problem does not arise in the general case where w(t) is nonzero). Typically a value $\bar{g}$ is assumed for g_0, and provided $0.5b_0 < \bar{g} < \infty$ the algorithm will still converge to give the correct control action. The control is then:

$$u(t) = - [\hat{f}_0 y(t) + \dots + \hat{g}_1 u(t-1) + \dots] / \bar{g} \qquad \dots (29)$$

and the corresponding model used in RLS estimation is:

$$y(t) - \bar{g} u(t-k) = f_0 y(t-k) + . + g_1 u(t-k-1) + . + \epsilon(t) \qquad \dots (30)$$

The operation of the self tuner is as follows. At each sample instant t:

1. From the current output y(t) form $y(t) - \bar{g} u(t-k)$ and update the parameters $\hat{f}$ and $\hat{g}$ using equations 26.

2. Given the parameter estimates, generate the control u(t) from eqn.29.

3. Shift all arrays along one place ready for the next sample.

4. Go to 1.

This self-tuner is called IMPLICIT as the estimated parameters are those of the feedback controller; an EXPLICIT self-tuner estimates the 'ordinary' plant model of eqn.4 and then performs a design calculation (e.g. solves a Diophantine equation). Predictive controllers lead naturally to implicit self-tuners; explicit self-tuners (such as pole-placement and GPC) require more calculations and are possibly more sensitive to parameter errors. In MRAC designs (Narendra and Monopoli (43)) the corresponding terms are DIRECT and INDIRECT.

Suppose two routines SAVE and MOVE are available, in which SAVE moves old data one place along an array and slots in the new data, whilst MOVE transfers data from one array to another. Then the basic implementation of the above self-tuner is shown in Table 2.

TABLE 2 A self-tuner for minimising $E\{y^2(t)\}$

```
<acquire data y(t) into Y>
YBAR = Y - GBAR*USAVE(K)              !g fixed
CALL MOVE (YSAVE(K),S,NY)            !old y into s(t)
CALL MOVE (USAVE(K+1),S(NY+1),NU)
CALL RLS (YBAR,S,THETA,NY+NU,FORGET)
U = -(SCAPRO(THETA,YSAVE,NY) +
+        SCAPRO(THETA(NY+1),USAVE,NU))/GBAR
IF (U .GT. UMAX) U = UMAX            !clip u(t)
IF (U .LT. UMIN) U = UMIN
CALL SAVE (U,USAVE,NU+K)             !save u(t)
CALL SAVE (Y,YSAVE,NY+K)             !and y(t)
```

Note that the estimation requires data in the USAVE and YSAVE vectors as far as NU+K and NY+K in the past. Hence the self-tuning algorithm should be sequenced to allow the data vectors to be filled with good data before estimation is allowed to start. The clipping of u(t) in the above coding is to cover the practical case where there are physical limits on the control action. Suppose the desired value of U is 200, but the actuator saturates at 100. Then if the above clipping is not done, the data in the USAVE array would not correspond to that sent to the plant, and the estimated parameters would deviate from the true values.

10.8.1 Operating The Self-tuner.

There are several choices to be made to set-up a self-tuner for best performance. Mostly these are easy to make and the closed-loop control is relatively insensitive to the values chosen. Considering the simple algorithm described above, the relevent parameters are:

1. The sample-time h. This is generally the most critical parameter. Typically 1/10 of the dominant plant time constant is taken, but if there is a significant dead-time h should be such that k is 2 to 3. If the control is bad, possibly due to nonminimum-phase zeros, h should be increased or a different law such as pole-placement or general predictive control used instead.

2. The model order n. A good choice here is 3. The F polynomial has degree n-1 and the G polynomial has degree n+k-1, so about 2+4 = 6 (or 7) parameters would need to be estimated. If a control weighting Q is used fewer parameters can still be effective.

3. The initial parameters $\hat{\theta}(0)$. A good choice is to make the start-up control a simple low-gain proportional law, or first to identify the plant in open-loop to obtain reasonable estimates.

4. The forgetting factor β. A data-window $\alpha = 1/(1-\beta)$ of 50 implies a 'rapidly' varying plant; if $\alpha = 1000$ the plant is taken to be reasonably steady. A variable forgetting factor can be used (Fortescue et al (23)), but the 'best' such algorithm is not clear. Generally P(0) is taken to be a large diagonal matrix, except when the initial parameters are well known (as in batch applications).

5. The fixed parameter $\bar{g}$. This can be done by making the corresponding element in P(0) zero; if the data is scaled properly (e.g. 0-100%) the value is quite easy to choose as it corresponds to the first nonzero value on the step response. This choice is not required in algorithms other than minimum-variance.

When the self-tuner has been set-up its operation can be

sequenced: first the data vectors should be filled (about n+2k samples), then the parameters should be estimated (about a further 2n samples), then closed-loop control asserted. During the start-up phase a test-signal could be used. It may be possible to close the loop immediately (in simulations anyway), though quite large control signals could result which naturally would give good parameter estimates. By various choices of $\hat{\theta}(0)$, $P(0)$ and constraints UMAX,UMIN a whole range of initial control actions are possible. For noncritical plant there is freedom to allow large controls, with correspondingly rapid tuning. Most commercial self-tuners provide for an initial test-signal phase.

The simple minimum-variance control law is not recommended for practical applications as it cannot cope with nonminimum-phase plant or with variable dead-time. Either the detuned model-following or the pole-placement algorithms are preferable; there is then a rather more complex initialisation but again there is no difficulty in finding good values provided some plant knowledge is used (such as the open-loop response time).

10.9 JACKETTING

In practice the parameter estimator is the critical part of a self-tuner: it should be switched on only if the data-vector is providing information which could improve the model. In many cases full adaptation during ·normal operating conditions would lead to a steady deterioration of the model as nonlinearities, quantization noise and load disturbances would predominate. 'Jacketting' software describes the process by which the estimator is monitored to allow or to inhibit estimation on the judgement of likely performance. At one extreme it could simply consist of a switch under the control of the operator, but in general the recursive estimator becomes:

$$\hat{\theta}(t) = \hat{\theta}(t-1) + j(t)\ a(t)\ M(t)\ s(t)\ \epsilon(t)$$

where $j(t)$ is either 1 (enabled) or 0 (inhibited). The switch $j(t)$ would be set to 0 depending on user-chosen estimation 'dead-bands' Dx:

   1. If $|\Delta u(t)| < Du$ the change of control might not affect the process (e.g. a valve with stiction).

   2. If $|\Delta y(t)| < Dy$ the change in output might be due to measurement or to quantization noise. For example a real change less than one transducer quantum would not be detected, and a naive estimator would deduce a process gain of zero.

3. If $|y(t)-w(t)| < D\epsilon$ the controlled plant could be in a quiescent state in which the data in unlikely to be exciting.

4. If $|\epsilon(t)| > D\epsilon$ the prediction error could be due to a load-disturbance.

Note that even if j(t) is set to 0 by any of the above conditions being true, the data vector s(t) should still be updated in case estimation resumes later. An important feature of dead-bands is that the estimator and the control law become __robust__ against 'unmodelled dynamics': ignored high-frequency poles do not significantly affect the adaptive performance.

Although self-tuning is derived assuming that the currently estimated model is used in the controller, this need not be the case. In practice __several__ models can be used:

1. θ_a = 'currently estimated model', updated when j=1.

2. θ_b = 'best model' used in the control design.

3. θ_c = 'initial model', etc.

A possible 'jacket' could consist of:

1. __if__ good-data __then__ ⟨update θ_a⟩

2. __if__ θ_a predicts better than θ_b for a while __then__ ⟨transfer θ_a to θ_b and update the controller parameters⟩

Another possibility is to monitor the predictions of all the models to detect a significant plant change, and if so to start self-tuning again from its initial conditions. Limits on individual parameters or on the controller gain can be included, as with the Foxboro PID tuner. With Turnbull's device the user is informed of suggested new PID settings together with a 'confidence' factor, and these settings may be transferred on demand to the PID controller which otherwise operates with fixed parameters.

10.9.1 Offsets.

An offset arises when the average value of y(t) does not equal the set-point w(t), which is assumed to be constant. Some self-tuning methods can lead to offsets if precautions are not taken: a discussion is in Clarke et al (17),(50). Inspection of eqn.21 shows that with a general predictive controller both Q(1) and E(1) should be 0, and P(1) should be 1. This implies that Q must contain a __differencing__ element and hence the associated cost-function J_1 weights __changes__ in control u(t)-u(t-1).

One way to avoid offsets is to augment the plant with an integrator and to ensure that the gain associated with w(t) equals that of the gain of the feedback path. It is better, however, to use an incremental predictor as this also ensures that the data in s(t) has zero mean — a precondition for good estimation performance. With simple assumptions about the plant (such as a Brownian motion disturbance x(t)) it is found that self-tuning PID control laws arise naturally (Gawthrop (27), Proudfoot et al (45)).

10.10   APPLICATIONS

A self-tuning controller is simple enough to be mounted on an 8-bit microcomputer (Clarke and Gawthrop (16)), even though floating-point arithmetic is required for the estimation as the parameters have a potentially wide dynamic range. Hence there are several implementation options:

o   as a stand-alone controller for critical single-loop applications (45); Andreiev(2); see also ASEA's NOVATUNE, Foxboro's 'Exact' controller and Turnbull's 6355 Autotuner

o   as part of a DDC package with a more powerful CPU in which several loops are tuned simultaneously (Fjeld and Wilhelm (22))

o   as part of a distributed control system (Halme et al (30))

o   as a self-tuned predictor in which the operator closes the loop.

Practical experience with self-tuning has been growing rapidly over the last few years — more so than with any other control idea. Several thousand loops are now being adaptively tuned. This is because that the methods are not difficult to implement (though care with jacketting is required) and appear to meet a real industrial need. There are many cases in which self-tuning has been in routine operation on an industrial plant, and surveys appear in (5) (7) (9) (20) (21) (22) (32) (34) (41) (42) (43) etc. These cover a wide range of processes from paper-making and ship-steering to batch-reactor and distillation column regulation. Simple self-tuners are being considered for applications such as heating and air-conditioning systems and diesel engine control. With the appearance of commercial self-tuners adaptive control is now being routinely applied in industry.

Self-tuners do not remove the need for the control engineer's skill, but instead he has to consider the real control objective of the plant so that an appropriate design method can be chosen. Moreover the intelligence built into a self-tuner will enable automatic plant diagnosis to be included in a DDC package. Self-tuning theory itself is being extended to consider important questions of robustness (Gawthrop and Lim (28)) and new algorithms such as LQG (Lam (36)) and GPC (Clarke et al (17)). The problems of convergence and of good jacketting (Schumann et al (48)) are also of central importance, particularly for processes with rapidly varying parameters (as in aerospace) or with significantly variable dead-time.

REFERENCES

1.   Allidina,A.Y., Hughes,F.M. and Tye,C., 1981,
        Proc.IEE, Vol.128, Pt.D, No.6, pp.283-291.

2.   Andreiev,N., 1981, Control Engineering, August.

3.   Astrom,K.J.", 1980, in 'Applications of Adaptive
        Control', (ed.) Narendra and Monopoli,
        Academic Press.

4.   Astrom,K.J., 'Ziegler-Nichols auto-tuners', 1982,
        Report LUDFD2/(TFRT-3167)/01-025/(1982),
        Lund Institute of Technology.

5.   Astrom,K.J., 1983, Automatica,
        Vol.19, No.5, pp.471-486.

6.   Astrom,K.J. and Wittenmark,B., 1973,
        Automatica, Vol.9, No.2, pp.185-199.

7.   Belanger,P.R., 1980, in 'Applications of Adaptive
        Control', (ed.) Narendra and Monopoli,
        Academic Press.

8.   Bierman,G.J., 1977, 'Factorization methods for
        discrete system estimation', Academic Press.

9.   Cegrell,T. and Hedqvist,T., 1975, Automatica,
        Vol.11, No.1, pp.53-59.

10.   Clarke,D.W., 1980, in 'Numerical techniques for
        stochastic systems' (ed.) Archetti and
        Cugiani, North-Holland.

11.   Clarke,D.W., 1981, in 'Self-tuning and adaptive
        control' (ed.) Harris and Billings,
        Peter Perigrinus.

12.   Clarke,D.W., 1982, Opt.Control App. and Methods,
        Vol.3, pp.323-335.

13.   Clarke,D.W., 1982, Trans.Inst.M.C.,
        Vol.5, No.2, pp.59-69.

14.   Clarke,D.W. and Gawthrop,P.J., 1975, Proc.IEE,
        Vol.122, No.9, pp.929-934.

15.   Clarke,D.W. and Gawthrop,P.J., 1979, Proc.IEE,
          Vol.126, No.6, pp.633-640.

16.   Clarke,D.W. and Gawthrop,P.J., 1981, Automatica,
          Vol.17, No.1, pp.233-244.

17.   Clarke,D.W., Mohtadi,C. and Tuffs,P.S., 1984,
          OUEL reports 1555,1557.

18.   De Keyser,R.M.C. and van Cauwenberghe,A.R., 1979,
          'A self-tuning predictor as operator guide',
          IFAC Symposium on Identification and System
          Parameter Estimation, Darmstadt, FRG.

19.   De Keyser,R.M.C. and van Cauwenberghe,A.R., 1981,
          Automatica, Vol.17, No.1, pp.167-174.

20.   Dexter,A.L., 1981, Automatica,
          Vol.17, No.3, pp.483-492.

21.   Dumont,G.A. and Belanger,P.R., 1978,
          IEEE Trans.Autom.Control, Vol.AC-23,
          No.4, pp.532-538.

22.   Fjeld,M. and Wilhelm,R.G., 1981, Control
          Engineering,   October, pp.99-102.

23.   Fortescue,T.R., Kershenbaum,L.S. and Ydstie,B.E.,
          1981, Automatica, Vol.17, No.6, pp.831-835.

24.   Franklin,G.F. and Powell,J.D., 1980, 'Digital
          control of dynamic systems', Addison-Wesley.

25.   Gawthrop,P.J., 1977, Proc.IEE,
          Vol.124, No.10, pp.889-894.

26.   Gawthrop,P.J., 1980, Proc.IEE,
          Vol.127, Pt.D, No.5, pp.229-236.

27.   Gawthrop,P.J., 1982, 'Self-tuning PI and PID
          controllers', IEEE Conference on Appns.
          of Adaptive and Mutivariable Control, Hull.

28.   Gawthrop,P.J. and Lim,K.W., 1982, Proc.IEE,
          Vol.129, Pt.D, No.1, pp.21-29.

29.   Grimble,M.J., 1981, Int.J.Control,
          Vol.33, No.4, pp.751-762.

30.   Halme,A., Ahava,O., Karjalainen,T., Torvikoski,T.
          and Savolainen,V., 1981, 'Implementing and
          testing of some advanced control schemes in
          a microprocessor-based process
          instrumentation system', IFAC Congress.

31.   Hodgson,A.J.F., 1982, 'Problems of integrity in
          applications of adaptive controllers',

OUEL report 1436/82.

32. Hodgson,A.J.F. and Clarke,D.W., 1982, 'Self-tuning applied to batch reactors', IEEE Conference on Applications of Adaptive and Multivariable Control, Hull.

33. Isermann,R., 1981, 'Digital control systems', Springer-Verlag.

34. Kallstrom,C.G. and Astrom,K.J., 1978, 'Adaptive autopilots for large tankers', IFAC Congress, Helsinki.

35. Kurz,H., Isermann,R. and Schumann,R., 1980, Automatica, Vol.16, No.2, pp.117-133.

36. Lam,K.P., 1980, 'Implicit and explicit self-tuning controllers', OUEL report 1134/80.

37. Latawiec, K. and Chyra, M., 1983, Automatica, Vol.19, No.4, pp.419-424.

38. Lim,K.W., 1982, 'Robustness of self-tuning controllers', OUEL report 1422/82.

39. Ljung,L., 1977, IEEE Trans.Autom.Control, Vol.AC-22, No.4, pp.551-575.

40. Ljung,L., 1978, IEEE Trans.Autom.Control, Vol.AC-23, No.5, pp.770-783.

41. Moden,P.E. and Nybrant,T.,1980, 'Adaptive control of rotary drum driers', IFAC Symposium on Digital Computer Applications to Process Control, Dusseldorf, FRG.

42. Morris,A.J., Nazer,Y., Wood,R.K. and Lieuson,H., 1980, 'Evaluation of self-tuning controllers for distillation column control', IFAC Symposium on Digital Computer Applications to Process Control Dusseldorf, FRG.

43. Narendra,K.S. and Monopoli,R.V. (ed.), 1980, 'Applications of adaptive control', Academic Press.

44. Peterka,V., 1975, Kybernetika, Vol.11, No.1, pp.53-67.

45. Proudfoot,C.G., Gawthrop,P.J. and Jacobs, O.L.R., 1983, Proc.IEE, Vol.130, Pt.D., No.5, pp.267-272.

46. Richalet,J., Rault,A., Testud,J.L. and Papon,J., 1978, Automatica, Vol.14, No.5, pp.413-428

47.   Sandoz,D.J. and Swanick,B.H., 1972, Int.J.Control,
         Vol.16, No.2, pp.243-260.

48.   Schumann,R., Lachmann,K.H. and Isermann,R., 1981,
         'Towards applicability of parameter adaptive
         control algorithms', IFAC Congress, Kyoto.

49.   Smith,O.J.M., 1959, Instr.Soc. of America Journal,
         Vol.6, No.2, pp.28-33.

50.   Tuffs,P.S. and Clarke,D.W., 1985, Proc.IEE, Vol.132,
         Pt.D, No.3, pp.100-110.

51.   Wellstead,P.E., Edmunds,J.M., Prager,D. and
         Zanker,P., 1979, Int.J.Control, Vol.30,
         No.1, pp.1-26.

52.   Whittle,P., 1963, 'Prediction and regulation by
         linear least-squares methods,
         English Universities Press.

53.   Wittenmark,B., 1973, 'Self-tuning regulators',
         Report 7312, Division of Automatic Control,
         Lund Institute of Technology.

54.   Yaglom,A.M., 1973, 'An introduction to the theory of
         stationary random functions'
         (translated by R.A. Silverman), Dover.

55.   Zanker,P.M. and Wellstead,P.E., 1979, 'Practical
         features of self-tuning', IEE Conference on
         Trends in On-line Computer Control Systems.

56.   Ziegler,J.G. and Nichols,N.B., 1942, Trans.ASME,
         Vol.64, pp.759-768.

Chapter 11

Computer aided design for industrial control systems

D.J. Sandoz

11.1 INTRODUCTION

This chapter presents a review of particular techniques that may be employed for the modelling and analysis of industrial processes and for the design of control systems to regulate those processes. These techniques have been integrated into a package of computer hardware and software that is now being marketed by Vuman Ltd. VUMAN stands for the "Victoria University of Manchester" and the company is wholly owned by Manchester University. The package is termed the "Plant Analysis System" or PAS for short. It has been successfully applied to a number of processes in a variety of industries.

The 'Plant Analysis System' is a companion package to the Vuman 'Real Time System.' The latter is an online facility for the monitoring of plant data and for the implementation of advanced controllers. Data collected via the 'Real Time System' may be communicated to the 'Plant Analysis System' for subsequent processing.

The capabilities of the PAS divide into four main functions; Signal Processing, Identification, Control System Design and Simulation. These functions are integrated together within a Computer Aided Design framework as illustrated in fig. 11.1. This is necessary since the procedures do not give rise to absolute solutions. The results are always open to interpretation by the user who is, ideally, able to apply engineering judgement to make appropriate decisions. Thus the normal sequence of analysis is not a single pass through to give rise to an answer. It rather involves a number of intermediate assessment stages and the reprocessing of the various functions to improve results in a systematic fashion. The PAS has therefore been structured to provide an efficient means for accessing the various functions and for interlinking their operations.

With reference to fig. 11.1, plant data that is to be analysed may be accessed via the 'Real Time System' or via some other mechanism (for example by directly entering the results of laboratory analyses or via some alternative automatic computer collection procedure). Various editting facilities are available to prepare the data for subsequent analysis. In addition, filtering and a variety of other forms of Signal Processing may be implemented to carry out preliminary calculations upon the data. Correlation and Power Spectrum analyses may also be applied in order to characterise the time domain and frequency properties of the data.

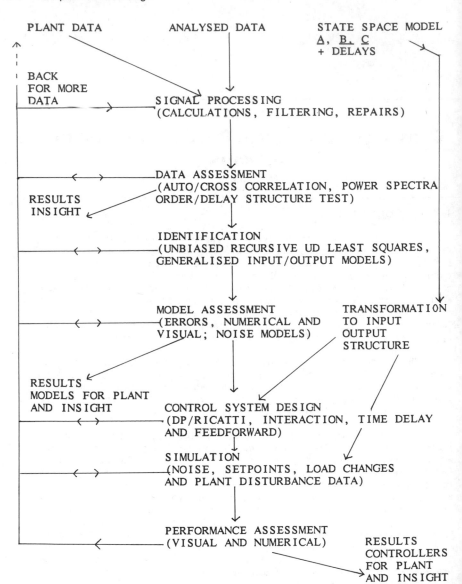

PLANT DATA   ANALYSED DATA   STATE SPACE MODEL
A, B, C
+ DELAYS

BACK
FOR MORE
DATA

SIGNAL PROCESSING
(CALCULATIONS, FILTERING, REPAIRS)

DATA ASSESSMENT
(AUTO/CROSS CORRELATION, POWER SPECTRA
RESULTS   ORDER/DELAY STRUCTURE TEST)
INSIGHT

IDENTIFICATION
(UNBIASED RECURSIVE UD LEAST SQUARES,
GENERALISED INPUT/OUTPUT MODELS)

MODEL ASSESSMENT   TRANSFORMATION
(ERRORS, NUMERICAL AND   TO INPUT
VISUAL; NOISE MODELS)   OUTPUT
STRUCTURE

RESULTS
MODELS FOR PLANT
AND INSIGHT   CONTROL SYSTEM DESIGN
(DP/RICATTI, INTERACTION, TIME DELAY
AND FEEDFORWARD)

SIMULATION
(NOISE, SETPOINTS, LOAD CHANGES
AND PLANT DISTURBANCE DATA)

PERFORMANCE ASSESSMENT
(VISUAL AND NUMERICAL)   RESULTS
CONTROLLERS
FOR PLANT
AND INSIGHT

Fig. 11.1

A "Model Identification" facility may be used to apply Least Squares procedures to build cause and effect models that approximate the dynamics that describe the variations in the data. These models have the capability to predict the behaviour of systems that involve multiple signals and may be used to simulate and investigate plant behaviour. In support of the Identification facility, there are techniques that may be employed to assess the most suitable model structures and to establish signal time delay characteristics. It is possible to directly enter linear dynamic models. These models take the standard state-space format and may be used as the basis for simulation and control system design studies.

Control System Design is achieved by carrying out optimisation calculations in association with an identified model to establish a controller that is able to regulate particular model output signals. The procedure is based upon a cost function minimisation criterion that uses Dynamic Programming/Ricatti Equation techniques and may give rise to a multivariable controller that incorporates both feedforward and time delay compensation. The derived controller has properties similar to those of a conventional industrial process control system, namely error feedback with integral action. It therefore provides a robust basis for industrial application.

A simulation facility provides a means for the assessment of the results of the identification and control system design procedures. Control systems may be simulated in application to derived models with disturbances applied that arise from real plant data. In this manner it is possible to evaluate the time response characteristics of a controller in order to assess if it is acceptable for implementation in real time.

The PAS operates in a computer that is based around the DEC 11/73 processor. An Acorn BBC computer is used to provide a high resolution low cost graphics terminal. System functions are accessed via an efficient menu-based command structure. Trend graphics provide the main means for presentation of information to the user. Many of the system facilities are automatically configured so that the user is not required to be too closely involved with housekeeping activities and may concentrate upon the engineering problems that are the subject of study.

This chapter concentrates upon two aspects in particular of the PAS.

i) the structure of models used for the representation of plant dynamics and the means for statistical identification of the parameters of such models (section 11.2); and

ii) the techniques for developing control systems from the models (section 11.3)

This chapter integrates with Chapter 21, which is concerned with particular case studies that relate to the material that is presented here. The two chapters must be taken together if a realistic appreciation of the techniques is to be obtained.

11.2 MODELLING OF PLANT DYNAMICS

This Section describes the techniques of the PAS that are available for determining mathematical models of plant dynamics to provide a basis for predicting process behaviour.   Models are established by applying statistical procedures to sampled information monitored from plant.   Such information is normally collected using the 'Real Time System'.   It is then transmitted to the PAS, via some disk storage medium, for subsequent analysis.   Data may also arise from other sources, such as from the results of laboratory analyses.

Section 11.2.1 presents the general characteristics of the models that are applied.   The model structure is discrete (ie it relates to sampled data) and may involve multiple plant input and output signals.   The structure is linear but can account for time delays that are associated with plant inputs.   The structure is attractive in that it may relate to simple single input and single output systems and also to complex systems involving many inputs and outputs.

Section 11.2.2 is concerned with constraints that must relate to the sampled data from plant in order for effective models to be established.   In particular, details relating to deliberate input signal perturbations for generating suitable plant data are considered.

Section 11.2.3 reviews the statistical procedures that are applied (ie Recursive Least Squares), giving an appreciation of the various figures of performance that arise to assist the user in judging the effectiveness of the results of analysis.

11.2.1 The Structure of Plant Models

Fig. 11.2 is a schematic illustration of the signals that may be associated with a model.

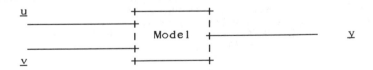

Fig. 11.2 Model Structure and Signals

The Signals are:

y,   a set of p measured values monitored from plant;

u,   a set of m independent actuation values applied to plant

and v a set of q disturbance values applied to plant.

The model predicts variations in y that arise because of variations in u and v.

Typically, signals associated with $\underline{y}$ will be temperatures, flows, densities etc. Those with $\underline{u}$ will be valve and pump settings or controller setpoints. Those with $\underline{v}$ will be a combination of both of the above, but will relate to external influences that are not affected by the $\underline{u}$ signals.

The model structure may be represented by a generalised equation (ie one that relates to all possible forms of models). This equation is detailed below(ie equation 11.1) and two examples are given afterwards for clarification.

Consider first the following definitions. These definitions arise as a result of extensive and detailed manipulation that commences with the standard State Space representation of linear multivariable systems (Sandoz and Wong (3), Sandoz (4))

i) $\underline{y}_k$ = value of $\underline{y}$ at instant k.

ii) The interval k to k+1 is T seconds.

iii)The Sampling Interval s = sT seconds, equivalent to the interval k to k+s.

iv) The Prediction/Control Interval c = cT seconds.

v)   The order of process dynamics is n, n$\geqslant$ 0.

vi)

$$\underline{Y}_k = \left|\begin{array}{c} \underline{y}_k \\ \underline{y}_{k-s} \\ : \\ \underline{y}_{k-(R-1)s} \end{array}\right| = \text{a } pR \times 1 \text{ vector of R samples } (R\geqslant 1)$$

of $\underline{y}$, sampled at instants sT seconds apart.   R is subject to the constraint n+p > p.(R-1).

vii) c> (R-1)s, ie. the total span of sampling within $\underline{Y}$ is less than the Prediction Interval.

viii) $\underline{Y1}_k = \underline{Y}_k$ if n +p $\geqslant$ p.R   otherwise $\underline{Y1}_k$ equals the first n+p elements of $\underline{Y}_k$.   The dimension of $\underline{Y1}_k$ is N1x1 with N1 = p.R if n+p $\geqslant$ p.R or N1 = n+p if n+p < p.R.

ix) $\underline{Y2}_k$ is compounded from the first n+p elements of

$$\left|\begin{array}{c} \underline{Y}_k \\ \underline{Y}_{k-c} \\ : \\ \underline{Y}_{k-(N-1)c} \end{array}\right| \quad \text{with } n+p > p.R(N-1) \text{ and } p.R \text{ } N\geqslant n+p.$$

x) $\underline{w} = \left|\begin{array}{c} \underline{u} \\ \underline{v} \end{array}\right| = \left|\begin{array}{c} w1 \\ : \\ wr \end{array}\right| = \text{an } r \times 1 \text{ vector} (r = m +q)$

compounded from both actuations and disturbances

xi) The signal wj, j = 1 to r, must either be constant across the interval k to k+c or its variation across this interval must be approximately linear (ie varying as a ramp). For the former case, ej = 0 and for the latter   ej = 1.

xii) dj is the minimum time delay associated with signal wj, rounded down to the nearest multiple of the prediction interval c.

xiii)

$$\underline{W}i_{k+c} = \left|\begin{array}{l} w_j{}_{k+(ej-dj)c} - w_j{}_{k-(ej-dj-1)c} \\ \qquad\qquad : \\ w_j{}_{k-(Rj-ej-1-dj)c} - w_j{}_{k-(Rj-ej-dj)c} \end{array}\right|$$

$\underline{W}j_{k+c}$ includes Rj samples of terms in wj. Each term represents the change in value of wj that has arisen across each Prediction Interval.

xiv) Rj = N+D$_j$ + e$_j$

D$_j$ is the "delay spread", D$_j \geqslant$ 0. If the minimum delay is correctly specified, then linear system analysis allows a delay spread of only 0 or 1. If 1, this accounts for the portion of delay discarded when rounding down to the minimum.

xv) If the term $\underline{Y}_{k-(N-1)c}$ in $\underline{Y}2_k$ involves only samples at the instant k-(N-1)c, after the truncation to n+p elements, then Rj is 1 less than specified above.

Given the above definitions, the generalised equation for plant models is

$$\underline{Y}1_{k+c} = \underline{\alpha} \ \ \underline{Y}2_k + \underline{\beta 1} \ \underline{W}1_{k+c} + \dots + \underline{\beta r} \ \underline{W}r_{k+c} \qquad\qquad 11.1$$

$\underline{\alpha}$ is the Transition Matrix and the matrices $\underline{\beta 1}$ to $\underline{\beta r}$ combine to form the Driving Matrix. The model predicts plant output values within $\underline{Y}1_{k+c}$, across the interval k to k+c, on the basis of a series of sampled output values within $\underline{Y}2_k$, sampled at instant k and earlier, and a series of changes in the various inputs $\underline{W}1$ to $\underline{W}r$. The latter changes relate to the interval k to k+c and a number of preceding intervals.

The model involves incremental terms in the input vectors $\underline{W}i$ (ie differenced values) and absolute terms in the measured vectors $\underline{Y}$. This format is used because it proves advantageous when the model is subsequently utilised to evaluate controllers, leading to error feedback with incremental actuation. The format requires that the Transition Matrix has a specific property such that in the steady state

$$\underline{Y}1_{ss} = \underline{\alpha} \ \underline{Y}2_{ss}$$

with $\underline{Y}1_{ss}$ and $\underline{Y}2_{ss}$ involving the same sampled vector $\underline{Y}_{ss}$ throughout.

Consider for example a first order single input and single output system. If y is the measured value, and u the actuation value, the model will have the form

$$y_{k+c} = (\alpha 1, \alpha 2) \begin{vmatrix} y_k \\ y_{k-c} \end{vmatrix} + \beta 1 (u_k - u_{k-c}) \quad 11.1a$$

if the delay spread is set to zero and if there is no time delay. It is assumed that u is constant across the interval k to k+c and that s =0.

Note that $\alpha 2 = 1 - \alpha 1$ in order to satisfy steady state requirements.

If $s \neq 0$, an alternative form is

$$\begin{vmatrix} y_{k+c} \\ y_{k+c-s} \end{vmatrix} = \begin{vmatrix} \alpha 11 & \alpha 12 \\ \alpha 21 & \alpha 22 \end{vmatrix} \cdot \begin{vmatrix} y_k \\ y_{k-s} \end{vmatrix} + \beta 1 (u_k - u_{k-c})$$

$$11.1b$$

Consider as a second example, a fourth order two input and two output system with c=4 and s=2. One input signal, u1, has a minimum delay of 0 with a delay spread of 2. The other signal, u2, has a minimum delay of 0, a delay spread of 1 and varies as a piecewise ramp.

The model has the form

$$\begin{vmatrix} y_{k+4} \\ y_{k+2} \end{vmatrix} = \underline{\alpha} \begin{vmatrix} y_k \\ y_{k-2} \\ y_{k-4} \end{vmatrix} + \underline{\beta 1} \begin{vmatrix} u1_{k-4} - u1_{k-8} \\ u1_{k-8} - u1_{k-12} \\ u1_{k-12} - u1_{k-16} \end{vmatrix}$$

$$+ \underline{\beta 2} \begin{vmatrix} u2_{k+4} - u2_k \\ u2_k - u2_{k-4} \\ u2_{k-4} - u2_{k-8} \end{vmatrix}$$

$$11.1c$$

Note that $\underline{\alpha}$ is a 4 x 6 matrix and $\underline{\beta 1}$ and $\underline{\beta 2}$ are 4 x 3 matrices.

The structure of the model is therefore determined by the following parameters:

the number of inputs r and outputs p;
the order of process dynamics n;
the Prediction/Control Interval c;
the Sampling Interval s;
and, for each input signal:
the minimum delay;
the delay spread; and
whether or not the signal varies as piecewise steps or ramps.

These parameters have to be specified by the engineer in order to define the structure of a model that is to be established by the statistical procedures.

11.2.2 Constraints on Sampled Data

The generalised model structure indicated above implies that certain constraints must apply to any data that is to be statistically analysed to establish a model.

With reference to section 11.2.1, the m inputs $\underline{u}$ may be varied independently. When collecting data from plant for the purpose of developing a model, it is usual to deliberately adjust these inputs to excite the plant and create data rich with information relating to plant dynamics. The most effective procedure is to apply random variations (eg Pseudo Random Binary Sequences) with a sequence of step changes being applied automatically. It is essential that these step changes are only implemented at what is considered to be a prediction instant of the model. The smallest interval between such changes represents the largest Prediction Interval (ie cT, see section 11.2.1) that can be utilised in generating a model from the resulting data. It is important that any interval between changes is a multiple of this "smallest interval," otherwise it will generally not be possible for the model to stay synchronised with the data (ie as the model progresses through the data, a step change will eventually arise that is not at a prediction instant).

If two or more inputs are being adjusted simultaneously, it is important that they are synchronised in that each relates to the same "smallest interval" or a multiple of it.

A further aspect of constraint with the sampled data relates to the dependent inputs $\underline{v}$ (see section 11.2.1). These inputs vary as they will. For modelling purposes it is necessary that the variations may be reasonably approximated as piecewise steps or ramps across each prediction interval, otherwise an established model is unlikely to be very representative. Continuously varying signals can often be closely approximated as a series of ramps provided the prediction interval is carefully chosen (ie not too large).

11.2.3 Identification of Model Coefficients

Given a defined model structure and a set of suitable plant data, Least Squares analysis may be applied to the data to establish the coefficients of the Transition and Driving matrices that characterise the model.

Equation 7.1 may be compacted to the form

$$\underline{Y1}_{k+c} = \Omega \ \underline{W}_k \hspace{2cm} 11.2$$

with $\Omega = (\alpha \ \underline{\beta1} \ \ldots \ \underline{\beta r})$ and $\underline{W}_k = \begin{vmatrix} \underline{Y2}_k \\ \underline{W1}_{k+c} \\ \vdots \\ \underline{Wr}_{k+c} \end{vmatrix}$

The identification problem is to determine the coefficients for $\underline{\Omega}$ that give rise to the most accurate prediction of $\underline{Y1}_{k+c}$ given $\underline{W}_k$, taking into account the complete span for which data is available for analysis (ie k=0, c, 2c, ...Rt.c, with Rt the total number of Prediction Intervals that are available within the span).

A Recursive Least Squares algorithm (Eykhoff, (2)) is used which has the general form

$$\hat{\underline{\Omega}}_{k+c} = \hat{\Omega}_k + (\underline{Y1}_{k+c} - \hat{\Omega}_k \hat{W}_k) \frac{\underline{P}_k \hat{W}_k T}{d_k} \qquad 11.3$$

$$\underline{P}_{k+c} = \frac{\underline{P}_k - \hat{W}_k \underline{P}_k \underline{P}_k \hat{W}_k{}^T}{d_k} \qquad 11.4$$

$$d_k = 1 + \hat{W}_k{}^T \underline{P}_k \hat{W}_k \qquad 11.5$$

$$\hat{Y1}_{k+c} = \hat{\Omega}_{k+c} \hat{W}_k \qquad 11.6$$

$$\hat{W}_{k+c} = \begin{vmatrix} \hat{Y1}_{k+c} \\ \underline{W1}_{k+2c} \\ \vdots \\ \underline{Wr}_{k+2c} \end{vmatrix} \qquad 11.7$$

k = 0, c, 2c, 3c, ...

$\hat{\Omega}_k$ is the estimate to $\underline{\Omega}$ that is available at instant k.

$\underline{P}_k$ is the symmetric covariance matrix evaluated at instant k.

$\hat{Y1}_k$ is the estimate to $\underline{Y1}$ at instant k, based upon $\hat{\Omega}_k$ and $\hat{W}_{k-c}$

$\hat{W}_k$ incorporates $\hat{Y1}_k$ rather than the $\underline{Y1}_k$ of $\underline{W}_k$.

The algorithm gives rise to unbiased estimates in the presence of coloured measurement noise.

The actual implementation of this algorithm is done in a 'robust' manner, utilising the upper diagonalisation concept(ie UD, Bierman (1)) This, in effect, takes account of the symmetry of $\underline{P}_k$ to ensure that $\underline{P}_k$ can never be other than positive definite.

To consider the manner in which the algorithm functions, presume that analysis commences with k =0. Equation 11.3 is to be utilised first. All parameters to the right in equation 11.3 must therefore be available. The following initial settings are therefore made

$$\hat{W}_O = \underline{W}_O = \begin{vmatrix} \underline{Y1}_O \\ \underline{W1}_C \\ : \\ \underline{Wr}_C \end{vmatrix} \qquad\qquad 11.8$$

$$\hat{\Omega}_O = \underline{0}$$

and $\qquad \underline{P}_O = b\,\underline{I}$

with b a large value and $\underline{I}$ a unit matrix.

Thus $\hat{W}_O$ is set with actual plant values rather than with estimated values. Initially, no knowledge of model coefficients is available, so they are simply set to 0. The covariance matrix $\underline{P}$ is set to a diagonal matrix with very large coefficients. This, in effect, indicates total lack of confidence in the current estimate $\hat{\Omega}_O$.

Following the application of equation 11.3, the first estimate $\hat{\Omega}_C$ is available. The covariance matrix $\underline{P}_C$, to be used at the next stage, is then computed (equations 11.4 and 11.5). Finally, $\hat{W}_C$, which includes the estimated $\hat{Y}1_C$, is computed (equations 11.6 and 11.7). The reference to the data then advances by c sample intervals and the complete procedure is repeated. This cycle progresses until the complete span of data that is to be analysed has been covered.

As the analysis proceeds through the data, there is normally a general convergence, with the estimated values $\hat{\Omega}$ tending to stabilise. Such convergence can be slow, particularly if the data is noisy. In these circumstances, to improve the convergence, it can be useful to cycle the data through the analysis a number of times. Each new cycle, $\underline{W}1_O$ would be reinitialised (equation 11.8), but $\hat{\Omega}_O$ and $\underline{P}_O$ would be retained equal to the estimates reached at the end of the previous cycle.

Two figures of merit may be obtained to give a measure of convergence. The first is the parameter d_k (equation 11.5). When the system is very close to having converged, d_k will be near to (but very slightly greater) than 1. Note that d_k should never be less than 1, such an occurrence is an indication of numerical problems with the calculations. A second figure of merit may be formulated from the elements of the covariance matrix $\underline{P}_k$. As convergence progresses these should become very small. Small elements in $\underline{P}_k$ provide a measure of confidence in the estimates $\hat{\Omega}_k$.

The plant data that is analysed using the Recursive Least Squares procedure is normalised before such use. This normalisation is based upon the mean and standard deviation of a signal across the span of data that is to be analysed. Thus each input and output signal processed is subject to the following normalisation

$$yn_k = (y_k - ym)/ysd$$

with y_k the signal value at instant k, ym the mean of the signal and ysd the standard derivation of the signal. yn_k is the normalised value that is utilised for the Least Squares analysis. This normalisation avoids numerical difficulties that can arise because of orders of magnitude of difference in the variation of different signals (for example, one measurement may vary by 100 units around 300 and another by 0.1 of a unit around 0.5). The analysis thus gives rise to a normalised model. The scaling factors y_{sd} may then be used to translate the model back to a form relating to engineering units. This normalised model is of use for the control system design procedures (see section 11.3.1) and is the form of model presented in relation to the case studies of Chapter 21.

Once a model has been established using the above procedures, the accuracy of the derived model has to be assessed. This is achieved by perturbing the model with the same input signal sequences that originally perturbed the plant. The model outputs may then be compared directly with the plant data to assess the degree to which the model follows plant variations. The differences between model and plant data are termed "Residual Errors", or Residuals for short. The residuals can be compounded to give a number of figures of merit that provide a measure of model accuracy. These are listed below.

Let y_k be an output monitored at instant k, and $\hat{y}_k$ be the value estimated by the model at the same instant and let Rt be the number of Prediction Intervals across which the figures of merit are to be evaluated, then:

figure of merit 1   (absolute mean error)

$$F1 = \frac{1}{Rt} \sum_{k=0}^{Rt.c} (y_k - \hat{y}_k) \qquad k = 0, c, 2c, \ldots$$

figure of merit 2   (absolute RMS error)

$$F2 = \frac{1}{Rt} \sqrt{\sum_{k=0}^{Rt.c} (y_k - \hat{y}_k)^2}$$

figure of merit 3   (incremental mean error)

$$F3 = \frac{1}{Rt} \sum_{k=c}^{Rt.c} [y_k - y_{k-c} - (\hat{y}_k - \hat{y}_{k-c})]$$

figure of merit 4   (incremental RMS error)

$$F4 = \frac{1}{Rt} \sqrt{\sum_{k=c}^{Rt.c} [y_k - y_{k-c} - (\hat{y}_k - \hat{y}_{k-c})]^2}$$

These figures of merit are best evaluated using normalised data values. The figures are then meaningful in the absolute sense, as a basis for interpretation of model accuracy. The values tend to restrict to within the range 0 to 1, with the smaller number indicating better accuracy.

F1 and F2 provide a measure of the extent to which data and model drift apart. F3 should always be close to 0. F4 provides a measure of the extent to which the shape of the plant and model outputs compare and is perhaps the most useful of the four figures.

The most effective way of assessing the accuracy of a model is visual, by direct comparison of graphs of plant data and residuals. These aspects are clarified in the case studies of Chapter 21.

11.3 CONTROL SYSTEM DESIGN

This section describes the facilities of the PAS that are available for designing control systems. The design procedures give rise to a set of gains for a sampled data control system. The design analysis requires that a model of the form already described (ie equation 11.1) has been established. The gains are determined for the control loops to which the model has been cross referenced. The structure of the control system (ie the number of gains, the signals to which the gains relate, and the sampling intervals) arises directly from the structure of the model.

The design analysis utilises the model to minimise a quadratic cost function that involves terms associated with errors (ie deviations from Setpoints) and the magnitude of actuator adjustments. Weightings that are associated with these quantities which may be selected to permit control system responses to be efficiently tailored to engineering requirements. Control systems may be designed to compensate for time delays, external disturbances and for interactions.

Section 11.3.1 reviews the concepts of design that are utilised and indicates the control system structure and the manner in which this relates to the model from which it is derived. The model of equation 11.1 must undergo certain transformations prior to the application of the design analysis. For completeness, a fuller description of the various transformations and of the minimisation procedures for control system design, is presented in Section 11.3.2.

11.3.1 Control System Design Concepts

Control system design and operation is based upon minimisation of a scalar quadratic cost function of the form

$$J = \sum_{i=1}^{M} \{ (\underline{y}_{(i+1)c} - \underline{y}_{ss})^T \cdot \underline{Q} \cdot (\underline{y}_{(i+i)c} - \underline{y}_{ss})$$
$$+ (\underline{u}_{i.c} - u_{(i-1)c})^T \cdot \underline{P} \cdot (\underline{u}_{i.c} - \underline{u}_{(i-1)c}) \} \quad 11.9$$

The design analysis makes use of the model (ie equation 11.1) to progress the minimisation. The interval i to i+1 is cT seconds (see section 11.1). The minimisation may be achieved using Dynamic Programming or Ricatti equation solution techniques. (Kuo (3), Silverman (6)).

$\underline{Q}$ and $\underline{P}$ are (pxp) and (mxm) diagonal weighting matrices respectively, which are constrained to have diagonal elements that are $\geqslant 0$(all off diagonal elements are fixed at zero).

$\underline{y}_{ss}$ is a vector of setpoints that defines the values to which the elements of the output vector $\underline{y}$ are to be set by the controller.

Each diagonal element of $\underline{Q}$ relates to a particular output and gives rise to an "error squared" term associated with that output. Thus, minimisation of J leads to minimisation of each such "error squared" term. The result is that when the designed controller is applied to plant each output approaches its setpoint as the minimisation progresses.

Each element of $\underline{P}$ relates to a particular control input (ie actuation) and gives rise to a term that involves the square of the changes in actuator setting that are to be applied. The minimisation invokes a constraint upon the degree to which an actuator may be adjusted during the course of control implementation.

For control system design, therefore, the facility is available to adjust the diagonal elements of $\underline{Q}$ and $\underline{P}$. An increase in the value of an element of $\underline{Q}$ implies that the associated error is to be reduced more urgently (ie that the associated output is to be driven more quickly to setpoint). An increase in the value of an element of $\underline{P}$ implies that the associated actuator is more constrained (ie that successive adjustments are smaller).

The minimisation analysis utilises the normalised model (see section 11.2.3). This has the advantage of simplifying the initial selection of weights for $\underline{Q}$ and $\underline{P}$. If all weights are initially set to unity, then in general, the analysis results in a sensible control system (ie one that gives rise to responses of a reasonable rate given the time constants of the process that is to be controlled). Thus an initial control system may be established without the need for any decisions from the engineer. Given an initial design, the engineer may then alter the weightings from unity to tailor the control system responses to be closer to those desired.

Note that if $p > m$, then only the first m of the p outputs of $\underline{y}$ may be driven to setpoints. In this circumstance, the engineer may only specify the first m diagonal elements of $\underline{Q}$, the remainder being set to zero.

The minimisation is carried out for M Prediction/control intervals (ie over a time span of M.cT seconds). At each stage of analysis (ie for each value of i, $i = 1$ to M) a set of control system gains is derived. As more stages are progressed, these gains will normally converge to a constant set of values. This converged set represents the control system that is to be utilised. One particular algorithm for minimisation is presented in Section 11.3.2.

The controller that arises from the minimisation has the structure

$$\underline{u}_k - \underline{u}_{k-c} = \underline{G1}.(\underline{Y2}_k - \underline{Y2}_{ss}) + \underline{G2}. \left| \begin{array}{c} \underline{v}_k - \underline{v}_{k-c} \\ \underline{V}_{k-c} \end{array} \right| \qquad 11.10$$

with
$$\underline{V}_{k-c} = \left| \begin{array}{cc} \underline{w}_{k-c} - \underline{w}_{k-2c} \\ \underline{w}_{k-2c} - \underline{w}_{k-3c} \\ . \\ . \\ \underline{w}_{k-Sc} - \underline{w}_{k-(S+1)c} \end{array} \right|$$

The matrix $\underline{G1}$, mx(n+p), comprises a set of gains that relate to error quantities. $\underline{Y2}_k$ and $\underline{Y2}_{ss}$ are defined in relation to equation 11.1. The controller calculates an amount by which the actuator values are to be adjusted. Its action is therefore incremental and is based upon error feedback.

The matrix $\underline{G2}$, mx(r.(s+q)) comprises a set of gains that relate to incremental actuation and disturbance quantities, see section 11.2.1, item x). The value of S depends upon the magnitudes of the various delays dj associated with the input Signals in $\underline{w}$, (see section 11.2.1, item xiii).

For any Signal wj, $j = 1$ to r, the number of samples required in $\underline{V}_{k-c}$ is

$S_j = R_j + d_j - 2$ if $d_j \geqslant 1$, $j = 1$ to m(actuation signal)
and if $d_j \geqslant 0$, $j = m+1$ to r(disturbance signal)

with R_j defined in section 11.2.1, item xiv).

If $d_j = 0$, then for an actuation signal (ie $j = 1$ to m) then S_j the number of samples required, is equal to R_j-1. In this circumstance, the actuation is presumed to be acting as a ramp input and the analysis requires one extra term.

Given the above, S is the maximum of the S_j for $j = 1$ to R. For any particular signal j, if $S > S_j$ then there will be more elements in $\underline{V}_{k-c}$ than required. In this case, the associated gains in $\underline{G2}$ are set to 0 so that the unwanted elements do not contribute to the controller.

The controller of equation 11.10 includes terms in $\underline{v}_{k-c}$ that are associated with previous actuation corrections. Such terms may be associated with high order process dynamics, but more usually provide for time delay compensation. The greater the time delay relative to the interval c, the greater will be the number of compensating terms in the controller. When the minimisation analysis is being carried out, M, the number of stages of minimisation, must imply an interval that is at least greater than the maximum time delay associated with the model. The convergence process will not initiate until the elapse of such an interval.

The controller of equation 11.10 also includes terms that are associated with the disturbance signals $\underline{v}$. These terms give rise to feedforward control. The controller will anticipate a disturbance and correct for it before its effect is apparent in the output. However, the feedforward terms that arise (together with the other terms) are based upon a common selection of the weighting matrix elements. There is not in this procedure an independent basis for defining the urgency of feedforward control relative to feedback control. To cater for a measure of independence, the engineer is provided with the means to select alternative weightings for the design of feedforward terms. A complete set of controller gains is established via this alternative but only those gains that relate to the disturbance terms are updated. The remainder are held at their previously designed settings.

Two examples of controller structures are given below. These relate directly to the examples of model structures presented in section 11.2.1. Consider first the example of equation 11.1a. A control system will result with the simple form

$$u_k - u_{k-c} = |\ G11\ \ G12\ |\ .\ \ \begin{array}{|c|} y_k - y_{ss} \\ y_{k-c} - y_{ss} \end{array}\ |\qquad 11.11$$

For the example of equation 11.1b, the control system is more complex and has the form

$$\begin{array}{|c|} u1_k - u1_{k-4} \\ u2_k - u2_{k-4} \end{array} = \begin{array}{|c c c|} \underline{G11} & \underline{G12} & \underline{G13} \\ \underline{G14} & \underline{G15} & \underline{G16} \end{array}\ .\ \begin{array}{|c|} \underline{v}_k - \underline{v}_{ss} \\ \underline{v}_{k-2} - \underline{v}_{ss} \\ \underline{v}_{k-4} - \underline{v}_{ss} \end{array}$$

$$+\ \begin{array}{|c c c|} \underline{G21} & \underline{G22} & \underline{G23} \\ \underline{G24} & \underline{G25} & \underline{G26} \end{array}\ .\ \begin{array}{|c|} \underline{u}_{k-4} - \underline{u}_{k-8} \\ \underline{u}_{k-8} - \underline{u}_{k-12} \\ \underline{u}_{k-12} - \underline{u}_{k-16} \end{array}$$

$$11.12$$

with $\underline{u} = \begin{array}{|c|} u1 \\ u2 \end{array}$. Note that the elements of $\underline{G23}$ and $\underline{G26}$ that relate to u2

are zero, since no gains are required for the term $u2_{k-12} - u2_{k-16}$.

One further facility is provided for the purpose of control system design. This is a mechanism to establish a controller that gives rise to a minimum time response. This form of controller will transfer the system from one steady-state to another with the minimum number of actuator corrections. The structure of such a controller satisfies the definitions above, although the values of the gains will be different. A different form of cost function is minimised to achieve the required design. This is

$$J_{MT} = (\underline{Y2}_{k+MT.c} - \underline{Yss})^T (\underline{Y2}_{k+MT.c} - \underline{Yss})$$
$$+ \underline{V1}^T{}_{k+(MT-1)c} \underline{V1}_{k+(MT-1)c}$$

$$\text{with } \underline{V1}_k = \begin{vmatrix} \underline{u}_k - \underline{u}_{k-c} \\ . \\ . \\ \underline{u}_{k-(s+1).c} - \underline{u}_{k-(s+2)c} \end{vmatrix}$$

The cost J_{MT} is minimised over the stages with i = 1 to MT. This is achieved by presuming $\underline{Q}$ and $\underline{P}$ are zero for i < MT and that they are unit matrices for i = MT. At the stage at which the new steady state is evident (ie when i = MT) under ideal conditions the cost J_{MT} will be exactly zero and the associated controller will give rise to the minimum time response. Thus, for a minimum time controller, as the design analysis proceeds, the engineer is presented with the cost J_{MT}. The engineer may decide at which stage this cost first attains the minimum. The decision is left to the user since numerical inaccuracy can make the break point somewhat fuzzy in certain situations.

A minimum time controller should always be treated with caution. It usually gives rise to fierce actuator corrections that are impractical for plant implementation. It is, however, quite useful for insight into plant behaviour when carrying out simulation studies.

11.3.2 Control System Design - technical details

This Section presents further technical information relating to the review of the PAS control system design technology presented above.

Equation 11.1 takes the form

$$\underline{Y1}_{k+c} = \underline{\alpha} \ \underline{Y2}_k + \underline{\beta1} \ \underline{W1}_{k+c} \ \cdots \ \underline{\beta1} \ \underline{Wr}_{k+c} \qquad 11.1$$

$$\text{with} \quad \underline{Wi}_{k+c} = \begin{vmatrix} \underline{Wi}_{k+(ej-dj)c} - \underline{Wi}_{k+(ej-dj-1)c} \\ : \\ \underline{Wi}_{k-(Rj-ej-1+dj)c} - \underline{Wi}_{k-(Rj-ej+dj)} \end{vmatrix}$$

This equation undergoes three forms of transformation, prior to the application of design analysis:

Transformation 1

This transformation is to standardise the dimensions of the input signal vectors and to eliminate the time delay specifications. Time delay representation is achieved by increasing the dimensions of the driving matrices and input signal vectors.
Let $Rmax$ be the largest of $Rj + dj$, $j = 1$ to r.

Equation 11.1 is then rewritten

$$\underline{Y1}_{k+c} = \underline{\alpha} \ \underline{Y2}_k + \underline{\mathcal{S}1} \ \underline{V1}_{k+c} + \ldots + \underline{\mathcal{S}r} \ \underline{Vr}_{k+c} \qquad 11.13$$

$$\text{with } \underline{Vi}_{k+c} = \begin{vmatrix} \underline{Wi}_{k+ejc} - \underline{Wi}_{k+(ej-1)c} \\ : \\ \underline{Wi}_{k-(Rmax-1+ej)c} - \underline{Wi}_{k-(Rmax+ej)c} \end{vmatrix}$$

Each vector $\underline{Vi}_{k+c}$ ($Rmax$ x 1) includes the vector $\underline{Wi}_{k+c}$ with, if appropriate, additional elements top and bottom. Each matrix $\underline{\mathcal{S}j}$ ($N1$ x $Rmax$) correspondingly includes the matrix $\underline{\beta j}$. The elements of $\underline{\mathcal{S}j}$ that relate to the additional elements in $\underline{Vj}$ are set to zero so that

$$\underline{\mathcal{S}j} \ \underline{Vi}_{k+c} = \underline{\beta i} \ \underline{Wi}_{k+c} \qquad , j = 1 \text{ to } r.$$

Transformation 2

If $p > m$, ie. there are more outputs than actuations, the excess outputs yi, $i = m+1$ to p, are expressed in incremental rather than absolute form. This is necessary since the excess outputs cannot be related to setpoints (only m outputs can be controlled to setpoints). Equation 11.13 resolves to

$$\underline{Y3}_{k+c} = \underline{\alpha 1} \ \underline{Y4}_k + \underline{\mathcal{S}1} \ \underline{V1}_{k+c} + \ldots + \underline{\mathcal{S}r} \ \underline{Vr}_{k+c} \qquad 11.14$$

$$\text{Let} \qquad \underline{y}_k = \begin{vmatrix} \underline{z1}_k \\ \underline{z2}_k \end{vmatrix} \text{ with } \underline{z1}_k = \begin{vmatrix} y1 \\ : \\ y_m \end{vmatrix} \text{ and } \underline{z2}_k = \begin{vmatrix} ym+1 \\ : \\ y_p \end{vmatrix}$$

$$\text{Then,} \qquad \underline{Y4}_k, \ [N-(p-m)] \times 1 = \begin{vmatrix} \underline{z1}_k \\ \underline{z1}_{k-s} \\ \underline{z2}_{k-s} - \underline{z2}_k \\ : \\ \underline{z1}_{k-(R-1)s} \\ \underline{z2}_{k-(R-1)s} - \underline{z2}_k \\ : \end{vmatrix}$$

$\underline{Y3}_{k+c}$ = the first N1-(p-m) rows of

$$
\begin{vmatrix}
\underline{z1}_{k+c} & \\
\underline{z1}_{k+c-s} & \\
\underline{z2}_{k+c-s} \; {}^{-}\underline{z2}_{k+c} & \\
\vdots & \\
\underline{z1}_{k+c-(R-1)s} & \\
\underline{z2}_{k+c-(R-1)s}{}^{-}\underline{z2}_{k+c}
\end{vmatrix}
$$

$\underline{\alpha 1}$ is an [N1-(p-m)]x[N-(p-m)] matrix derived by manipulation of the coefficents of $\underline{\alpha}$. The matrices $\underline{\zeta j}$, j=1 to r, are similarly derived by manipulation of the coefficients of the matrices $\underline{\beta j}$.
This transformation is achieved in two stages.

Firstly, $\underline{Y1}_{k+c}$ =
$$
\begin{vmatrix}
& \begin{vmatrix} \underline{z1}_{k+c} \\ \underline{z2}_{k+c} \end{vmatrix} & \\
& & \\
& \begin{vmatrix} \underline{z1}_{k+c-s} \\ \underline{z2}_{k+c-s} \end{vmatrix} & \\
& \vdots &
\end{vmatrix}
$$

The set of p-m rows of equation 11.14 that comprise rows m+1 to p (ie that relate to $\underline{z2}_{k+c}$ on the lhs) may be subtracted from the successive sets of rows that comprise rows p+m+1 to 2p, 2p+m+1 to 3p etc. (ie that relate to $\underline{z2}_{k+c-s}$, $\underline{z2}_{k+c-2s}$ etc.). The rows m+1 to p may then be eliminated from the equation and the form of $\underline{Y3}_{k+c}$ prevails in the lhs. The matrices $\underline{\zeta 1}$ to $\underline{\zeta r}$ are also established by this mechanism, although not $\underline{\alpha 1}$ at this stage.

The second stage of the transformation relates to the columns of $\underline{\alpha}$. Now

$$
\underline{Y4}_k \quad = \quad
\begin{vmatrix}
& \begin{vmatrix} \underline{z1}_k \\ \underline{z2}_k \end{vmatrix} & \\
& & \\
& \begin{vmatrix} \underline{z1}_{k-s} \\ \underline{z2}_{k-s} \end{vmatrix} & \\
& \vdots &
\end{vmatrix}
$$

The set of p-m columns of $\underline{\alpha}$ (or the equivalent matrix arising from the first stage of transformation) that comprise columns m+1 to p and relate to $\underline{z2}_k$ on the rhs, may be subtracted from the successive sets of columns that comprise columns p+m+1 to 2p, 2p+m+1 to 3p etc. (ie that relate to $\underline{z2}_{k-s}$, $\underline{z2}_{k-2s}$ etc.). The columns m+1 to p may then be eliminated and the form of $\underline{Y4}_k$ then prevails, with the matrix $\underline{\alpha 1}$ established.

Transformation 3

This transformation is to express equation 11.14 in a standard matrix format that is suitable for the iterations required by the design analysis. Equation 11.14 resolves to

$$Y5_{k+c} = \alpha2 \; Y5_k + \Omega \; v2_{k+c}$$

$$\text{with} \qquad Y5_k = \begin{vmatrix} Y4_k \\ v3_{k+c} \\ v1_k \\ \vdots \\ v1_{k-(Rmax-2)c} \end{vmatrix}$$

with $v1_{k+c} = \begin{vmatrix} v2_{k+c} \\ v3_{k+c} \end{vmatrix}$, $v2_{k+c} = \begin{vmatrix} w1_{k+e1.c} & - & w1_{k+(e1-1)c} \\ & \vdots & \\ wm_{k+em.c} & - & w1_{k+(em-1)c} \end{vmatrix}$

and

$$v3_{k+c} = \begin{vmatrix} w\delta_{k+e\delta.c} & - & w\delta_{k+(e\delta-1)c} \\ & \vdots & \\ wr_{k+er.c} & - & wr_{k+(er-1)c} \end{vmatrix} \text{with } \delta = m+1.$$

Let
$$\eta = N + Rmax.r - m \qquad \text{if} \quad m \geqslant p \quad \text{and}$$
$$= N - (p-m) + Rmax.r - m \qquad \text{if} \quad p > m.$$

Then $\alpha2$ has the dimension η^2 and Ω has the dimension $\eta x m$.

Extra rows are added to equation 11.14, consistent with the definition for $Y5_{k+c}$ and the matrix coefficients (for $\alpha2$ and Ω) associated with these rows are 0 or 1 as necessary to balance the equation. The elements of the rhs are shuffled to give rise to the defined format.

The vector $v2_{k+c}$ is the set of independent actuations that can be selected at instant k to carry out control corrections.

Control System Design involves the minimisation of the cost function

$$J = \sum_{i=1}^{M} (z1^T_{k+(M-i+1)c} \; Q1_i \; z1_{k+(M-i+1)c} + v2^T_{k+(M-i+1)c} \; P_i v2_{k+(M-i+1)c})$$

with $Q1_i$ and P_i diagonal matrices.

Dynamic programming may be applied to minimise this cost function. The minimisation algorithm takes the form

$$K_i = - (\Omega^T R_i \Omega + P_i)^{-1} \Omega^T R_{i-1} \alpha2 \qquad \text{and}$$

$$R_{i+1} = Q_{i+1} + \alpha2^T (\Omega K_i + I)^T R_i (\Omega K_i + I) \alpha2 + K_i^T P_i K_i,$$
$$i = 1 \text{ to } M$$

with $\underline{R}_1 = \underline{Q}_1$, $\underline{Q}_i = \underline{Q}$ and $\underline{P}_i = \underline{P}$.

The control system that arises is

$$\underline{v2}_{k+c} = \underline{K}_M \; \underline{Y5}_k.$$

Bringing in setpoints to the rhs of the control system, it may be written out more fully as

$$\underline{v2}_{k+c} \quad = \quad \underline{K}m \; \begin{vmatrix} \underline{z1}_k - \underline{z1}_{ss} \\ \quad | \; \underline{z1}_{k-s} - \underline{z1}ss| \\ \quad | \; \underline{z2}_{k-s} - \underline{z2}_k \; | \\ \quad : \\ \quad \underline{v3}_{k+c} \\ \quad \underline{v1}_k \\ \quad \underline{v1}_{k-(Rmax-2)c} \end{vmatrix}$$

Note that the term $\underline{v2}_{k+c}$ is evaluated at instant k. It may be that elements of the disturbance terms $\underline{v3}_{k+c}$ are not available at instant k (ie if particular e_i terms are 1 for ramp approximation). In this case the elements of $\underline{K}_m$ that relate to these are eliminated and the structure is reduced in dimension (in effect the particular terms of $\underline{v3}_{k+c}$ are presumed to be zero).

Finally, the equation is resolved to the more consistent form

$$\underline{v2}_{k+c} = \quad \underline{G} \; \begin{vmatrix} | \; \underline{z1}_k - \underline{z1}_{ss} \; | \\ | \; \underline{z2}_k \quad | \\ \\ | \; \underline{z1}_{k-s} - \underline{z1}_{ss}| \\ | \; \underline{z2}_{k-s} \quad | \\ \quad : \\ \quad \underline{v3}_{k+c} \\ \quad \underline{v1}_k \\ \quad : \end{vmatrix}$$

with $\underline{G} \quad = [\underline{G1} : \underline{K}m \;]$

and $\underline{G1}$ formulated by summating the elements of $\underline{K}_M$ that relate to $\underline{z2}$ (and negating the result).

For Minimum Time design, $\underline{Q}_i$ and $\underline{P}_i$ are set to unit matrices for i = 1 and are set to zero for i = 2 to M. The result is to determine the control system to set

$$\underline{v2}_{k+Mc} = 0 \; \text{with} \; \underline{Y5}_{k+Mc} = 0$$

which implies a steady state. The number of stages to reach this situation is apparent from inspection of $\underline{R}$, which should fall to zero at the point at which the steady state can be achieved.

11.4 DISCUSSION

This chapter has been concerned with the mathematical basis for the modelling and control system design techniques of the VUMAN 'Plant Analysis System.' Chapter 21 continues with the case studies that relate to the material presented above. In Chapter 21, two industrial systems are studied. The first is a simple single input and single output system that involves a measured steam flow rate and a valve in the steam pipeline. The second is more complex multivariable process, a multiple effect evaporator with various flow and density parameters. These two case studies bring out the significance of the mathematical material presented in this chapter.

The VUMAN 'Real Time System' package further enhances the capability of the technology reviewed in this chapter. The concepts of identification and control system design have there been integrated together to create a powerful online self-tuning facility. This provides the essential capability to maintain the complex control systems. It is straightforward for the engineer to re-establish control systems that may have deteriorated in performance because of the usual process reasons, even though they may involve multiple signals with time delays in both a feedforward and a feedback structure. This technology has now been proven in a number of arduous plant situations (eg cement kilns, spray driers) and has been shown to hold the potential for real cost benefit improvements in plant operations

11.5 References

1. Bierman G J 'Measurement updating using the U-D factorisation' Automatica, Vol. 12, 1976

2. Eykhoff P 'System Identification' (Wiley, 1974)

3. Kuo B.C: 'Discrete Data Control Systems' (Prentice Hall 1970)

4. Sandoz D J and Wong O 'Design of hierarchical computer control systems for industrial plant' PROC IEE, Vol. 125, No.11, 1978

5. Sandoz D J 'CAD for the design and evaluation of industrial control systems' PROC IEE, Vol. 131, No.4, 1984

6. Silverman L M 'Discrete Ricatti equations: alternative algorithms, a symptotic properties and systems theory interpretations' in G T Leondes (Ed.) Control Dynamic Systems, Vol. 12 Academic Press 1976.

Chapter 12

Reliability and redundancy in microprocessor controllers

P.A.L. Ham

12.1. INTRODUCTION

Increasingly, the "Reliability" of microprocessor controllers is an important feature of the design. This is particularly so where a potential hazard to life is involved or where the financial or other penalties of the plant being out of service are high. Examples of industries where these factors are important are: aerospace, power generation (conventional and nuclear), oil/gas production facilities and air/rail transport.

Whilst the steady improvement in component technology of itself leads to a general improvement in controller reliability, particular design techniques such as redundancy and self-checking strategies are necessary in the more demanding applications. However, no meaningful assessment of improvements in reliability can be properly made until definitions are formulated and ways in which they can be measured have been defined. Some workers in this field have used the term "Dependability" to cover aspects of reliability which are associated with three broad classes of system as follows:

12.1.1 High-Integrity or Safety Systems

A system of this kind is required to be "fail-safe", i.e. any failures should not lead to a greater hazard. It is usually also required that the mode of degradation for successive failures should be gradual rather than a sudden, total collapse.

12.1.2 High-Continuity Systems

With a system intended for continuous operation, it is of interest as to what proportion of its life it will be not available; alternatively, it may be of interest as to how likely it is to successfully complete a life of given duration.

12.1.3 Protective or Standby Systems

With systems which are only invoked in an emergency, it is of interest as to how likely it is that they will not operate when required.

Estimates of Reliability are made in the design stage using statistical estimates based upon historical component data; in absolute terms these are, of course, often subject to large errors, but comparative predictions based, say, on various system architectures can provide a sound basis for the design of more reliable systems.

12.2. RELIABILITY DEFINITIONS

Reliability theory is based on probabilistic concepts as outlined in the monograph by Humphries (1) and the standard definition of Reliability given by Smith (2) is "the probability that an item will perform a required function, under stated conditions, for a stated period of time".

12.2.1 Mean Time Between Failure (MTBF)

This is the reciprocal of the Failure Rate (λ) for a complete equipment based on summated data for all the component parts. The expression is used in the context of repairable systems, whilst Mean Time To Failure (MTTF) has the same meaning but applies to non-repairable systems.

For a constant Failure Rate, the reliability decreases exponentially with time so that we have the relationship of Fig. 12.1 where:

$$R = e^{-\lambda t} \qquad \ldots(1)$$

Where, R = Reliability (In the range 0 to 1)

λ = Failure Rate

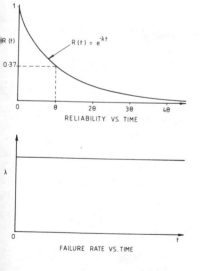

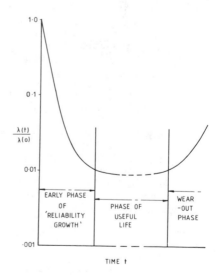

Fig.12.1 Standard Definition of Reliability using an exponential distribution.

Fig.12.2 Typical variation of Reliability as a function of time.

And $\quad \dfrac{1}{\lambda} = \theta = \quad$ Mean Time Between Failure

All equipments tend in practice to follow the "bath-tub" characteristic of Fig. 12.2. This illustrates that the "Reliability Growth" discussed by Crombe & Merril (3) may approach two orders of magnitude. The various factors have been codified in tables of data such as the IEE /Inspec publication (4). Some, such as temperature, are very significant as has been pointed out by Wright (5) and this is typically illustrated in the curve of Fig. 12.3. Additional factors allowed for in the data given in (4) are: Hermetic/Non-hermetic Device, Number of Gates, Application Code (environmental severity) and Number of Functional Pins.

The MTBF for a complete microprocessor controller may be found by summating the products of Failure Rate (λ) and numbers used of each individual component type. This is best carried out by adopting a tabular format, which may be extended to incorporate tabulation of designated failure modes (e.g. short circuit, etc.) following the standard practice of many manufacturers, as indicated by Zinniker (6).

12.2.2 Unavailability

In an equipment in which all failures are both revealed and repairable, the Unavailability, U, is that proportion of an equipment's life for which it is out of service. Sometimes expressed as the Forced Outage Rate in percentage terms, it is of particular value in the context of 12.1.2 above, and is determined by the overall Failure Rate, θ, and the MEAN TIME TO REPAIR (MTTR) represented by γ. It is given by:

$$U = \frac{\gamma}{\theta + \gamma} \approx \lambda \gamma \qquad \ldots(2)$$

It is also possible to define an expression for the Availability, A, noting that:

$$U + A = 1 \qquad \ldots(3)$$

12.2.3 (The probability of a) Failure On Demand (FOD)

A microprocessor-based supervisory or protection system is normally dormant until called upon to act; it is of interest to know what is the probability that it will then fail to perform. This is particularly relevant in the context of 12.1.3 above, and where the product of Failure Rate for UNREVEALED faults (λ') and the MEAN TIME TO TEST (MTTT, represented by T) is small, it can be shown that the probability of Failure On Demand is given by:

$$Df = \frac{\lambda' T}{2} \qquad \ldots(4)$$

In practice, no safety system is generally implemented in single-channel form, and several channels are passed through a MAJORITY-VOTING System. In such a system, for example, if two of the three outputs coincide, then these two prevail whether the output of the third channel is in accord or not.

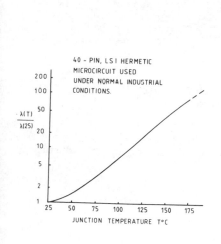

Fig.12.3 Typical variation of device Failure Rate with Temperature.

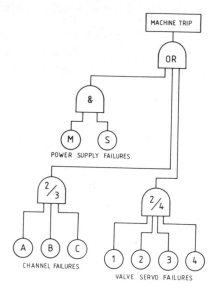

Fig.12.4 A "Fault Tree" diagram.

The FOD for such a system can be shown to be:

$$Df = \binom{n}{r} \frac{\left(\lambda' T\right)^r}{r + 1} \qquad \ldots(5)$$

Where: n = the number of redundant channels.
             r = the number to fail for a system failure.

12.3. REDUNDANCY AND FAULT TOLERANCE

12.3.1 The Concept of Fault Tolerance

In any system exhibiting a redundant structure, the MTBF will no longer represent a true picture of the reliability of the equipment as a whole, and the Availability or the FOR will become the dominant parameter. The very purpose of redundancy is, of course, to confer the ability to continue to perform the design functions (or as many of those functions as can in the circumstances be preserved) in the presence of the widest possible range of individual component failures. Many component failures will thus not be immediately revealed at all, or not lead to an observable performance degradation.

A system demonstrating these properties may be described as FAULT-TOLERANT. Original work was carried out in this field in the mid-1970's by Avizienis (7), but for an overview in the context of computer systems the reader is referred to the Review by Depledge (8). Whilst Redundancy is generally an essential feature of Fault-Tolerant systems, the use of software methods such as, for example, a repeated execution may also be important.

12.3.2 Summary of Methods of Analysis for Redundant Systems

Methods of analysis for complex system architectures are based on various forms of diagram; probably the most familiar is the Fault Tree, in which logic symbols similar to AND & OR gates are used, as in Fig. 12.4. Other forms of diagram which may be used have been listed (6) as follows:

> The State Diagram.
> The Reliability Block Diagram.
> A Cause/Effect Diagram.
> A Truth Table.

Expressions are available to allow prediction for various redundant configurations (1). These differ considerably according to the rules by which the channels are combined, as follows:

A system divided into three sections, any one of which can individually cause failure, will have an Availability defined by:

$$A(sys) = A_1 A_2 A_3 \qquad \ldots(6)$$

If, as is usually the case, the Unavailabilities are small, then:

$$U(sys) = U_1 + U_2 + U_3 \qquad \ldots(7)$$

The Failure Rate for such a system is given:

$$\lambda(sys) = \lambda_1 + \lambda_2 + \lambda_3 \qquad \ldots(8)$$

Where we have a system which is so interconnected that only a single section operating correctly will lead to successful system operation, then:

$$A(sys) = 1 - U(sys) \qquad \ldots(9)$$

And:

$$U(sys) = U_1 U_2 U_3 \qquad \ldots(10)$$

Assuming that the sections are identical, the MTTF for the complete system is given by the expression:

$$\theta(sys) = \frac{1}{3 \lambda_1^3 \gamma^2} \qquad \ldots(11)$$

For a 2-from-3 Majority-Vote, it can be shown that the Availability of the system is a function of the individual channel availabilities as follows:

$$A(sys) = A_1 A_2 + A_1 A_3 + A_2 A_3 - 2 A_1 A_2 A_3 \qquad \ldots(12)$$

The Unavailability of the system is similarly given as follows:

$$U(sys) = U_1 U_2 + U_1 U_3 + U_2 U_3 - 2 U_1 U_2 U_3 \qquad \ldots(13)$$

If, as is common, the three channels are identical, then:

$$A(sys) = 3 A_1^2 - 2 A_1^3 \qquad \ldots(14)$$

And:

$$U(sys) = 3 U_1^2 - 2 U_1^3 \qquad \ldots(15)$$

$$= 3 U_1^2 \quad \text{if the value of U is small.}$$

12.3.3 Improvements to be gained by Redundancy

The degree of improvement which may be anticipated with various levels and forms of Redundancy has been estimated, on a comparative basis, in the context of safety systems (5) and is summarised in the Table below; notes on the reading of these figures follow:

TYPE OF SYSTEM	FAILURE PROBABILITY		
Single-Channel System:	1.0	to	0.01
Single-Channel System With Fault-Detection:	0.3	to	0.003
Redundant System with Similar Channels:	0.05	to	0.0005
Redundant System with Partial Diversity:	0.01	to	0.0001
Redundant System with Full Diversity:	0.001	to	?

12.3.1.1 The Use of Diversity.

Diversity may cover both dissimilar hardware and software, i.e. either or both may be obtained from independant design and/or manufacturing sources, in order to reduce common errors (see 12.3.1.4 below). There are in practice more examples of diverse software than hardware, mainly in the fields of Civil Aviation, Military Avionics, Nuclear Reactor Protection systems and most notably, in the U.S. Space Shuttle.

12.3.1.2 Assumptions Relating to the Forms of Redundancy.

It is assumed in the above that all redundant channels are fully active and, for 3-channel or higher systems, have equal authority. Two-channel systems are a special case, one usually being in control, and the second being run as a HOT STANDBY (i.e. powered-up and with all data kept updated in memory).

Note that it is not generally possible, in two-channel systems, to compare the outputs and, without additional information, to determine which is in error. The changeover operation must therefore rely on other means to identify a failure, such as a WATCHDOG TIMER, which detects a failure of the program to complete execution in a given time; other independent fault-detection means may also be involved if they can be devised. Systems with three or more channels, however, have the advantage of being able additionally to utilise Majority-Voting techniques.

12.3.1.3 Implications Relating to the MTBF.

The improved performance figures quoted above only apply to the overall system, and reflect the success in masking individual channel failures. The total number of failures actually experienced, however, will be at least n-times greater with an n-channel system. It will thus be clear that multi-channel systems carry a significant overhead in terms of total failures; for this reason they are only used in well-defined contexts and are most often configured as Triplex systems. In situations where rapid repair is difficult, additional Hot or Cold (i.e. not powered-up) Standby channels may be provided by way of in-situ spares, to be switched in automatically when required.

12.3.1.4 Avoidance of Common-Mode Failures. Any improvement on the scale indicated will only be achieved if the system has NO COMMON MODE FAILURES, i.e. single faults which affect, or can propagate to, all channels simultaneously. Defences against this type of situation are largely centred on procedural rigour in design, manufacture, test and maintenance as discussed by Bourne et. al. (9).

12.3.1.5 The Need for Maintenance. It should also be noted that the continued achievement of a satisfactory performance relies upon effective maintenance procedures being carried out, to ensure that any faulty channel is repaired promptly, and hence operation with the full degree of redundancy restored.

12.4.   THE MINIMISATION OF FAULTS DUE TO SOFTWARE ERRORS

The avoidance of errors due to software is an important consideration; there are a number of disparate aspects to this question, involving how the software is stored and in what ways it may be corrupted, as well as how it is specified, prepared, validated and maintained.

12.4.1 The Effects of Transient Phenomena

Power-Supply-borne transients or Radio Frequency Interference (RFI) can corrupt readings from peripherals, interfere with the operation of the system bus, and directly or indirectly lead to corruption of data held in Random Access Memory (RAM). Any high-reliability application will require the provision of safeguards in the form of RFI protection, Mains Input Filters, and the like. It is possible that a limited amount of corruption arising in this way may be successfully cleared by the use of ERROR CORRECTING CODES (see 12.4.3.1) but these are generally unable to cope with more than one or two bits in error in any one data word.

It is advantageous to prepare software in such a way that all data held in RAM is automatically updated, or re-validated, at regular intervals, so that random errors from any cause are automatically cleared. Where possible, vital data such as Interrupt Vectors are best located in Read Only Memory (ROM), and it is generally considered preferable to aquire input data from peripherals by polling rather than raising an Interrupt upon a change of state. It is also good practice to associate any Interrupts which are necessary with a timeout procedure, so that a reversion to a pre-determined state can be initiated if the Interrupt does not arrive as expected.

12.4.2 Hard Memory Errors

"Hard" errors are caused by irreversible defects arising in the memory component, causing malfunction every time a faulty cell (carrying one bit of information) is accessed. The effect on the running of the processor will not in general be predictable, but it must be assumed that a proportion of such defects will lead to a complete system failure. It is thus advisable to incorporate a separate Watchdog Timer in each system or channel to initiate a controlled shut-down in this eventuality.

The incidence of such errors is significantly less than that of "Soft" errors (see below), and it was reported in 1982 (10) that some designs of 64K RAMs had a "hard" failure rate of approx. 0.02% per 1000Hrs. at 55 deg. C. Such errors can be detected by one of the memory test pattern routines such as, for example the "Walking-Ones" pattern. Whilst this may well be carried out as a production procedure, it may more usefully be incorporated as part of the start-up or Reset procedure of a microprocessor controller, so that there is a possibility of on-line error detection and reporting.

Errors of this type can also arise in ROM, but there is scant published data on the subject. It is, however, relatively easy to carry out a simple test on ROM by means of a CHECKSUM, which is an algorithm for adding, with rotation, the words in a particular area of memory, and comparing this with an expected value. Such a procedure may also form part of a start-up or Reset procedure as above; a typical arrangement is illustrated in the Flowchart of Fig. 12.5.

12.4.3 Soft Memory Errors

"Soft" errors are caused by the spontaneous change of a bit from a

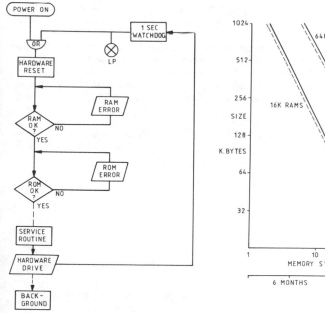

Fig.12.5 The use of a Watchdog Timer and memory tests on start-up.

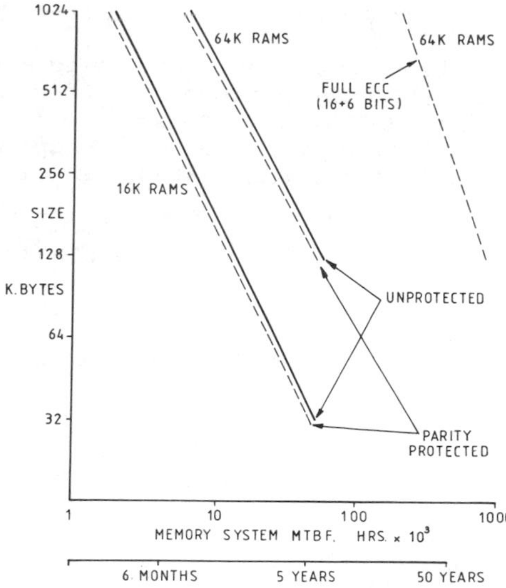

Fig.12.6 Example of Failure Rates for various types of memory.

high to a low state or vice-versa. Such occurrences can be caused by alpha-particles (helium nuclei) generated by the packaging material of the chip itself, and at one stage was significant enough to be a source of some concern. More recently, however, packaging materials have been improved, designs have been changed to increase the storage charge level, and protective coatings have been developed to better absorb the energy of alpha particles. It was also reported in 1982 (10) that the best current designs of 64K dynamic RAMs exhibited a "Soft" failure rate of 0.1% per 1000Hrs. at 55 deg. C. and that, due to design improvements, 64K RAMs were, on a per-Kilobit basis, 3.5 times as reliable as 16K RAMs, as is illustrated in Fig. 12.6.

This trend may be expected to continue; nevertheless, in systems employing a large amount of semiconductor RAM and where reliability is of some concern, various approaches have been adopted in an attempt to safeguard against random bit errors. These are summarised as follows:

12.**4.3.1** Error-Correcting Codes (ECC). This is sometimes known as Error Detection and Correction (EDC). Extra bits in each word of memory allow a form of encryption which is able in the commonest embodiment, the HAMMING CODE, to provide double-error detection and single-error correction, described in the book by Peterson (11). Hardware systems are nowadays available in which the operation of such systems are transparent to the user, and an MTBF improvement of between 20 and 100 times may be expected.

The simplest form of error detection is that of PARITY-CHECKING, i.e. adding the bits in a word together, to see if the Least Significant Bit (LSB) in the resultant is "1" or "0". However, this cannot lead to an error correction, but only a re-try, re-start or shut-down procedure.

The effects of ECC are beneficial overall only because the improvements due to error correction outweigh the penalties due to additional hardware and increased access times; ECC may well not be justified in systems of less than 512K byte capacity (10). The increased "overhead" as the data word gets smaller, is illustrated in the Table below:

DATA WORD LENGTH	SINGLE ERROR CORRECTION		SINGLE ERROR CORRECTION /DOUBLE ERROR DETECTION	
	Extra Bits	Increase %	Extra Bits	Increase %
4	3	75	4	100
8	4	50	5	62.5
16	5	31.25	6	37.5
32	5	15.6	6	18.75
64	6	10	7	10.9

12.**4.3.2** Error Recovery Techniques. Methods of recovering system operation after a fault have been investigated for asynchronous systems and theoretical concepts have been given by Campbell in the 1983 IFAC Proceedings (12), which incidentally includes a number of other references relevant to this Chapter. The concept of Recovery Blocks in software was first stated by Randell (13), and has been applied to small, Real-Time Microprocessor Controllers by Stewart (14).

The latter involves a multi-level approach which uses, amongst other things, program copies or alternative algorithms in memory. However, it should be observed that since the holding of replicated copies in a single memory space could be construed as a potential Common Mode Fault, a fully redundant structure is in principle superior.

12.4.4 The Minimisation of Design and Coding Errors in Software

All Programmers, however skillful, appear to generate code which contains a small proportion of latent errors or "Bugs"; tracing and recifying these at a later date is a major cost involvement quite apart from the special hazards of systems required to be reliable. It is now commonly agreed that not only must these initial faults be minimised, but the code should be written in such a way and with such documentation that it is easy to correct later.

Approaches to the problem have developed significantly during the period since 1982, and have evolved from separate viewpoints, notably those of Computer Science on the one hand, and Management Control on the other. Significant advances have also been made in the hardware and software tools which are now available to assist the Designer. This general area of discipline is part of the field now known as "Software Engineering" described in the book by Depledge [Ed.] (15). Some significant points on which emphasis is now laid are briefly discussed in the following paragraphs.

12.4.4.1 The Use of Better Coding Methods.

Emphasis is now increasingly laid upon order and "Structure" in the generation of Code; programs written in this way are "Modular" i.e. they are divided into sections which are self-contained and have clearly-defined entry and exit conditions. The use of "Pseudocode" to specify software is gaining acceptance, and certain forms of "Structured Programming" have been advanced by George (16) in which flowcharts are much reduced.

A figure quoted by Barnes (17) for the error-rate in coding when the greatest possible care is taken, is less than 1 error per 1000 lines of source code, whereas the Industry average is considered to be 3 to 5. Since the rate is thought to be independent of language, this tends to argue the case for High-level Languages rather than Assembler, but in real-time Industrial systems this will always be subordinate to considerations of program run-time.

It is as well also to remember that a Compiler itself is yet another program which, until it has been fully proven with the target processor, may be assumed to contain errors which could in turn lead to the production of faulty code.

12.4.4.2 Advances in Design Methodology.

Better constructed and more error-free programs can result from applying structured analysis and design methodologies. These are particularly valuable in large software projects and are assisted by the use of an inherently-structured language such as Pascal; however, the principles have been shown by Mills (18) to be equally valid for use with Assembler. The best known methods of Structured Analysis and Design are probably those due to De Marco (19) and Yourdon (20).

Some attempt has been made towards the the mathematically formal specification of software requirements in such a way that a precise verification of correctness may be attempted; this is difficult, however, and it seems very unlikely that the approach will be of use in the context of Industrial control systems into the forseeable future.

12.4.4.3 Management Procedures; Quality Assurance. Whether an established Methodology is used or not, it is always possible that errors (including those of omission) may be made, or that errors of interpretation may subsequently arise. In a typical Industrial situation, it is also likely that requirements will change or that there will be enhancements to the original concept. There is therefore a need for organisational ways of coping with change on a continuous basis, whilst maintaining a consistent check on what has been done.

The management control procedures covering these activities have come to be known as SOFTWARE QUALITY ASSURANCE, and a summary of the principles can be found in Reference 2. It is a cornerstone of this approach that objectives are defined and shown to have been met at each stage, and records are kept to demonstrate that the correct procedures have been adhered to.

For large software projects, it may be of benefit to adopt procedures such as those epitomised in the IEEE Software QA Plans (21). These will define all aspects, including the ways in which the REQUIREMENTS and SOFTWARE SPECIFICATIONS are to be drawn up and how the end result is to be VERIFIED (to check that all levels of code match each other) and VALIDATED (to check that the Specification has been met overall). There are numerous other procedures which may also be used, such as formalised Design Reviews, Software Audits, Reliability Assessments, and Document Substantiation, preferably carried out by independant teams.

An essential activity is CONFIGURATION MANAGEMENT, which is concerned with controlling which Versions of Software Modules are in use at all stages, including in the field. An objective much to be sought is the formation of ASSURED CODE LIBRARIES, which will allow the design overheads to be shared amongst a number of applications.

In Industrial control systems, particularly in a Reliability context, nothing assumes greater importance than that of INTEGRATION TESTS, in which rigorous final testing takes place with the target hardware. In all systems of the type considered in this Chapter, there is likely to be a significant component of inter-dependancy between software and hardware, and the most appropriate trade-off between the two needs to be determined. A noteworthy example of a procedural document which can not only take account of such trade-offs, but also categorises software according to the degree of hazard associated with its malfunction, is that relating to Software in Civil Aviation Airborne Applications (22).

12.4.4.5 Development Aids and Software Tools. Attempts have been made to develop software tools with a view to blocking as many initial coding errors as possible (17), but none have reached the stage of general

acceptance. On a rather more practical level, the increased sophistica-
tion of Microprocessor Development Systems generally is probably the
most significant contribution. Such Systems not only permit a choice of
Languages, but also provide a comprehensive environment of editing,
linking, modifying with record-keeping, In-Circuit Emulation and general
documentation facilities. Systems are becoming available which use
colour-graphics to extend this range of support back into Structured
Analysis using the techniques of Yourdon and De Marco.

12.4.4.6 Other Approaches to Software Reliability.

Much research work
has taken place on means of obtaining more reliable software and on
methods of quantifying the reliability; it is only possible to allude
briefly to to examples. One approach to generating "error-free" code due
to Geller (23) involves the use of "Dual Coding" in which a "Reference
Language" (in this case AP/L) is used in addition to another high-level
"Target Language" to arrive at the final code.

A number of models have been tried in an attempt to generate predictions for software reliability in an analogous manner to that for
hardware. One such due to Musa (24) is based upon an "Execution Time"
model, in which early observations of software failures during the
development phase of a program are used to predict future behaviour.
Methods for relating the results to real time are very complex, and it
would seem that such techniques are only of benefit in very large
software projects where the number of such failures is large enough to
have statistical significance.

12.5. NOTES ON THE IMPLEMENTATION OF RELIABLE SYSTEMS

12.5.1 Levels and Types of Redundancy

In implementing a reliable system involving redundancy, it is
essential to adopt a hierarchical approach and provide redundancy only
where it is of most value. In the control of Steam Turbines as described
by Ham (25), for example, the control of Speed and certain protective
functions are implemented in TMR form, but less vital features such as
variable rate-of-change are only implemented in a higher-level Interface
processor which communicates with all three channels. In general, many
forms of multiple-processor system may be envisaged, but in a reliability context the following are of most value:

12.5.1.1 Distributed Processing.

In this configuration, the tasks are
subdivided between processors such that failure of one leaves an adequate range of total system functions still available; this ensures a
gradual degradation of performance rather than a sudden total system
failure. An Interface processor combined with a TMR section (as above)
illustrates this type of redundancy in one of its many possible forms.

12.5.1.2 Modular Redundancy.

In this configuration, each system or pro-
cessor performs (or is able to perform) all the possible tasks in parallel with its fellows; the ultimate control action is obtained by some
form of Majority-Voting.

12.5.2 The Provision of Automatic Fault-Detection

In order to maintain the integrity of a redundant system at the initial level, it is essential that it be maintained in good repair; such a system will, by its basic nature, also mask faults so that it becomes essential to provide comprehensive automatic fault-detection or diagnostics so that maintenenance can be initiated in a timely manner. The following techniques are widely used in such systems:

12.5.2.1 Validity-Checking. This covers Transducer out-of-range checks and excessive rate-of change of variables, for example, and may be hardware or software orientated.

12.5.2.2 Cross-Channel Comparison. This is especially valuable in TMR or higher-order systems, and enables a faulty channel to be identified and appropriate measures taken without, however, of itself identifying the causative defect.

12.5.2.3 On-Line Proof Testing. This is particularly appropriate to processor-based systems, where test routines can be run in apparent concurrency with the main control program.

12.5.3 Distribution of Functions Between Hardware and Software

Many of the considerations described above are not new, and have been applied over a number of years in analogue electronic systems in a number of separate fields. It is now possible with the introduction of processor-based systems to implement these various reliability-enhancing features in either hardware or software and a fine measure of engineering judgement is required as to the most effective assignment. As a general rule, ultimate safety is best assured by enclosing software-based systems with an overriding hardware trap, such as a Majority-Voter or a Watchdog-Timer.

Since both software as well as hardware will exhibit some failure-rate (even if it may not be possible to quantify it) then it follows that a limit should be placed on the amount of additional software or hardware which is specifically devoted to fault-detection or diagnostics, compared to the amount already used to implement the main control function. A small fraction -say 10%- could be considered an effective design whereas 100% or more would by most standards be classed as unsatisfactory.

12.6. MAJORITY-VOTING TECHNIQUES

In designing reliable systems using Redundancy, the need to avoid Common- Mode faults has already been mentioned. A desirable feature which helps towards that aim is to maintain separate channel integrity for input data; in many cases this requires separate transducers. It will then be necessary to apply Majority-Voting to the system outputs.

In some cases, it may also be necessary to apply Majority-Voting to intermediate stages in the process; this may be particularly useful, for example, where the reliability of input transducers is not of the best, or where the structure of the software makes it convenient to sectionalise the process. In some systems where the channels are not wholly independent, such that shared memory is involved; for example, intermediate Majority-Voting may be carried out on a number of signals such as may constitute a parallel data bus, as described by Yaacob (26).

Depending upon the requirements and the structure of the particular system therefore, there may exist several discrete areas where Voting techniques may advantageously be applied, and where different criteria for the choice of method between the possible alternatives, both in hardware or software, may be appropriate. These are discussed in more detail below.

12.6.1 Types of Signal to which Voting is Applied

Two kinds of signal have normally to be considered, namely, digital, or two-state signals and modulating signals representing a continuous variable. With microprocessor- based systems, the modulating signal is generally represented by a 2's complement signal held in a register or memory space. In nearly all cases, there is a hardware equivalent to a software technique, and vice-versa.

12.6.2 The Treatment of Digital Signals

The following procedures may be be employed in particular contexts in redundant systems:

12.6.2.1 Output True if any single input High. This is an OR Function as shown in Fig. 12.7(a). A duplex safety system might well operate upon this principle.

12.6.2.2 Output True if all inputs High. This is an AND Function as shown in Fig. 12.7(b), and may represent an inverse-logic version of 12.7(a). Thus, for example, a turbine may require to shut down if both oil-pumps have failed.

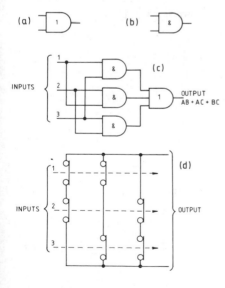

Fig.12.7(a) An OR-function
       (b) An AND-function
       (c) A 2-from-3 in logic
       (d) A 2-from-3 with relays.

Fig. 12.8 Flowchart for a 2-from-3 majority-vote in software for digital signals.

12.6.2.3 Output True if a Majority of inputs High. An m-from-n Function
in the general case, this is shown for a 2-from-3 embodiment in Fig.
12.7(c). This form is now common in protective systems such as turbine
overspeed (25), where it may be implemented by hard-wired relays, as in
Fig. 12.7(d). A software-based equivalent can be formed using the
flowchart shown in Fig. 12.8.

12.6.3   The Treatment of Modulating Signals

There is no "best" method of arbitrating between groups of modulating signals without reference to the particular application.It will be
clear that there are, however, relative trade-offs involving different
amounts of hardware or software associated with the different strategies. In systems where Integrity is important, for example, it may be
better to carry out the voting in hardware after the signal has passed
through A/D Conversion. Hence the following Sections show some of the
available methods and the equivalence between the various hardware and
software techniques.

12.6.3.1   A Single-selection procedure (i.e. SELECT-HIGH or SELECT-LOW).
Such a procedure may be used where it is desired to limit a control
variable to a particular maximum or minimum value. Single-selection can
be achieved by means of biassed diodes, as shown in Fig. 12.9. Where
software methods are involved, the need can be satisfied by the simple
routine of Fig. 12.10.

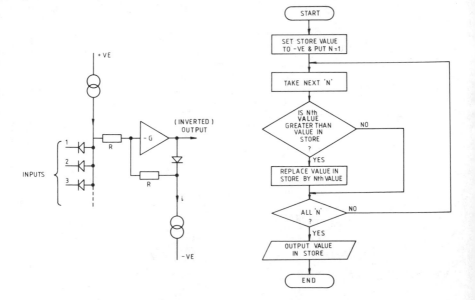

Fig.12.9   Circuit for an analogue   Fig.12.10   Flowchart to select the
           select-low function.   highest of a number of
                                                      modulating signals.

12.6.3.2 A Simple Average or Mean Value. This is simple to implement and is satisfactory where the signals do not deviate from each other to a significant degree or where the number of signals is large; in a TMR system the effect of a "hard-over" failure of one will be significant.

12.6.3.3 An Average with Automatic Self-Purging. It is possible to improve the performance of the Simple Average procedure described above by providing additional means to disconnect one incoming signal completely when it deviates from the average value by more than a prescribed amount, as in the method described by Ham (27), shown in Fig. 12.11. This procedure leads to a transient change in output upon operation.

12.6.3.4 Addition by Variable Weighting. A possible solution to the transient problem is to employ signal addition with variable weighting so that the contribution of a particular signal is reduced progressively; this is illustrated in principle in Fig. 12.12. A quadratic decision algorithm has been described by Broen (28), and for a TMR system the weighting factors are given by the following expressions:

$$w1 = \frac{z^2 + p^2}{D}, \quad w2 = \frac{z^2 + q^2}{D}, \quad w3 = \frac{z^2 + r^2}{D} \quad \ldots(16)$$

Where:
$$D = 3z^2 + p^2 + q^2 + r^2 \quad \ldots(17)$$

And:
- $w1$ = the weighting factor for channel 1, etc.
- z = the deviation threshold.
- p = the deviation between channels 2 and 3.
- q = the deviation between channels 1 and 3.
- r = the deviation between channels 1 and 2.

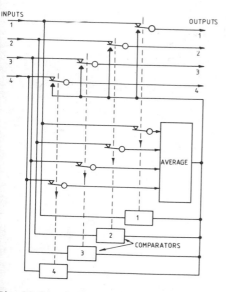

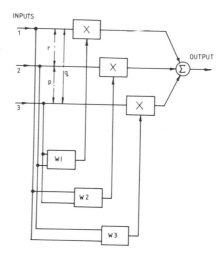

Fig.12.11 Relay-type voter for 4-channel systems using "self-purging".

Fig.12.12 Voting by the method of summation with variable weighting.

Note that: $w1 + w2 + w3 = 1$ ...(18)

And: p, q, r and z lie in the range 0 to 1.

It can readily be shown that, for small values of z and large discrepancies in input signals, the output signal approximates to the median (see below). Due to complexity, the arrangement is probably restricted in its field of application to the more powerful processor-based systems; the basic form of a flowchart to obtain such a result is shown in Fig. 12.13.

12.6.3.5 A Median-Selection Procedure. An alternative to forming a mean value from a number of signals is to select the median value; this is simple and avoids switching transients. A selection routine is shown in Fig. 12.14(a), which can comprise arrays of diodes as described by Klaassen (29), or groups of circuits such as that illustrated in Fig. 12.9. The equivalent procedure in software is shown in Fig. 12.14(b).

12.6.3.6 A Reduced-Median or Reduced-Mean. For safety reasons it may be desirable to incorporate additional means to disconnect, or restore to a safe value, any inputs which can be identified as having failed.

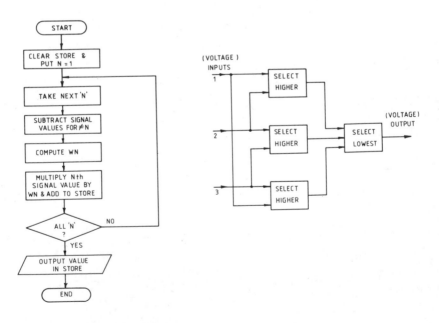

Fig.12.13 Flowchart for voting by   Fig.12.14(a) Selection of the media
          summation with variable   from three modulating
          weighting.   signals.

12.**6.3.7** A Blocked-Averaging Procedure. A median signal does not represent the best use of the information contained by the whole population of signals; it could thus be advantageous to form a mean value using all the available signals provided they all lay within a certain acceptance band. The analogue circuit due to Ham (30) and shown in Fig. 12.16 can be used following an A/D conversion. It derives an average from a number of approximately equal inputs, but for large discrepancies automatically reverts to a median-selection.

Neglecting the effects of diode voltage-drop, a transition takes place and a particular input signal becomes isolated from the formation of the mean when:

$$\frac{2Vin}{3R} = \frac{v}{r} = -\frac{2Vout}{R} \qquad \ldots(19)$$

Where:

Vin = the voltage of the channel under consideration relative to the other two channels (assumed the same).
Vout = the mean/median output of the circuit.
R = the input/feedback resistor value.
v = the diode quad half-supply voltage.
r = the diode quad half-series resistor.

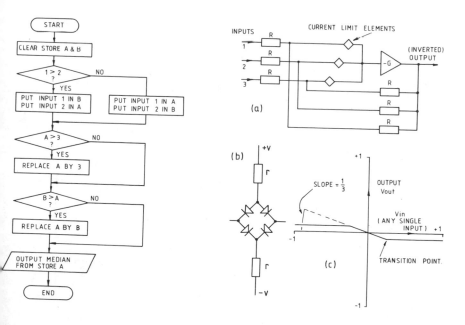

Fig.12.14(b) Flowchart to select the median from three modulating signals.

Fig.12.15(a) Blocked-averaging voter
(b) Current-limiting element
(c) Idealised characteristic.

12.7. REQUIREMENTS FOR THE PARALLEL OPERATION OF CHANNELS

12.7.1   The Minimisation of Divergence and Need for Equalisation

It is to be expected that parallel, redundant systems will show errors; some of these errors will be constant with time and represent variations in input data or the like, and can up to a point be tolerated.

In systems such as guidance and navigation systems as described by Padinha (31), and which incorporate an integral term, or alternatively those in which the synchronisation of time in redundant computers is not fully implemented, a slow divergenge of control signal values can occur. This is not, of itself, a fault and total disconnection may not be the most appropriate strategy - some form of correction or CHANNEL EQUALISA-TION will be needed.

A common method of Equalisation (31) is the feedback, generally with integral action, of the difference between a channel's output command and the controlling mean or median, as in Fig. 12.16. The flowchart for an arrangement used in a TMR Processor-based Turbine Control System has been described by Ham (32) and is shown in Fig. 12.17.

12.7.2   Inter-Channel Communication and Initial Harmonisation

In order to implement Majority-Voting in software, carry out Channel Equalisation or simply to perform comparisons for fault-detection purposes, it is necessary to provide a means for inter-channel communication. Whilst parallel bus structures or byte-serial methods such as

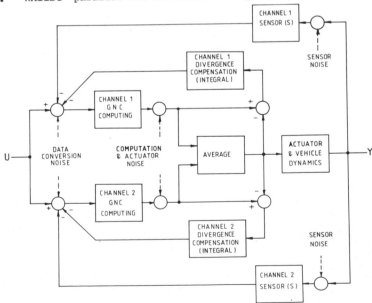

Fig. 12.16   General form of divergence minimisation using integral feedback.

the IEEE 488 interface could be employed, many problems such as those of synchronisation may be avoided by the use of asynchronous bit-serial links described by Cox (33). Some of the problems inherent in the synchronisation and matching in Fault-Tolerant systems have been studied theoretically and are reviewed by Moore & Haynes (34).

It will be necessary to include means to set all reference values (e.g. Setpoints) to safe initial values on start-up. Routines must also be included to harmonise all reference values in a channel just returned to service, by checking first that the running channels are in agreement, and then forcing the values in the returned channel to comply with these before the channel is re-instated as part of the TMR system.

2.8. CASE STUDY 1; A TMR OVERSPEED TRIP SYSTEM

An example of a Safety System is a device intended to shut down a piece of rotating machinery if the speed exceeds a pre-set value, as shown in Fig. 12.18. It consists of three entirely separate probes operating with a toothed-wheel, each feeding to a Speed Detection Module.

The Speed Detection Modules are hardware-based, and are all interfaced with a monitoring Processor. This enables an accurate time-versus-tooth count to be transferred to the the Processor at 10 msec. intervals, so enabling three speed measurement values to be computed for both monitoring and display purposes. The Speed Modules incorporate individual hardware level-detection systems, based upon a pre-set tooth/time count to generate an independent Trip output for transfer to

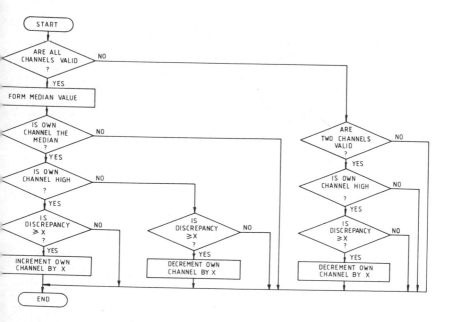

Fig. 12.17 Flowchart for a form of divergence minimisation as used in a turbine-control system.

a relay-type Majority-Voter.

The core of the system, wherein the maximum reliability is needed, comprises only a part of the Speed Detection Modules, and the single Processor is devoted to continuous on-line diagnostics and manually-initiated but automatically-run on-line testing and calibration. This leads to a very small number of unrevealed faults and, as a result, the following performance parameters are obtained:

Assumed Manual Test Interval: 10 weeks

Assumed Speed Module Unrevealed Faults: 30%

Predicted FOD: 5.9×10^{-6}

The equivalent FOD figure for a single-channel system, assuming that some means could be devised to carry out on-line testing, is:

$$1.72 \times 10^{-3}$$

In this case, therefore, a TMR system has improved the failure probability by over three orders.

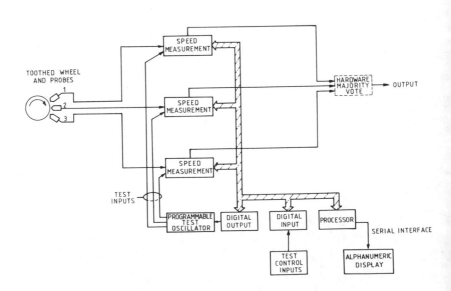

Fig. 12.18 Case Study 1: A TMR Overspeed Trip System.

12.9. CASE STUDY 2: A DIGITAL ELECTROHYDRAULIC GOVERNING SYSTEM

This system comprises a full TMR control system for Steam Turbines (25) and is shown in principle in Fig. 12.19. It incorporates not only a speed controller operating down to almost zero, but also controls the synchronised Load, allows programmed Speed and Load changes, and can automatically reduce the Speed or Load in response to incorrect operating conditions. Various linearising functions are performed, particularly that relating to Steam Mass Flow.

A number of Control Options are available by the addition of standard Peripherals and Software Modules. These include the ability to control the opening sequences of all the Steam Valves, in various Stages, either as functions of steam pressure or to suit operational needs, and to implement various control strategies designed to enhance Power System stability.

In accordance with the principles described, the more vital functions, including a Manual back-up control, are resident in each of the three channels, whilst those of lower priority, such as the control of Loading Rate for example, are resident in the single Interface Processor. Safety functions, such as signals to trip the machine in the event of certain fault conditions, are brought out on separate channel outputs and majority- voted in hardware. The Steam Valve controllers, of which there may be up to ten, are provided with individual Majority-Voting systems, fed from individual A/D Convertors, so minimising the possibility of Common- Mode faults.

Inter-Channel Communication is provided by a total of six bi-directional serial links, three between the Channel Processors, and three between the Interface Processor and the Channel Processors. This allows the input signals from pressure trandsducers, for example, to be exchanged between channels so that Majority-Voting can be carried out

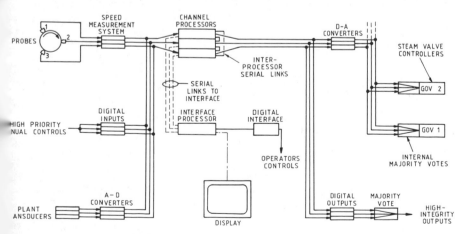

Fig. 12.19 Case Study 2: A Digital Electrohydraulic Governing System for steam turbines.

and errors due to dissimilar data can be minimised; all the necessary routines for Divergence Minimisation and Harmonisation are also carried out by this means.

The system has the ability to run for short periods with one main Channel out of service, and one or more Steam Valves closed. It also exhibits the characteristic of a controlled degradation in performance following a series of faults and is, for all forseeable faults Fail-Safe, i.e. it will shut down rather than overspeed the Turbine.

ACKNOWLEDGEMENTS

The Author thanks the Directors of NEI Parsons Ltd. for permission to publish the material contained in this Chapter.

REFERENCES

1. Humphries, M., 1982: "The Application of Reliability Techniques to Instrumentation Systems". SRS/GR/55.

2. David J. Smith, 1985: "Reliability and Maintainablity in Perspective" (2nd. Edition), Macmillan.

3. Crombe, R. C. & Merril, R. A., 1981: "Industrial Process Control Instrumentation Systems' Reliability Assurance Program, Reliability Growth and Field Results". Joint Organisers NCSR, Culcheth, Warrington & IQA, 53 Princes Gate, London.

4. "Electronic Reliability Data", 1981: IEE Inspec/NCSR.

5. Wright, R. I., 1982: "Microprocessor Hardware Reliability". SRS/GR/60.

6. Zinniker, P., 1984: "Predicting the Reliability of Electronic Units" Brown-Boveri Review, 6/7-84, 256-266.

7. Avizienis, A., 1976: "Fault Tolerant Systems", IEEE Trans., C-25, 1304-1312.

8. Depledge, P. G., 1981: "Fault Tolerant Computer Systems", IEE Proc., Vol. 128, Pt. A, No. 4, 257-272.

9. Bourne, A. J., et. al., 1981: "Defences against Common-Mode Failures in Redundancy Systems", SRD/R/196.

10. Lucy, D., 1982: "Choose the Right Level of Memory Error Protection", Electronic Design, Feb. 18, 37-42.

11. Peterson, W. W., 1961: "Error Correcting Codes", MIT Press, Camb., Mass.

12. Campbell, R. H., et. al., 1983: "Practical Fault Tolerant Software for Asynchronous Systems", IFAC Safecomp '83, Cambridge. (Pergamon) 59-65.

13. Randell, B., 1975: "System Structure for Software Fault Tolerance", IEEE Trans. on Software Engineering, SE-1, June 1975, pp. 220-232.

14. Stewart, T. R., & Preece, C., 1982: "Fault Tolerance as a Feature of Digital Governing", Proc. 17th. UPEC Conf., UMIST 1-6.

15. Depledge, P., Ed., 1984: "Software Engineering for Microprocessor Systems", Peter Peregrinus Ltd.

16. George, R., 1977: "Eliminate Flowchart Drawings", SOFTWARE-PRACTICE AND EXPERIENCE, Wiley-Interscience, Vol. 7, pp. 727-732.

17. Barnes, D, 1983: "Software Reliability", Electronic Design, April 14, 172-180.

18. Mills, B. E., 1980: "Good Programming in Assembly Language", IEE Proc., Vol. 127, Pt. E, No. 6, pp. 241-248.

19. De Marco, T., 1978: "Structured Analysis and System Specification", Yourdon Press, N.Y., ISBN: 0-917072-07-03.

20. Yourdon, E. & Constantine, L. L., 1978: "Structured Design", Yourdon Press, N.Y., ISBN: 0-917072-11-1.

21. IEEE Standard 730-1982: "Software Quality Assurance Plans", IEEE Service Center, Piscataway, N. J.

22. RTCA Doc. No. DO-178A, 1982: "Software Considerations in Airborne Systems and Equipment Certification", Radio Technical Commission for Aeronautics, One McPherson Square, 1425K Street N.W., Suite 500, Washington D. C. 20005.

23. Geller, G. P., 1983: "Coding in Two Languages Boosts Program Reliability", Electronic Design, March 31, 161-170.

24. Musa, J. D., 1980: "The Measurement and Management of Software Reliability", Proc. IEEE, 68 No. 9, 1131-1143.

25. Ham, P. A. L., 1982: "The Application of Redundancy in Controllers for High Capital Cost or High Integrity Plant", IEE Conf. "Trends in On-Line Computer Control Systems". Pub. No. 208, pp. 181-191.

26. Yaacob, M. et. al., 1983: "Operational Fault Tolerant Microcomputer for Very High Reliability", IEE Proc., Vol. 130, Pt. E, No. 3, 90-94.

27. Ham, P. A. L., 1966: Brit. Pat. No. 52932/66, "Improvements in or relating to electric circuits".

28. Broen, R. B., 1975: "New voters for redundant systems", Trans. ASME, J. Dyn. Syst. Meas. Contr., 97, 41-45.

29.   Klaassen, K. B., 1984: "Reliability of Analogue Electronic Systems", Elsevier; Studies in Electrical and Electronic Engineering, Vol. 13.

30.   Ham, P. A. L., 1975: Brit. Pat. No. 1555123, "Improvements in and Relating to Logic Circuits and Control Systems Incorporating such Circuits".

31.   Padinha, H. A., 1972, "Divergence in Redundant Guidance, Navigation and Control Systems', IBM No. SSE-01-006, IBM Electronics Systems Center, Owego, N.Y.

32.   Ham. P. A. L., 1985: "Arbitration Between Signals in Redundantly-Structured Control Systems", IEE Conf. "Control '85", Pub. No. 252, Vol. 1, pp. 60-65.

33.   Cox, H., 1979: "The Advantages of Pure Bit Serial Interfaces for Microcomputers and Other Controllers", IEE Conf. "Trends in On-Line Computer Control Systems", Pub. No. 172, pp. 26-28.

34.   Moore, W. R., & Haynes, N. A., 1985: "A Review of Synchronisation and Matching in Fault-Tolerant Systems", IEE Proc. Vol. 131, Pt. E, No. 4, July, pp. 119-124.

Chapter 13

Software and hardware aspects of industrial controller implementations

M.G. Morrish

13.1   INTRODUCTION

In the industrial control field the CEGB, like many other industries, are now using computers as a replacement for our traditional tailor-made electronic and pneumatic control systems. The reduction of costs of microprocessors means that it is now effective to use several such distributed processors to perform the overall control function, with obvious benefits in terms of reliability and limited degradation of performance due to any one failure. The concepts of such a distributed system with each processor working semi-independently with a minimal amount of communication between them is now well established. As an alternative to previous large single computer systems which had to be designed with considerable back-up features, the present solution shows considerable benefits both in terms of cost and performance. Costs of equipment are not yet at the level where we use one microprocessor per control loop but our current philosophy tends to apply one microprocessor to control each plant area. As examples we can quote one processor controlling all the superheater temperatures controls on a major generating unit or one microprocessor controlling the combustion controls on a large oil fired power station. Small applications include items such as the control of fuel oil pumps for a generating unit performed by one microprocessor, or sequence control of a coal handling tower performed by one microprocessor.

The microprocessor has not solved, and will not solve, our technical problems overnight. The same basic engineering principles and requirements have to be considered in the design, construction, commissioning and maintenance phases to ensure successful control operation. Digital systems operate in fundamentally different manner from previous continuous analogue systems, the consequences of which should not be underestimated.

I would like to consider the engineering aspects of such control installations in two main sections, namely hardware and software.

13.2   SOFTWARE

13.2.1   CUTLASS.

One of the major benefits of using microprocessor based control systems is in their flexibility of operation obtained by making changes to the software, or control programs. The ease with which this can be done depends on the software system being used. At an early stage the CEGB decided to operate using high level software systems. That is, software systems which were easily understood and operable by Station Engineers. Having surveyed the languages that were available from commercial manufacturers, the Board decided to develop its own system due to lack of a suitable commercial product. This language and operating system is known at CUTLASS, and is really a combination of two previous languages developed within the Board, DDACS and SWEPSPEED.

The CUTLASS software system is designed to be a high level application language combined with a sophisticated operating and compiler system for use by Engineers rather than programmers. The reason for this is that the Board decided that Engineers were more cognizant of plant problems, they then would be the best people to perform this function. In order to provide the sophisticated compiler which can check application programs for sense before they are actually run, the design calls for a host computer through which all the reprogramming of the control computers is performed. The high level language combined with this sophisticated compiler means that much of the logical manipulation normally required in control programs does not need to be included by the Engineer. Such items as auto manual changeover and balancing between auto and manual states are automatically catered for within the software.

In order to simplify the application programming further, the programs themselves are sub-divided into schemes and tasks. Each scheme is designed to perform a particular function such as generation of a display, or control of a particular plant item. The schemes themselves are divided into tasks, each one of which may be run on a regular basis depending on its clock connection rate. Thus CUTLASS is a true real-time, multi-task, application and operating software system (Fig 13.1).

13.2.2   Features of CUTLASS.

The high level language components of CUTLASS are divided into several instruction sets. Each scheme within the processor is defined as being one of four types. These are, DDC, display, sequence and general purpose. The high level aspects of the language means that it is possible to assemble a control task with a very small number of calls. For instance, in order to implement a standard PID type controller, it is only necessary to write one statement. The Engineer specifies as arguments within this statement the proportional gain, the integral time, the derivative

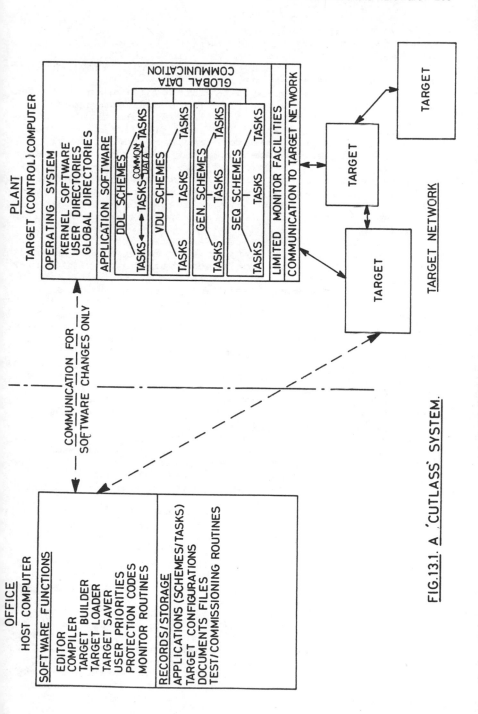

FIG.13.1. A 'CUTLASS' SYSTEM.

time and the derivative roll off function, if desired. The
CUTLASS operating system then ensures at compilation time
that the PID algorithm is correctly implemented and that
sufficient data is stored as history to enable the correct
functioning of the algorithm.

Bad data is automatically catered for with suitable
safe responses from control algorithms and from plant
driving systems. Bad data is defined as data which is
either out of range or which has been flagged as bad from
the analogue scanner. A feature known as bad data
propagation, is incorporated within the system. This means
that if any bad data is involved in a calculation, the
result is flagged as bad also. In this way, the engineer
who is programming the system has to perform a minimal
number of checks on the outputs from his control programs,
and can ensure that suitable safe responses are obtained in
all conditions.

Auto/manual transfer is also catered for
automatically in those situations where multi- or cascaded
loops are involved. It is only necessary to specify either
the auto or the manual condition in one task of any scheme
and the operating system itself will then ensure this
information is transmitted to all the other tasks in that
scheme. In addition, if a change is made from manual to
auto the system automatically enters an intermediate mode
known as match. In this mode sufficient executions of the
control and data gathering programs are executed to ensure
that the transients pertaining on auto/manual changeover are
overcome before true auto control is selected. In the same
way if a system trips to manual through acquisition of bad
data for instance, the match mode is also entered in order
to re-initialise the systems. In this way much of the
initialisation procedures that are normally to be found in
control programs are catered for automatically. However,
the flexibility of pgoramming in a high level language is
not lost because there exists an explicit mode in which the
automatic matching between auto and manual can be
overwritten. This is sometimes useful in parallel actuator
situations.

Power fail recovery is also automatically catered for
by the setting of known flags at the beginning of each
scheme in the event of a power fail and subsequent recovery.
Use is made of the match mode when power fail recovery takes
place and once again the programmer does not have to worry
about details of initialisation of control programs. The
exact response of the system to a power fail can be set up
during the build-up of the software for the system.
Facilities exist for automatically reloading the software
from a storage medium, or relying on recovery from a storage
medium, or relying on recovery from a battery backed memory
in which case all the schemes are returned in manual mode.

One of the aims of using a Board standard software
system is that eventually standard approaches to problems
and standard solutions may be implemented using widely
different equipment. The biggest variation in equipment

used comes in the input/output hardware which is connected to the various microprocessors.

This variation in equipment and the problem of finding software to drive this hardware has been overcome by the use of what is known as generalised input/output calls. As far as the engineer/programmer is concerned any analogue input, analogue output, digital input or digital output call looks identical for any particular piece of hardware. The ability to drive different manufacturers' hardware is obtained at software configuration time, when the system is set up and the appropriate routines are included in the operating system and compiler. Thus it is possible to run basically the same control programs in different types of computer.

13.3   APPLICATION SOFTWARE DESIGN

There are aspects of the application software design which are independent of the language or operating system used, but which have fundamental implications on the ease of commissioning/maintaining the system and on the overall performance. Some of the features which we have found important through experience on several systems installed to date are detailed below.

A multi-task software system such as CUTLASS can be used to advantage to ease commissioning and maintenance of the system if a suitable split of software functions is chosen. Referring to CUTLASS schemes and tasks as detailed above, it is possible for individual schemes to communicate certain data items between themselves (this is achieved by loading the data into what is known as global variables). It is only possible for one scheme to write data items into a specified global variable, but any scheme can read the results. Thus if one scheme writes the results of the scanning of analogue signals into global variables the remaining schemes in the system can read the analogue data. In addition, if a scheme is stopped from running, it is then possible to artificially load data items into the global variables. In this way it is possible to test the reading schemes for correct operation by allowing them to respond to dummy data.

To illustrate this concept, let us look at a typical example of a small system. Software is built up from six basic schemes and these are listed below together with their basic functions.

Typical Schemes

SCAN   The scan scheme sets up a crude database for both analogue and digital input data. The analogue data is digitised and perhaps scaled to engineering values and the results loaded into global variables. The digital input data is scanned directly and the results loaded into a global array.

DESK   The desk scheme handled all the operator inputs
                for controlling the plant. The results of these
                operator actions such as auto manual transfers or
                set point changes also load into globals for
                subsequent reading by other schemes.

CONTROLS   The controls scheme or schemes performs the basic
                DDC operation. The tasks read global input data
                and operator inputs and respond to these globals
                to perform normal control. The outputs from the
                control algorithms are usually loaded into global
                variables for subsequent transmission to actuator
                drive scheme.

DISPLAY   The display scheme or schemes read global data
                from SCAN, from DISK and from ALARM. All this
                information can then be processed and arranged as
                suitable VDU formats with regular data updates
                from SCAN and ALARM.

WATCHDOG   The watchdog scheme is the overall system monitor.
                It is usually arranged to be capable of monitoring
                the correct operation of all the other schemes on
                the system and stopping certain schemes or plant
                outputs if software failures occur. Once again
                the watchdog scheme may well read global data
                coming from all the other schemes not to determine
                whether the schemes are running correctly or not.

        In addition, for more complex systems there may be
additional schemes such as system start-up, power fail
recovery and alarm overrides. These additional schemes all
work in a similar manner using global data for inter-scheme
communication.
        If we now take the example of attempting to
commission the control scheme software, this can be achieved
in isolation by stopping both the SCAN and the DESK scheme
and loading appropriate dummy data into the globals read by
the control scheme. It is thus possible to put a test
software harness around the control schemes and investigate
their responses by reading global outputs. The next stage
on from this level of commissioning is to allow the DESK
scheme to run and to attempt to influence the control scheme
by normal operator actions. Once again the results of the
control scheme can be monitored by looking at the global
data output and the conditions for control can be set up by
loading dummy data into the SCAN global variables. The
final level of commissioning before actually driving the
real plant is to set the SCAN scheme running and use dummy
signal inputs and investigate the response of the control
scheme by measuring the physical signal outputs. Whilst all
this activity is taking place it is obviously also possible
to set the ALARM and DISPLAY schemes running and observe
correct operation and writing of alarm messages for example
        For complex systems, it is possible to install a
plant model within the control computer. In this case

rather than using a test software harness, the plant model is used to generate suitable inputs in place of the SCAN scheme. Normal operator controls can then be performed from the desk to ensure that the control system logic will control the plant model in the correct fashion. This is also very useful for operator training and the general acceptance of advanced systems.

NOTE: In this case, it is not necessary to have a plant model which exactly mimics the real plant, because in most cases the only items it is designed to check are the logical decisions made by the control systems rather than their absolute control performance. Thus in most cases a relatively simple plant model can be used.

One can regard this type of commissioning as the straightforward substitution of a plant model for the scan scheme.

Thus it can be seen from the above that by use of a suitable multi-task system, the split of software functions can be chosen appropriately and the interactions between the software functions can be well defined in such a manner as to make commissioning and maintenance a very straightforward procedure. Our experience to date shows it is well worthwhile splitting up the software in this type of manner rather than merely writing one scheme for each control loop for instance.

Any control system is only as good as the data being input to it. Thus it is well worthwhile putting in some additional checks on the data in the scan schemes. These checks usually consist of items such as:

    Rate limits
    Data in range
    Good data
    Discrepancy checks
    Scaling of data to engineering values
    Digital filtering of critical data

All these items may be built into the scan scheme with the overall desire to come out with suitable values loaded into the global data arrays. If a fault occurs in the data input, it is usually desirable to load some default value into the global array which ensures suitable control output in response to the bad data. As an example of this, let us look at the operation of a rate limit. If two successive scans of a particular analogue item differ by more than the rate limited amount then the data loaded into the global variable corresponds to the original item. Thus if a spike occurs on the incoming analogue data this will be ignored by the control system since it would merely receive the same value as the previous valve. Such spikes are usually of very short duration and usually the third scan will result in a good data value within the rate limit to which the control system is allowed to respond.

Protection of plant from control system malfunctions is obviously of prime importance. Once again a multi-task software system allows this protection to be applied in a systematic manner. This is normally done by setting up two levels of protection, the first being a system monitor

scheme which reads data items relating to software functions
and takes appropriate action by stopping specific schemes
if malfunctions such as schemes overrunning their allotted
run time, take place. Thus only those schemes related to
a particular malfunction are stopped and the rest of the
multiple tasks can carry on as normal. The second level of
protection is watchdog system which is normally installed
to monitor overall processor activity. If the watchdog task
does not receive regular updates, it can be made to time out
and inhibit all control activity in that processor. In
addition, once this trip situation has occurred, it is
possible to send a suitable signal to other processors in
the system to indicate a particular failure and to possibly
indicate the need for a back-up control.
      NOTE: These two levels of protection are used in
addition to the basic methods employed in the design of the
control algorithms. Normally the control algorithm itself
will check to ensure that it is operating on good data and
that the plant status' are appropriate to an auto control
situation. These are normally known as availability checks
and if the availability check does not give rise to a good
result then it is not possible to select auto control, or
the system will trip from auto control to manual control.
In this case, no software malfunction has occurred but it
is highly likely that a hardware malfunction has occurred
and the loss of availability can be flagged as a suitable
global data item to the displays or alarm schemes.

13.4   DOCUMENTATION AND TRAINING

      The use of a high level software system which is
modular in design tends to lead to a self-documenting
listing. The level of comments often included in the basic
running software is often insufficient for maintenance
purposes by relatively unskilled staff, however, in this
case listings are prepared with virtually one comment per
line of coding. Comments at this level do not waste
processor time and space since they are ignored by the
compiler and do not appear in the running software (Fig 13.2
These documented listings are amplified by a description of
the function of each scheme which is regarded as a heading
to the documentation listings. The final piece of
documentation associated with each scheme is the signal
input/output schedule. This can consist of lists of
physical signal inputting to a scan scheme for example or
it may only include a lists of global variables indicating
data flowing in or out of a particular scheme. Inclusion
of these three items together with the normal type of syste
instruction manual is usually found to be adequate for
Station documentation.
      The use of a host computer for programming the targe
machines implies the retention of documentation of system
software as files on the host machine. Thus the
documentation of the system software as opposed to the
application software is normally held as files on some type
of magnetic storage medium. Hard copy listings of these

```
VMIN:=R11FA[5]
VMAX:=R11FA[6]
SC:=R11FA[7]
RATIO:=R11FA[8]
FCOP:=R11FA[9]
MMCOP:=R11FA[10]
CAL:=R11FA[11]
KFUEL:=R11FA[12]
NMIN:=I11FA[1]
NMAX:=I11FA[2]

; SET AUTO AVAILABLE IF MILL RUNNING.AND N2 GOOD.N2 IS GOOD ONLY
; IF SIGNALS AND GAINS GOOD
; HOWEVER, FEEDER REG. IS HELD WITH RTS CONDITIONS UNTIL
; FEEDER MOTOR IS INSERVICE.

OK1:=MMC>MMCOP
OK:=OK1 AND GOODINT(N2)

AUTO:=(OK AND AMB)                       ;AUTO IF AVAIL AND BUTT

AUTOMAN AUTO

;RETURN TO START ON ACTUATOR IF FEEDER TRIPS

RTS:=FTRIP OR ((FC<FCOP) AND GOODREAL(FC))
IF RTS AND GOODLOGIC(RTS) THEN
VTOP:=VMIN
ELSE
VTOP:=VMAX
ENDIF


E:=RATIO*PAD-MD

DX0:=INCPID(E,K,TI,TD,TF)                ;FEEDER CONTROL


DX:=DX0+REM                              ;TOTAL % MOVEMENT WANTED THIS TIC
VDOWN:=VMIN-V                            ;DISTANCE TO BOTTOM
VUP:=VTOP-V                              ;DISTANCE TO TOP
LIMIT OUTPUT DX1,LO,HI INPUT DX,VDOWN,VUP          ;LIMIT MOVEMENT THIS TIC

INCS OUTPUT N2,DX2,REM INPUT DX1,SC,NMIN,NMAX      ;CONVERT TO PULSES
N:=N2
FLIPFLOP M

;THE PA CONTROL IS INHIBITED FROM GOING TO MPC DRIVEN AUTO UNLESS THE
;FEEDER IS ON AUTO.HENCE SET GLOBAL STATUS FLAG FAUTO TO INDICATE FEEDER
;STATUS.FOR TOTAL FUEL LOOP IN MPC WE NEED TO KNOW THE FUEL OUTPUT
;FROM EACH MILL.THIS (FUEL) IS CALCULATED IN MW BY NOTING DISTANCE
;OF MILL DIFF FROM CLEAN AIR LINE

FAUTOA:=AUTO AND NOT(RTS)
IF OK1 AND GOODLOGIC(OK1) THEN
FUELA:=(MD-CAL*PAD)*KFUEL
ELSE
FUELA:=0.0
ENDIF
IGOUT INPUT OK CARD DO1,B4 CLEAR,SET,CLEAR         ;OUTPUT TO AM STATION
IGOUT INPUT OK CARD DO2,B7 SET,CLEAR,SET           ;ALARM

I11FTA[1]:=MD
I11FTA[2]:=PAD
```

FIG. 13.2. DOCUMENTED CUTLASS LISTING.

files are not normally retained since this information is not normally used on a day to day basis by the Station operating staff.

One aspect of training that we have found particularly useful is to assemble a set of Station spares into a running computer and then allowing Station staff to use this machine to develop their own non-critical pieces of software. The multi-task software system allows trainee programmers or engineers to produce small sections software, for example VDU displays and gain hands-on experience of using the system. Similarly, the modular design of the software allows a test harness to be applied to the small sections of development programs in order to allow off-line debugging before being applied to the real plant. The CUTLASS operating system also have provision for several levels of protection to be applied to various modules of the software. This is useful during training in that it allows trainees to investigate what is happening within a machine but does not allow them to change any of the control codes. As users become more competent in using the system, their individual protection codes can be increased to allow them greater access to the various monitor and editing commands.

An example of a typical CUTLASS scheme listing is shown in Figure 13.2 illustrating samples of a documented listing appropriate to this system.

13.5   HARDWARE

With the wide variety of computer control hardware available and the diversity of plant to which it may be applied, it is difficult to describe specific hardware details for general applications. However, our experiences to date have proved the worth of certain design philosophies which are described below. In order to show the scope of hardware required, a typical processor/plant schematic is shown in Figure 13.3. This indicates the hardware connected to a typical plant processor or target machine. In this example, the processor was being used to control both pressure and temperature of a fuel oil system.

13.5.1   Auto Manual Interface.

The detailed design of the auto manual interface is fundamental to the successful and reliable operation of any DDC system. It is essential that manual control is always maintained whatever the state of the control computers and their associated hardware. A schematic of a typical general system is shown in Figures 13.4. The basic element in this system is a latching auto manual changeover relay, which is wired such that breaking the latch or general loss of excitation voltage, causes the manual mode to be selected. Thus operation of the manual pushbutton breaks the latch to cause a trip to manual, and opening of either the availability relay or the watchdog relay causes loss of excitation volts, which also causes a trip to manual. Selection of auto control can only be made by outputting a

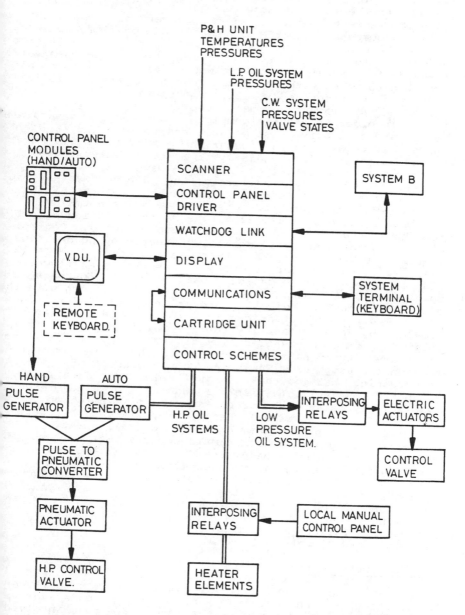

FIG.13.3. TYPICAL PROCESSOR / PLANT SCHEMATIC (SYSTEM A)

signal from the computer to the latching relay. Thus auto
control may only be selected if both the availability and
watchdog relays are energised, thus providing excitation
voltage to the latching auto manual relay. The fact that
the auto request signal to the changeover relay emanates
from the computer output means that it is possible to check
within the computer that auto control is possible. If auto
selection had been made available directly to the changeover
relay as for manual control, it would be possible to select
auto control even with the computer not running. This is
obviously an erroneous situation and therefore the
philosophy as drawn is normally used.

In order to provide complete redundancy, two separate
manual and auto drive systems are used to feed the
changeover relay. In particular the manual drive system
should be powered independently from the computer thus
guaranteeing that manual control is always available even
in the case of complete failure of the computer system. We
find it preferable to use pulse drives to plant actuators
and therefore the auto and manual drive units normally
consist of pulse or oscillator type devices feeding directly
into the contacts of the manual changeover relay. Control
of these pulsing units is normally directly by pushbuttons
in the manual drive case and by raise and lower signals in
the case of the auto drive units. These raise and lower
signals for auto drive may be a pair of digital outputs, or
merely a number transmitting to the auto drive unit with the
sign of the number indicating the raise or lower direction.

The protection for each control loop is normally
provided by the availability relay. The control of this
relay depends on an availability digital output which is set
or cleared depending on the success of checks made within
the control computer. These checks may consist of
ascertaining that good data is available, that data is in
range and that suitable plant is running. Opening of the
availability relay contacts causes loss of excitation
voltage on the auto manual changeover relay and therefore
an automatic trip to manual control. Overall plant
protection is provided by the main computer watchdog relay.
This relay is normally held closed as long as the computer
watchdog receives regular updates from the DDC control
programs. If these updates are not received the watchdog
hardware times out causing the loss of field volts on to the
main busbar which provides excitation voltage for all the
auto manual changeover relays associated with that
processor. This is shown diagrammatically in Figure 13.4
with auto manual relays 2 and 3.

13.5.2   Output Drives.

There are several ways of communicating raise/lower
information to plant actuators, but of the methods available
we prefer sending pulses over the long cable runs involved
This method has several significant advantages, the main
ones being security of signal transmission, bumpless auto
manual transfer and positive resolution. If a cable shoul

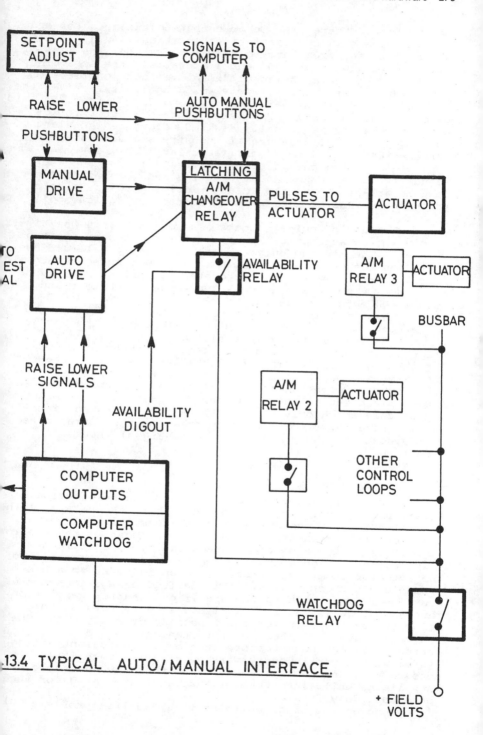

13.4 TYPICAL AUTO/MANUAL INTERFACE.

become broken there will be no output to the actuator and
it will freeze in its last driven position. In a similar
way on transfer from auto to manual operation, the loss of
pulses during the auto to manual operation, the loss of
pulses during the auto manual changeover has no effect on
the plant actuator which merely waits until it receives the
next train of pulses, which can emanate from either the auto
or manual drive units. Thus signalsare only sent to the
actuator when it is actually desired to change the plant
control parameter. The use of industry standard standing
current levels of 4 to 20 milliamps invariably involves some
type of balancing hardware so that there is no change of
signal during an auto manual transfer and loss of an output
line invariably means the actuator driving to one stop or
the other unless it has specific protection devices built
into it.

The use of pulse outputs also means that a positive
resolution is automatically achieved for actuator position
changes. The minimum pulse length is usually well-defined
as is the response to a minimum raise/lower pulse. Thus it
is relatively simple to incorporate appropriate dead bands
into the system to avoid excessive actuator chatter and
associated actuator wear.

The simplest way of decoding actuator pulses is to
use stepper motors at the actuator head end of the hardware
chain. Maximum security is achieved by using 4 phase
stepper motors which do however require rather more cores
for the signal path. For new installations this increased
security is thought to be well worthwhile. However, there
are many situations where control system refurbishments are
taking place and in this instance two phase stepper motors
may be used which only require three cores for the signal
path. These three cores usually exist as the normal three
wire system for the previous analogue signal input
actuators.

13.5.3   Actuators.

Virtually all types of actuators can be interfaced
to all DDC control systems, however, it is the nature of the
digital outputs which drive the actuators which can give
rise to problems on hardware selection. Most of the control
algorithms used are of the incremental type, that is a
calculation is made as to the desired change in position of
the actuator from its current position in order to achieve
suitable plant control. If this desired change in actuator
position is greater than the specified actuator resolution
then an output is made. This leads to actuators either
being turned fully on for short periods or fully off. The
bang bang nature of these outputs tends to put a more
onerous duty on the actuators than on an equivalent analogu
system with a continuously varying signal. Thus it is
necessary to ensure that actuators either have suitable dea
band limits built into them or that they are rated for such
continuous duty.

One of the areas where many reliability problems

occur is in the use of limit switches or local position loops on actuators. This situation can easily be avoided in DDC control systems by providing a good local position measurement and feeding this signal back to the main control computers. Any local position loops or limiting functions can then be performed within the processor with no fear of reliability problems due to sticking contacts etc. In terms of improving control systems reliability this is one of the most beneficial design changes that can be made over the conventional analogue type system.

13.5.4   Operator Interface.

The operator interface consists of a control panel through which an operator may communicate to and receive information from the main control computers. This is usually done via switches or lamps, pushbuttons, meters and VDU type displays. The use of a DDC system implies great flexibility in the implementation of control strategies and algorithms. In order to match this degree of flexibility with the operator interface it is convenient to manufacture the desk or panel as a series of standard size interlocking units. This is known as a DIN grid type of panel and an example is shown in Figure 13.5. Each module is made up of a standard size square plastic tile (72 mm x 72 mm square). These tiles may contain switches, lamps or small strip meters and interchangeable to any position on the grid. In addition, it is possible to have double or quadruple size modules which are large enough to show higher resolution strip displays or small charge recorders. A standard range of such modules is available and if a new loop is implemented it is merely a case of inserting appropriate tiles in appropriate sections of the grid.

All the modules are designed to be entirely passive, that is they consist purely of switches, lamps or meters. Each module is wired in a standard way to a standard plug and socket at the rear. Thus modules can easily be interchanged or replaced in the event of failure. In order to communicate information back to the operator it is normal to use colour VDU displays. These are usually of the high resolution type, so that accurate graphical information can be presented directly to the operator and the control of which displays are presented is normally achieved by direct pushbutton actions on the control desk. The rates of changing displays are important and it is found that operators usually require a new picture to be completely drawn in under two seconds to be satisfactory. The rate at which detailed information on each display is updated can, however, be much greater than this and will depend on the type of parameter being monitored. In a case of fast control looops, such as feedwater, this update may be of the order of one or two seconds, however, for metal temperature monitoring situations the refresh time may be of the order of 15 to 30 seconds.

With the advances in colour display controllers that have recently occurred there is great temptation to produce

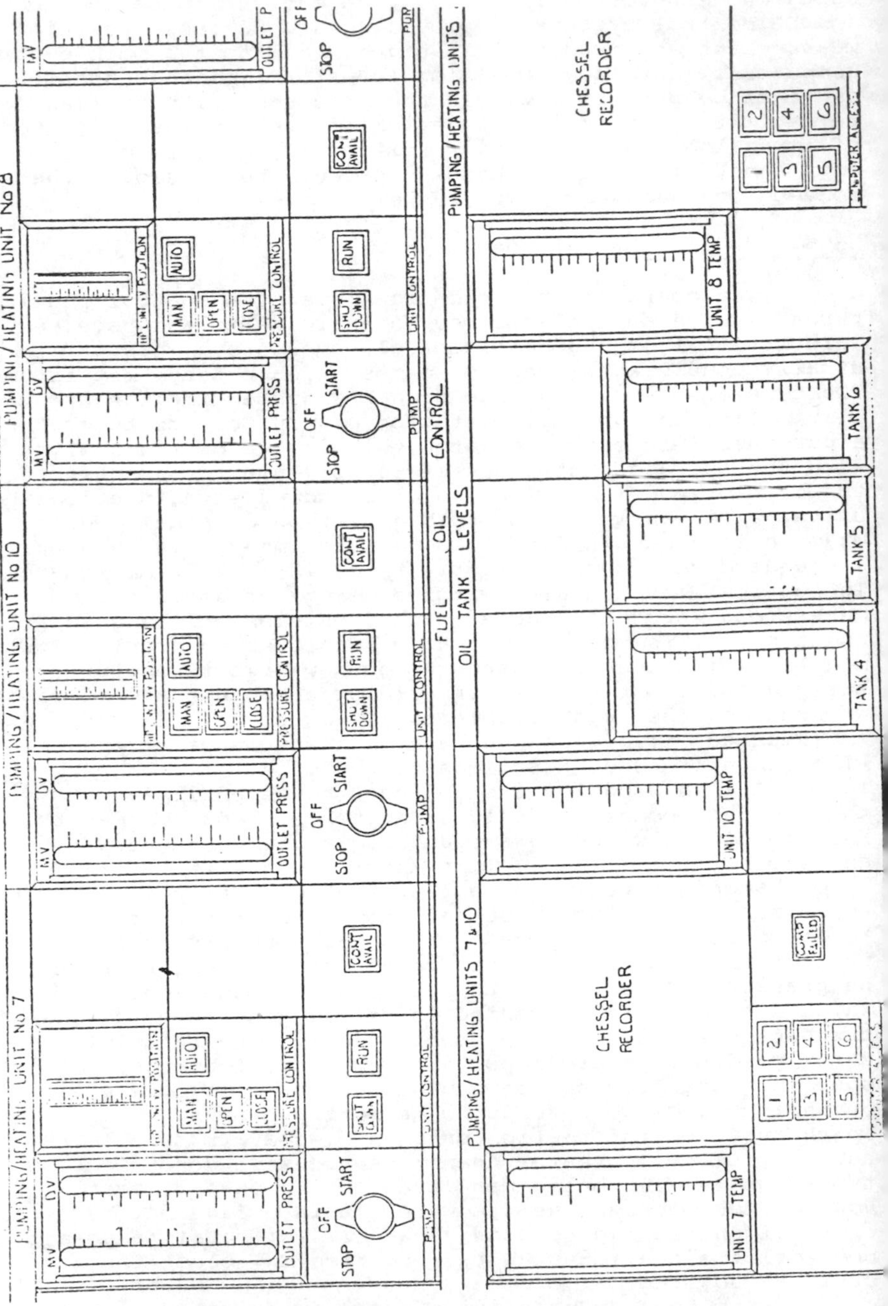

FIG.13.5. TYPICAL MODULAR OPERATORS PANEL.

highly complex colour displays containing much detailed information. In practice, this has been found to be unsatisfactory and very simple plant mimics are preferred by operators with secondary displays being used to amplify the basic information on the primary formats. In addition, the use of interactive displays has been found to be beneficial, for instance set point adjustment made on the control desk are displayed directly at suitable points on a plant mimic for the operator to very quickly appreciate exactly what is going on on the plant. Examples of typical colour displays are shown in Figure 13.6.

13.5.5   Computer Selection.

The choice of computer for real time control applications depends mainly on the data sampling rates required and the number of loops being implemented in each machine. It is not difficult to see that the most suitable sampling period is in some way related to the time constants of the plant to be controlled. In practice, we have found that sampling intervals of between 1/5th and 1/10th of the basic plant time constants are the minimum necessary for reasonable control response. For control loops on nuclear and fossil fired power stations the sampling period used would typically be in the range 1 to 10 seconds, with alarm and operator interface functions rather faster, of the order 200 milliseconds. Thus, whilst the cost of computer core storage is falling and becoming less significant, with small microprocessors computation time is often critical when considering real time operation. Therefore, it is often a case of selecting the maximum sajpling period that will be adequate. When multiplication and division operations are required it is frequently beneficial to add a hardware arithmetic unit in order to speed up the processing rate. Where very fast sampling rates are required in order to guarantee the speed of response of a particular control loop to say limit switches, use can be made of interrupting digital inputs in order to achieve a saving in computer loading and execution times.

The other main area of computer operation which determines the type of machine to be used is that of communications. Communication of information to or from the control computer via standard serial lines can be extremely time consuming if even relatively small amounts of data are to be regularly updated around a typical network of target machines. This particular problem can become so acute as to require the necessity of machines dedicated solely to communications. Usually, however, a compromise is made where critical data is duplicated in each machine requiring the information and suitable slower speed communications are made in orer to provide a back-up facility.

Currently the type of machines that are typically used consist of Digital Equipment Corporation (DEC) LSI/11/23 machines, with from 32K to 128K words of memory.

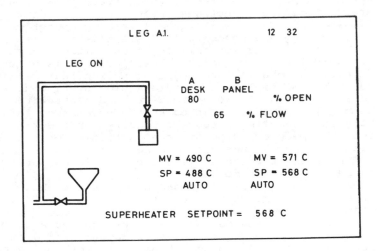

LOAD			SUPERHEAT		12	32

LOADING AUTO			LEGS	A1	A2	B1	B2
			OUTLET TEMP MV	571	565	567	564
LOAD	455	MW	OUTLET TEMP SP	568	568	568	568
TARGET	450	MW	LOOP STATUS	AUTO	AUTO	AUTO	MANUAL
FREQUENCY	50, 01	HZ	ATTEMP TEMP MV	490	485	491	475
VALVE RATIO	78	%	ATTEMP TEMP SP	488	487	489	472
			LOOP STATUS	AUTO	AUTO	AUTO	AUTO
FIRING AUTO			VALVE POSN A %	80	78	70	1
PRESSURE MW	164	BAR	VALVE POSN B %	0	2	0	76
PRESSURE SP	165	BAR	SPRAY FLOW %	65	63	50	59
DEADBAND	5	BAR	LEG STATUS	LEG ON	LEG ON	LEG ON	LEG ON
PA AUTO	24, 5	MB	LEG INHIBIT				
PA MANUAL	5, 4	MB					
PA TOTAL	29, 9	MB	SAT. TEMPERATURE		364		

LEG A.1.   12   32

LEG ON

               A   B
            DESK   PANEL
              80   % OPEN
                   65   % FLOW

            MV = 490 C   MV = 571 C
            SP = 488 C   SP = 568 C
               AUTO   AUTO

         SUPERHEATER SETPOINT =   568 C

FIG.13·6. VDU DISPLAYS

13.5.6   Environmental Considerations.

The ability to function in hostile environments is a factor which has to be taken into consideration when making the choice of computer control equipment. Parameters such as ambient temperature, humidity, dust levels, vibration, power supply variations and electrical interference must be considered carefully. In the CEGB we have standard specification categories for all these items and the requirements must be carefully checked against manufacturers' specification and tests performed to demonstrate the equipment performance. The power station environment tends to be more severe, especially in terms of electrical interference and power supply variations, than most manufacturers contemplate when specifying their equipment. For example, plus or minus 10% power supply voltage variations are commonly specified by manufacturers whereas we have experienced plus or minus 20% variations on site.

Such variations are normally the most difficult items to deal with and use has to be made of constant voltage transformers, radio frequency filters and other similar items in order to guarantee smooth supplies to the computer. It is normally not possible to install break-free supplies especially on refit type installations and therefore some form of either reloading the computer or maintaining the memory in the case of power supply failure is required. Currently many of our systems use battery back-up facilities to guarantee resistance to power supply failures, however I feel that the use of PROM for storing software is a better solution to this problem. Current technical advances in this field mean that we should very soon be able to hold all the application software in PROM (Programmable Rear Only Memory).

NOTE: That even with systems installed to guarantee a retention of memory, it is normal to install a back-up memory storage device such as a magnetic tape cassette unit, in order to be able to reload software at will.

13.6   CONCLUSIONS

There will be a substantial increase in the number of computer based control systems used throughout industry over the next decade, particularly with regard to the refurbishment or enhancement of existing plant. In many of these areas the trend of the application will be towards distributed systems in order to gain the technical and economical advantages made possible by the advent of low cost mini and microcomputers. To take full advantage of the possibilities of these systems it is necessary to consider carefully the differences between digital and the previous analogue type of controllers. I have tried to outline some of these possibilities in this paper and I hope that in the future there will be an adoption of common user requirements and standard hardware and software effort required for individual projects. One of the main areas where the Board

has tried to achieve this is in the use of its standard
software system, CUTLASS.

Standard approaches to hardware design will be more
difficult to achieve considering the wide variation of plant
to be controlled. However, the use of consistent design
philosophies as outlined in this paper should enable maximum
benefit to be obtained from the use of digital control
systems.

Chapter 14

Application of distributed digital control algorithms to power stations

J.N. Wallace

14.1 INTRODUCTION

It has been evident for many years that the enormous advances in digital estimation and control theory offered the potential for improving control system performance. However, hardware restrictions placed severe implementation constraints on process control engineers. Despite the difficulties the CEGB recognised the benefits of digital control and pursued the application of process control computers to complex power station problems. For example, the second generation of nuclear power stations (the AGR's) such as Hinkley 'B' incorporate reactor and some boiler/testing control functions within the computer function.

However, during the last ten years the rate of development and application of digital control to large CEGB power stations has accelerated as a result of:-

a) The development of reliable, low cost mini and micro computer equipment.

b) The development of efficient software systems supporting easy to use, general purpose language facilities.

c) The rapid escalation in oil and coal prices necessitating more flexible operation of large fossil fired units and greater attention to maximising output from operating nuclear stations.

The particular combination of fossil and nuclear power stations in the South West Region of the CEGB has resulted in Regional C & I specialist being heavily involved in this fundamental change in power station control. The first major commercial distributed digital control system were installed at Pembroke (4 x 500 MW oil fired) and Didcot (4 x 500 MW coal fired) in 1979 (Wallace (1)). Steady expansion has occurred on a number of sites since that time. The purpose of this paper is to identify and describe the various aspects of the application of digital control algorithms to power stations in the region.

14.2 CONTROL SYSTEM MODES, STRUCTURES AND LOOP CONCEPTS

Before dealing with the more technical aspects of digital control algorithms it is worth briefly commenting on the overall organisation problems that large scale, high performance digital control systems can pose. The complexities of the overall power plant control problem may be judged by reference to figure 2.1 showing the schematic structure of a Pembroke boiler-turbine unit.

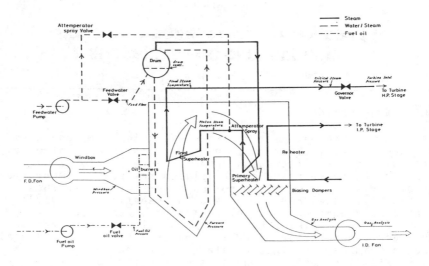

Figure 2.1   Boiler/turbine schematic

Traditionally, although primary inputs and outputs on reactors, boilers and turbines are interdependent, the CEGB has designed and implemented its control systems on a loop by loop (multi loop or cascade) basis for reasons of security and simplicity of implementation. The author and his colleagues have retained the loop concept rather than designing one very large complex multivariable control system. Thus a hierarchical approach has been adopted with the loops broadly falling into one of three categories:
LEVEL 1 - Primary sub loops for actuator modulation.

LEVEL 2 - Main Plant Area loops which determine set points for Primary loops.

LEVEL 3 - Supervisory loops which determine set points for the main Plan Area loops.
Using this functionally distributed approach it has been possible to exploit the potential of the flexible digital computer whilst providing gentle degradation in the case of partial measurement, computer or actuator failure. The use of primary sub loops is also recommended as good

practice in terms of containing local disturbances and ensuring that overall unit response to supervisory or optimisation commands is predictable.

Figures 2.2 and 2.3 illustrate the original 1977 and current 1986 processor configurations. Although the communications technology has changed from point to point serial links to an Ethernet bus, the control loop hierarchy remains unchanged (Fig.2.4).

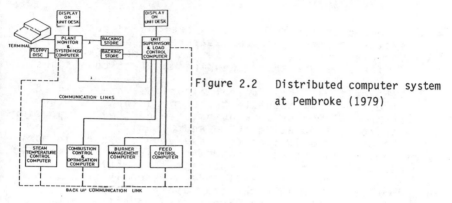

Figure 2.2 Distributed computer system at Pembroke (1979)

Figure 2.3 Augmented distributed computer system at Pembroke (1986)

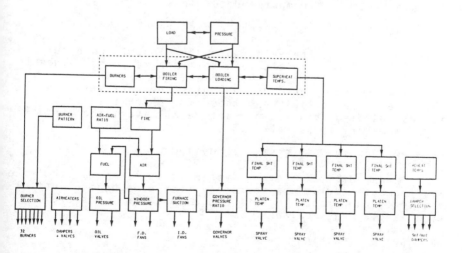

Figure 2.4 Control system hierarchy at Pembroke

The retention of a loop structure does, however, pose some organisational problems both in terms of how the operator perceives the system and in terms of software organisation. Difficulties arise when:
a) More than one plant actuator is used to control one process variable. (e.g. two oil valves, two FD fans and two ID fans are employed at Pembroke to control oil pressure, air flow and furnace suction respectively).

b) A loop operates in more than one mode depending on the status of other loops or measurement availability (e.g. The air-fuel ratio optimiser at Pembroke executes different algorithms depending on plant conditions).

c) True multivariable control algorithms are employed to simultaneously modulate two or more inputs to the control of two or more outputs (e.g. the load-pressure control system at Pembroke and Dicot).

The correct solutions to these problems are, in many cases, application specific and the subject of qualitative judgement. However, it is important that whatever convention is adopted is applied consistently throughout the computer system in order that operators are not confused. At Pembroke and Didcot the convention adopted is based on the concept of operator responsibility. Thus, each loop has a loop status code consisting of:

Standby Manual   -   Operator responsibility to determine the action manually.

Computer Manual -   Operator responsibility to determine control action, which is initiated through the computer.

Computer Auto   -   Computer determines control action.

Tracking   -   No control action required since controlled variable is determined by behaviour of other loops. Loop set points made to track measured values to ensure bumpless transfer.

The operator can thus determine the activities for which he is responsible and those for which the computer is responsible, even if interactive multivariable control algorithms are in operation.

14.3 MEASUREMENT PROCESSING ALGORITHMS

In any digital control system it is essential that the measurement samples are correctly processed before control action is computed. The principle considerations are:
a) Ranging and Scaling

b) Noise smoothing

c) Failure detection.

With real arithmetic, <u>ranging and scaling</u> of measured data is a straight forward operation. However, where integer arithmetic is employed reference must be made to the control algorithm and actuator resolution in order to avoid quantisation noise or loss of resolution.

<u>Noise smoothing</u> represents more of a problem since it impacts on control loop stability, response to plant disturbances and frequency of control action. The simplest method of smoothing measurements is to use the discrete equivalent of a first order lag, the required algorithm being presented in Table 1.

TABLE 3.1 Measurement Filtering Algorithm

Problem
$\underline{x}_{k+1} = F\ \underline{x}_k + G\ \underline{u}_k\ + \text{noise}$
$y_{k+1} = H\ \underline{x}_{k+1}$
$z_{k+1} = y_{k+1} + \text{noise}$

Lag filter
$\hat{y}_{k+1} = \hat{y}_k + K^f\ (z_{k+1} - \hat{y}_k)$

Kalman filter
$\hat{\underline{x}}_{k+1} = F\ \hat{\underline{x}}_k\ + G\ \underline{u}_k$
$\hat{\underline{x}}_{k+1} \equiv \hat{\underline{x}}_{k+1} + K^f\ (z_{k+1} - H\ \hat{\underline{x}}_{k+1})$
$\hat{y}_{k+1} = H\ \hat{\underline{x}}_{k+1}$

As an example, this form of algorithm has been successfully employed on the primary draught plant loops (oil, FD and ID fans) at Pembroke. Such filtering can substantially reduce measurement noise and it is possible to detune the controller in order to account for the additional lag introduced into the loop. The main snag is that the loop response to plant disturbances is degraded and a compromise must be struck which, in some cases, is not entirely acceptable.

Variations such as second order filtering can be introduced which may improve the design compromise. However, a more systematic method of smoothing measurements is to use a Kalman filter. This uses a discrete state-space model of the plant to predict dynamical changes in the plant measurements and then correct the state estimate as a function of the prediction error. The algorithm possesses a general form, illustrated in Table 3.1 for the scalar measurement case. The filter gain matrix K^T which represents the feedback gain from prediction error to model state is computed from a matrix riccati equation. For a more detailed explanation the reader is referred to Sorenson (2). This book is devoted entirely to the theory and application of Kalman filtering. Such a filter is used at Pembroke to process the pressure and steam flow measurements used in the load-pressure control system. The state equations for this example are contained in Table 3.2. In simplified form it has also been used at Oldbury to predict boiler conditions ahead of real time and hence allows improved control. (See Section 14.5.4).

From a control point of view, the major advantage is that it can provide excellent noise smoothing without interposing a lag into the control loop. Experience from Pembroke and Oldbury suggests that the approach can work in practice and can yield additional useful information regarding plant behaviour. The major disadvantages are that additional design effort is required and that a reasonable mathematical model of the plant is a prerequisite of the approach. However, in cases where simple lag filtering compromises the control loop design, the Kalman filter offers a practical alternative. The resulting 'state' estimates and capability of predicting future behaviour can also be useful in plant control.

The final, and probably most important, topic regarding measurement processing is that of failure detection. In a control environment it is essential that every care is taken to detect failures in the measurements. The commonly used checks are amplitude and rate of change of measurement (Table 3.3). The latter has proved particularly effective at throwing up gross measurement failures such as transducer faults. However, plant disturbances can also produce rapid changes in measurements and hence rate limits have to be set with this in mind.

Unfortunately, measurement systems (transducer, transmitter and analogue scanning systems) are capable of a whole host of unpredictable failures. For example, measurement 'freeze', cross channel interference in multiplexed scanners and step or ramp changes in calibration levels have all been personally witnessed by the author. Interpretation of measurement data under transducer, transmitter or scanner overload also needs careful attention.

TABLE 3.2. State Equations for Kalman Filter at Pembroke

Continuous Plant Model

$$\underline{\dot{x}} = A\underline{x} + B\underline{u}$$

$$\underline{y} = H\underline{x}$$

where

$$A = \begin{bmatrix} -\hat{a}_1 & \hat{a}_1 & a_1 & 0 & a_1 \\ a_2 & -(a_2+\hat{a}_2) & 0 & 1 & 0 \\ 0 & 0 & -a_3 & 0 & 0 \\ 0 & 0 & 0 & -a_4 & 0 \\ 0 & 0 & 0 & 0 & -a_5 \end{bmatrix}$$

$$B' = \begin{bmatrix} 0 & 0 & a_3 & 0 & 0 \\ 0 & 0 & 0 & a_6 & 0 \end{bmatrix}$$

$$H = \begin{bmatrix} 0 & 1 & 0 & 0 & 0 \\ 0 & \hat{a}_2 & 0 & a_7 & 0 \end{bmatrix}$$

Discrete Filter Model

$$\underline{x}_{k+1} = F_k\,\underline{x}_k + G_k\,\underline{u}_k + \underline{w}_k$$

$$\underline{z}_{k+1} = H_{k+1} + \underline{v}_{k+1}$$

where $\underline{a}' = (a_1\ a_2\ a_3\ a_4\ a_5\ a_6\ a_7)$ Calculable from $\underline{x}$

$\underline{\hat{a}}' = (\hat{a}_1\ \hat{a}_2)$ Estimated

$\underline{w}_k, \underline{v}_k$ White noise sources with covariance Q, R

$$F_k = (I + A\ \Delta t)^4$$

$$G_k = \sum_{i=0}^{3} (I + A\ \Delta t)^{3-i}\ B\ \Delta t$$

TABLE 3.3.   Measurement Failure Criteria

BASIC MEASUREMENT CHECKS	ALGORITHM
Max limit	$z_k > z_k^{max}$
Min limit	$z_k < z_k^{min}$
Gross rate of change limit	$\left| z_k - z_{k-1} \right| > \lambda_1 \cdot \Delta z_k^{lim}$
Cumulative rate of change limit	$\left| z_k - z_{k-1} \right| > \Delta z_k^{lim}$ N times in J samples
KALMAN FILTER RESIDUAL CHECKS	**ALGORITHM**
Gross rate of change limit	$\left| \tilde{z}_k \right| > \lambda_1 \cdot \sigma_k$
Cumulative rate of change limit	$\left| \tilde{z}_k \right| > \lambda_2 \cdot \sigma_k$ N times in J samples
Variance max limit	$(\rho_k)^2 > \lambda_3 \cdot (\sigma_k)^2$
Variance min limit	$(\rho_k)^2 < \dfrac{1}{\lambda_3} \cdot (\sigma_k)^2$

A systematic approach to detecting measurement failure is the examination of the Kalman Filter prediction residuals. Using one at a time measurement processing (Bierman (3)), the prediction residual for the ith measurement at t_k is:

$$\tilde{z}^i_k = z^i_k - H^i_k \hat{\underline{x}}_k$$

If the measurement noise variance is R^i, then the variance estimate for this residual is given by

$$(\sigma^i_k)^2 = (H^i_k)' P_k H^i_k + R^i$$

Whilst the actual Kalman filter residual variance $(\rho^j_k)^2$ can be computed using algorithms such as:

$$(\rho^j_k)^2 = \sum_{j=1}^{n} \frac{(\tilde{\underline{z}}^i_{k-j})^2}{n}$$

One can then apply the rate of change and variance checks tabulated in Table 3.3 to the prediction residual $z^1{}_k$. The checks can be more severe than the conventional ones applied to $z^1{}_k$ (Table 3.3), even under transient conditions and also reveal problems such as measurement freeze.

Such tests were incorporated into the Pembroke load/pressure control system (Wallace et al (4)) but only for alarm purposes. Commitment to other projects prevented a full evaluation of this technique. However, it was clear that unmodelled disturbances and modelling errors both disturbed residual behaviour. It proved difficult to set check limits which were significantly tighter than the normal rate checks without generating spurious alarms.

In many cases a single measurement noise spike is not indicative of measurement failure. Thus in general, only if a measurement fails to satisfy the various check 'N' times in 'J' samples is the measurement designated 'bad'. The plant consequences of failure determine the selection of N and J.

14.4 SINGLE LOOP CONTROL ALGORITHMS

14.4.1 Three Term Controllers

The digital equivalent of analogue PI and PID controller are commonly used at the primary sub loop level where the dynamics are fast (i.e. time constants of a few seconds) or of low order. They are usually implemented in incremental form since this simplifies bumpless-transfer problems and operation close to constraint boundaries. Typical examples are the oil pressure, windbox pressure, furnace suction loops and governor valve pressure ratio loop at Pembroke. Derivative action on set point changes is usually omitted to avoid unwanted kicks in control loop output.

Digital PID controllers have generally been found to be more successful than their analogue equivalents because of features such as drift-free long term-constant integral terms and programmable constraint handling and/or detection.

14.4.2. Time Delay Compensation

Whilst the PI and PID algorithms are satisfactory for single loop with low order dynamics, when the process exhibits significant time delay it is difficult to achieve tight control with such a simple controller. Detuning the PID controller yields stability at the expense of slow response to distrubances, particularly if the derivative term is constrained by measurement noise sensitivity. Typical examples are the measured combustion product response to air-fuel ratio response and the measured attemperator temperature change in response to spray flow perturbations.

With the programmability built in to the CEGB's real time software systems it is possible to compensate for time delay and hence achieve an acceptable quality of control over a range of operating conditions. Table 4.1 presents two compensation algorithms that have been employed:

TABLE 4.1   Time Delay Compensation

Series compensator for 'n' sample delay	$e_k = (z_k^{set} - z_k)$
	$\lrcorner u_k = f^c(e_k)$ normal control algorithm
Typical $\lambda = 0.5$	$\lrcorner u_k = u_k - \lambda u_{k-n}$
Smith predictor for 'n' sample delay	$e_k = z_k^{set} - z_k$
	$\hat{x}_k = \sum b_i\, u_{k-i} - \sum a_i\, \hat{x}_{k-i}$
	$e^*_k = e_k - (\hat{x}_k - \hat{x}_{k-n})$
	$u_k = f^c(e^*_k)$ normal control algorithm

The series compensator in Table 4.1 was applied to the superheater attemperator back-up control at Pembroke. Although simple to implement in a single board PROM based actuator control/drive card (1) it proved an effective addition to the basic PI control.
The Smith predictor (Marshall (5)) was applied to the combustion CO trim loop at Pembroke to compensate for the propagation delay between combustion chamber and stack where CO was sensed. The application was successful but the improvement in performance was not as great as desired due to the variability of the process dynamics.

14.4.3 Z - Transform Controllers

The particular algorithms described in 14.4.1 and 14.4.2 are particular cases of the more general 'Z' transform controller algorithms. As design techniques improve these controllers are being slowly introduced where control is unusually difficult. The mill control and Didcot P.S. are an example. The possibility of using self-lining packages to initially commission and periodically maintain the algorithms is also being explored.

14.5 MULTIVARIABLE CONTROL ALGORITHMS

14.5.1 Multi Input/Output Feedback Control

14.5.1.1 Problem Definition. Although one may design primary and plant area control loops by neglecting the effects on other plant areas there is every reason to tackle such multivariable problems at the supervisory level. Essentially, with multivariable control one manipulates multiple control inputs to multiple plant outputs in some coordinated fashion. One such problem which has received considerable attention is that of integrated unit load-boiler pressure control. The task, illustrated in Fig. 5.1, is to manipulate the governor valve and total boiler heat input such that the pressure and steam flow (load) remain close to set points which can rapidly change under certain operational conditions. A major constraint; particularly at Pembroke is that disturbances to the energy balance should be minimised in order to avoid plant damage.

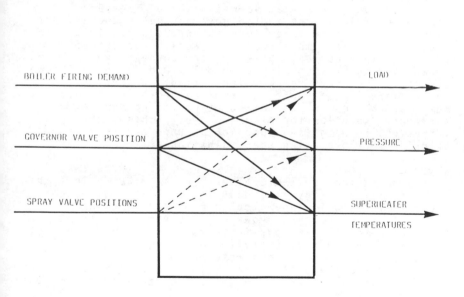

Figure 5.1   Boiler load/pressure/temperature dynamics

Coordinated control schemes were developed by the author and his colleagues for each of the four units at Didcot P.S. and Pembroke P.S. The schemes evolved using the same overall strategy but differing techniques because of significantly different plant characteristics and operating regimes. It is of interest to compare the structures of the two schemes using the plant model structure in Table 3.2. This model (incorporated into a Kalman filter) represents Pembroke's boiler dynamics. However, if one makes the simplifying assumption that steam flow and load are synonymous and the mill transfer function is a first order lag then the model serves as a useful reference for control structure comparison.

14.5.1.2 A multi-loop approach.   The approach adopted at Didcot was to combine a series of single loop concepts into a coordinated framework. This coordination process was based on previous CEGB experience with the single loops, an understanding of their limitations and of the particular response characteristics of the Didcot mills and boilers. Thus, after careful analysis of simulation results and plant trials the multivariable control algorithm, represented in figure 5.2 in state space form, was evolved. The augmentation of the state vector to achieve integral control is a well known technique. However, if state rather than output feedback had been employed more gain terms would be none zero. A noticeable feature of the algorithm is the time varying gain structure. As pressure errors (z_2 set $- z_2$) increase, the relative weighting of pressure and load error feedback on to the governor valve changes until eventually the load error term is suppressed. This gradual change of emphasis in the control action has proved effective in practice. However, the concept evolved through understanding of a particular problem and in general a more systematic method of generating variable weighting controllers is desirable. Work at Pembroke on the optimal multivariable conrol was aimed at providing such a systematic solution for other control problems through variations in weighting functions. Unfortunately higher priority work prevented the evaluation of this concept.

14.5.1.3 A discrete optimal control approach.   The load-
-pressure control system evolved for Pembroke follows the same general strategy as Didcot in terms of providing accurate load following which is modified if plant constraints are reached. However, the requirements at Pembroke were more arduous in terms of range and speed of operation and plant sensitivity to firing and steam flow perturbations. Hence a discrete optimal control algorithm has been used, the state information being derived from an adaptive Kalman filter. The optimal controller minimises a performance function of the form:

$$ J = \sum_{k=0}^{\infty} \underline{x}'_k Q^c \underline{x}_k + \Delta\underline{u}'_k R^c \underline{u}_k + \Delta\underline{u}'_k S^c \Delta\underline{u}_k $$

where

$$ \Delta\underline{u}_k = \underline{u}_k - \underline{u}_{k-1} $$

The weighting on rate of change of input was included by augmenting the state equations as illustrated in Fig.5.3. The F and G matrices are constantly updated by the

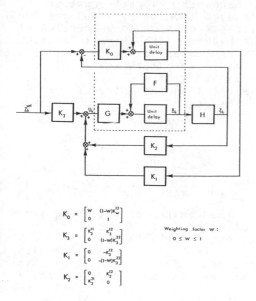

$$K_0 = \begin{bmatrix} w & (1-w)K_w^{12} \\ 0 & 1 \end{bmatrix}$$

$$K_3 = \begin{bmatrix} K_3^{11} & K_3^{12} \\ 0 & (1-w)K_3^{22} \end{bmatrix}$$

$$K_1 = \begin{bmatrix} 0 & -K_3^{12} \\ 0 & -(1-w)K_3^{22} \end{bmatrix}$$

$$K_2 = \begin{bmatrix} 0 & K_2^{12} \\ K_2^{21} & 0 \end{bmatrix}$$

Weighting factor W :

$0 \leq w \leq 1$

Figure 5.2   Load control structure at Didcot

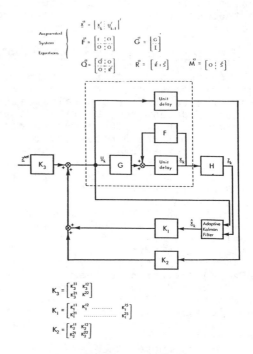

$$\underline{x}^a = \left[\underline{x}_k^t \mid \underline{u}_{k-1}^t \right]^t$$

Augmented System Equations

$$\tilde{F}^a = \begin{bmatrix} \Gamma & 0 \\ 0 & 0 \end{bmatrix} \qquad \tilde{G}^a = \begin{bmatrix} G \\ I \end{bmatrix}$$

$$\tilde{Q}^a = \begin{bmatrix} \tilde{q} & 0 \\ 0 & \tilde{q}' \end{bmatrix} \qquad \tilde{R}^a = \left[\tilde{R} \, , \, \tilde{S} \right] \qquad \tilde{M}^a = \left[0 \mid \tilde{S} \right]$$

$$K_3 = \begin{bmatrix} K_3^{11} & K_3^{12} \\ K_3^{21} & K_3^{22} \end{bmatrix}$$

$$K_1 = \begin{bmatrix} K_1^{11} & K_1^{12} & \cdots & K_1^{15} \\ K_1^{21} & & \cdots & K_1^{25} \end{bmatrix}$$

$$K_2 = \begin{bmatrix} K_2^{11} & K_2^{12} \\ K_2^{21} & K_2^{22} \end{bmatrix}$$

Figure 5.3   Load control structure at Pembroke

adaptive Kalman filter and hence the Riccati equation is also recursively updated at each time step. Thus continual re-optimistation of the feedback control as a function of current plant dynamics is achieved. The overall feedback control algorithm is also illustrated in figure 5.3. Since, in the long term, the Kalman filter adapts the model parameter such that states x_4 and x_5 are driven to zero, the model statics match the plant and no integral term is required in the controller. Set point control is achieved by chosing K_3 such that:

$$\bar{z} = \bar{\hat{z}} = H \bar{\hat{x}} = z^{set}$$

where

$$\bar{x} = (\hat{F} + \hat{G}(I-K_2)^{-1}K_1)\bar{\hat{x}} + (\hat{G}(I-K_2)^{-1}K_3)z^{set}$$

At present the system uses fixed values of weighting matrices Q^C, R^C and S^C. However, following on from the Didcot experiences, on-line adjustments of Q^C, R^C and S^C are planned in order to change the emphasis of control action in accordance with plant conditions. The importance of this is that it provides a systematic method (with proven stability characteristics) of modifying control action not only in load-pressure control systems but in other equally complex problems. The Q^C, R^C, S^C weighting factors also potentially provide less expert station staff with a simple method of re-tuning multivariable control systems without recourse to sophisticated design methods.

14.5.2 Measured Performance Index Minimisation.

The control algorithms described in section 14.5.1 follow in general terms, classical feedback control concepts. However, combustion control in fossil fired power stations requires an entirely different approach in order to achieve optimum control. The approach described below exploits the potential of the digital computer to perform complex calculations in real-time, thus emphasising the value of a digital approach to process control. The approach was pioneered at Aberthaw'A' a 120 MW unit oil fired station, in 1975 and later at Pembroke P.S. in 1981. Currently work is focussed on Didcot P.S.

The requirement is to compute the optimum ratio between combustion air and fuel (known as AFR) in the face of conflicting constraints. Each of the combustion products is related to AFR in the general fashion presented in figure 5.4. From an efficiency point of view there is an optimum AFR and deviations either side of that value produce excess CO or O_2 and result in thermal efficiency losses. However, the optimum AFR from a thermal efficiency point of view does not necessarily correspond to minimum production of corrosive acid smuts, black smoke, slagging or furnace corrosion.

Furthermore, the relationship between these various parameters changes with plant conditions and time. For example at Pembroke the optimum AFR at 500 MW is approximately three times that required for optimum performance at 100 MW whilst a two or three percent change in AFR at low load can result in a dramatic increase in CO and smoke production.

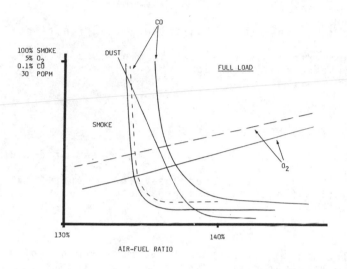

Figure 5.4   Combustion product relationships

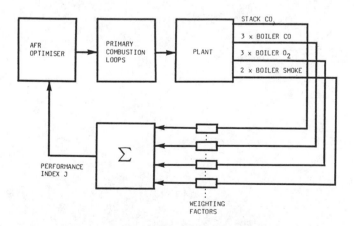

Figure 5.5   Combustion optimiser at Pembroke

In order to optimise the combustion process, a performance index J is computed from the various measurements of combustion products as follows:

$$J = \sum_{i=1}^{n} W_i \, e_i$$

where e_i is the error between the ith measurement and its set point

W_i is the corresponding weighting function.

The performance index can be minimised by using a 'steadfast descent' approach, which minimised $\int_{o}^{\infty} J^{-2} \, dt$.

i.e. $\dfrac{da}{dt} = K \, J \, \dfrac{dJ}{da}$

where a is the air fuel ratio

K is a constant determining the speed of descent.

At Pembroke, it proved possible (by appropriate choice of weighting functions) to approximate the above by the relationship.

$$\frac{da}{dt} = \frac{K}{\frac{dJ}{da}} \, \frac{dJ}{da}$$

and calculate $\dfrac{dJ}{da}$ from successive measurements of J.

The air-fuel ratio adjustment algorithm is thus:

$$(AFR)_{k+1} = (AFR)_k + K \, (J_k - J_{k-1}) \, / \, ((AFR)_k - (AFR)_{k-1})$$

whilst the overall combustion control scheme takes the form presented in figure 5.5.

The changes in AFR set point are rapidly translated into changes in AFR by the primary control loops. However, the optimiser sampling-update rate is restricted to 4 minutes by large delays in part of the combustion product measurement system. In order to compute how changes in AFR have affected J, the algorithm must wait until the measurement system responds. Thus, during rapid load changing the AFR is controlled using scheduling and stack CO (high load) or smoke (low load) measurment both of which are avalable from relatively quick response instruments. (See section 14.8)

At Didcot it was found that the rising and peaking nature of the combustion process prevented sensible evaluation of the performance index gradient. A more satisfactory approach was proven to be the off-line determination of the average value of each gradient term from PRBS/correlation data. The steepest descent algorithm was therefore approximated by the relationship

$$\frac{da}{dt} = K \sum_{i=1}^{n} W_i \cdot \overline{\frac{de_i}{dt}} \cdot e_i$$

The digital optimise algorithm thus takes the simplified form

$$\dot{a}(k) = \sum_{i=1}^{n} b_i \cdot e_i(k)$$

$$(AFR)_{k+1} = (AFR)_k = \dot{a}(k) \cdot (t_k - t_{k-1})$$

where

$$b_i = K W_i \left. \frac{de_i}{da} \right|_i$$

This, in effect, is a PI type algorithm using the weighted sum of combustion product errors. It has undoubtedly proved successful at Didcot but does suffer from one disadvantage compared with the Pembroke type algorithm in that measurement biases affect the operating point. The algorithm in figure 5.5 filters out such effects by of the gradient calculation.

14.5.3 Predictive Control

In some complex plant situations where a combination of time delay, spread of time constants and severe non-linearities occur, the normal methods of digital control offer only limited performance. In such cases, the use of a model based predictive control algorithm can be considered.

One such case occurred at Oldbury-upon-Severn Power Station which has two reactors, each currently rated at 217 ME(e). The reactors use Magnox fuel, and carbon dioxide gas cooling. The reactor is encased in a cylindrical concrete pressure vessel. To reduce the number of penetrations in the concrete, once-through boilers are used. The coolant gas is circulated by steam-driven gas circulators. There are four boilers, four gas circulators and one turbo-alternator per reactor.

The design of the Oldbury boilers is somewhat complex. A diagram of the feed system is given in Fig.5.6.

The feed water flow in each boiler is passed through three separate tube banks, which share a common gas flow. High pressure feed water passes through economiser, evaporator and superheater regions, before passing to a small back-pressure turbine. This turbine is directly coupled to the gas circulator. The exhaust steam from this turbine passes back to the reheat section of the boiler, where it is reheated, then mixed with steam from the other three reheater boilers and passed to the main turbine-alternator. In addition, low pressure feed water is passed through economiser, evaporator and superheater sections, before joining the reheat steam and passing to the main turbine-alternator.

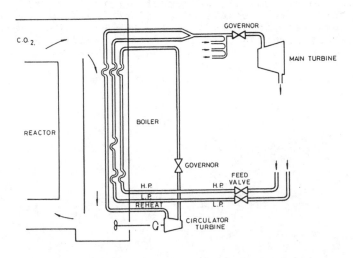

Figure 5.6   Oldbury reactor/boiler schematic

The running costs of Oldbury are amongst the lowest in the country, and the Station is run for maximum electrical output. In addition, the Station achieves very high levels of plant avalability. These factors, along with the ever-increasing cost of generation at fossil-fuel fired power stations, mean that there is a very great incentive to meet, and if possible, exceed the generation targets for the Station.

The reactor is operated with the aim of maximising power output without infringing the constraints defined by safety and economic factors. A governor valve on the gas circulator turbine is used to control the circulator speed to its desired value. The HP feed valve is selected to control the HP steam pressure to its set point, which is chosen to be as low as possible, consistent with maintaining the HP governor valves in a regulating position. In order to minimise throttling losses.

The LP feed valve is used to control the boiler gas exist temperature (T1). This is chosen to be as low as possible, in order to put the maximum amount of feed through the LP boiler.

The original design of the Station was for a boiler gas inlet temperature (T2) of 411 deg.C and the boilers were oversurfaced, with steam temperatures close to gas temperature. Automatic control of HP steam temperature was not necessary.

The Station was downrated in 1969, to reduce gas side corrosion of the boilers, the boiler gas inlet temperature has been reduced to less than 370 deg.C, and now the HP steam temperatures ae below the gas temperature. This means that disturbances can cause fluctuations in HP steam temperature.

The HP steam temperature cannot be allowed to fall indefinitely, because of the possibility of carryover of water droplets into the HP turbine. The HP feed is tripped if the HP steam temperature falls below a pre-set margin.

The HP steam temperature therefore becomes one of the power-limiting factors for the boilers, and had to be controlled by means of manual adjustments to the LP feed flow via the T1 set point.

The boiler behaviour has been investigated in the past, and some general conclusions may be drawn. The HP steam temperature varies as the LP feed flow rate is varied, however the relationship is highly non-linear (Fig.5.7), because of the need to run the gas circulator turbine at constant power. Furthermore, although the HP steam temperature is affected by LP feed flow, it is much more sensitive to the boiler gas inlet temperature (T2), and the HP feed conditions.

The HP steam temperature can be controlled manually, by adjusting the T1 set point to get the steam temperature to the required margin above thetrip level. However, the boiler is still very sensitive to disturbances. If the HP steam temperature is subject to a downward perturbation from this steady state, the circulator controls will increase

the steam flow to keep the circulator speed constant. This
will lead to a further drop in steam temperature, and the
process will continue at an ever-increasing rate until
corrective manual action is taken. Figure 5.8 shows plant
data illustrating the difficulty of manual control.

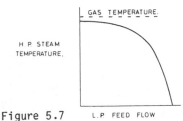

Figure 5.7   L.P FEED FLOW

HP Temperature/Flow relationship

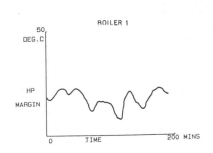

Figure 5.8

Manual temperature control

The use of a model-based control system is suggested by the
fact that the major non-linearity and cause of instability
in the boiler is well understood and predictable. This is
the mechanism whereby a falling steam temperature will lead
to an increasing flow demand from the feed controls and
hence a lower steam temperature. This effect is included
in the on-line model.

The control algorithm is based on the use of the on-
line model as a predictor of boiler behaviour. The model
is used in two modes. A multi-step prediction is made to
obtain an estimate of future states of the boiler. This
prediction is used to derive an error signal which drives
a simple proportional plus integral controller.

In addition, the model used to make a single step
prediction, which is stored and compared with the plant
measurements when they become available. The error between
model and plant is passed through a set of filter gains and
used to drive the model to match the plant. The gains are
derived by running the model against recorded plant data
in the laboratory, and using a Kalman filter to produce a
set of gains. It is found that these gains will converge
to a steady-state value, which can then be used on the plant
with a consequent saving in processor time and memory
requirements.

The structure of the model, controller and filter is
shown in Fig.5.9.

Each quadrant has its own 13th order model which is
updated from plant measurements every 5 seconds. Every 5
seconds the model is also used to predict HP steam
temperature 10 minutes ahead and control action evaluated
using the PI algorithm evaluated.

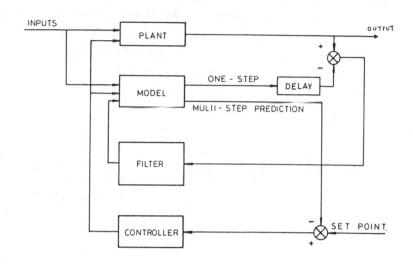

Figure 5.9   Predictive controller structure at Oldbury

The control system trials were pursued over a period of time, beginning with steady-state trials, and proceeding to more onerous plant transients as experience was gained. In genral, the most severe transients are those caused by the reactor, such as sector trimming or refuelling.

The long term performance of the control system is illustrated by Fig.5.10. The relative frequency of occurrence of steam temperature margins is plotted for each boiler. It can be seen that three of the boilers are controlled to the set point of 20 degrees with a maximum deviation of no more than three degrees, and a variance of no more than one degree. The other boiler, Boiler 1, was on manual control. During this period. a normal range of plant operations were carried out, including reactor trimming and refuelling.

The attainment of a steady 20 degrees C. margin reduces the number of occasions when output falls below target. A test was conducted which demonstrated that moving from a margin of 25 degrees to 20 degrees led to a gain of 2-3 MW(e) per reactor.

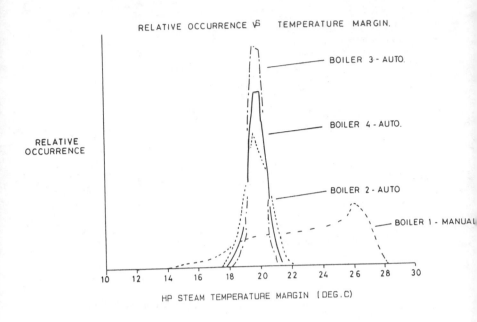

Figure 5.10 Long term performance comparisons

14.6 FEEDFORWARD CONTROL

        The general concept of feedback control is that one
senses errors between plant output set points and measured
values and then manipulates the plant inputs to reduce or
eliminate such errors. This is a well proven concept but
it has one major snag; namely that output errors are
required to generate changes in the control inputs.
        In many CEGB plant situations the disturbances can
influence the plant outputs more rapidly than either one
can measure the outputs or regulate them using control
inputs. In such circumstances the complementary concept
of feedforward control is employed. Essentially one
ascertains that a particular disturbance event has taken
place and attemps to modulate the control inputs such that
the effect of the disturbance are cancelled out. Such
techniques have been used beneficially at Pembroke, Didcot
and Oldbury. Two methods of using disturbance feedforward
information are currently in used. These are:
a)   Disturbance feedforward into control inputs.

b)   Disturbance feedback into a Kalman filter containing
      a plant mode with control action based on filter
      output.
        The first method relies on a calculation of the change
in input needed to offset the change in output resulting
from the disturbance. Typical examples are the load demand
feedforward to firing demand at Didcot and the scheduled

changes in air-fuel ratio at Pembroke as a function of firing transients.

In situations such as those cited, feed forward can be effective. The problem in applying it is that there is no inherent feedback and if one miscalculates the effect of the disturbance (or the effects are unpredictable) then one is likely to over-correct and not achieve the required result.

The second method of applying feedforward (b) above) partly overcomes the problem of imprecise feedforward data. By imputting the disturbance information to the model and comparing the prediction with actual plant measurements, some negative feedback is applied to the feedforward disturbance estimate. The Oldbury predictive control scheme described in section 14.5.3. makes good use of this technique since small changes in reactor T2 conditions constitute amajor disturbance to boiler temperatures. By feeding T2 data into the model the predictor is able to forecast the future temperature disturbance and adjust LP feed flow accordingly.

A variation in this theme which has only been tried experimentally on the Pembroke superheater controls (Wallace et al (6)) is to augment the Kalman filter state vector with a state space model of the plant disturbances. State feedback of both plant states and disturbances is employed and feedforward disturbance information is fed into the Kalman filter. Simulation results suggest that the resulting improvement in control performance is much less dependent on the accuracy of the information than if the latter is fed directly to the control input.

14.7 ADAPTIVE CONTROL PROCEDURES

14.7.1 The Need for Adaptive Control Procedures

It is no exaggeration to suggest that most items of plant, large or small, change their characteristics either with operating conditions or time. If the control algorithms do not reflect these changes then the inevitable result is degraded performance through loop instabilities or detuned control action. This has been recognised in the academic literature where there is much emphasis on self-tuning controllers. However, in practice, many of the plant changes are well understood and calculable from plant data. Modified control algorithm coefficients can thus be scheduled without recourse to more complex on-line identification alogrithms. This is not always the case and hence there are effectively two techniques to consider, parameter scheduling and parameter estimation. Both techniques have been used at Pembroke and Didcot and each is briefly considered below.

14.7.2 Parameter Scheduling

Many of the primary sub loops use parameter scheduling because actuators such as FD fans, ID fans, attemperator

spray water valves and governor valves exhibit highly
variable incremental gain variations. A five or ten fold
change in loop gain is not uncommon and cannot be ignored.
In some cases the gain change is reasonably predictable
given information regarding actuator position. In such
cases the controller gain can be adjusted to yield constant
overall loop gain. In other cases the process dynamics
change as a function of plant operating conditions. A
typical example is the final superheater at both Pembroke
and Didcot where the overall process gain varies as a
function of pressure (which determines the temperature-
enthalpy relationship) whilst the response time varies
inversely with steam flow (= load). The temperature
controller must reflect these changes by way of modified
controller settings if good, wide range control is to be
achieved.

A difficulty that can arise with scheduling is that
additional measurements not otherwise utilised, are
sometimes required. From a reliability point of view this
is undesirable. However, a little ingenuity can often
overcome such problems. For example, if on considers the
attemperator temperature control loop at Didcot, the
incremental process gain relating changes in spray valve
position to temperature is highly dependent on the
differential pressure across the valve. This varies
significantly but a direct measurement can be avoided by
noting that the combination of differential pressure and
valve opening is determined by the spray flow requirement.
This is well correlated with steam flow (= Load) which is
measured for other reasons. Least squares regression
analysis of plant data demonstrated at Didcot that an
adequate (from a control loop point of view) estimate of
process gain could be determined using a quadratic
polynomial in valve position, load and the product of valve
position and load.

14.7.3 Parameter Estimation

In some situations it is either convenient or necessary to
estimate plant parameters which affect control algorithm
settings. An example of this is the load-pressure control
system at Pembroke. The Kalman filter estimate of the state
vector is used to calculate parameters such as saturated
steam enthalpy and superheater flow resistance. These
parameters are used to update the transition matrix (F) both
for the filter and the optimal controller Ricatti equations.
In addition, other important parameters such as governor
valve position and feed enthalpy are estimated from
regression on the disturbance estimates (Jazwinski (7)).
This technique has the advantage that it decouples the
convergence rates for the Kalman filter from the rate at
which the parameter estimates are updated. The resulting
two stage estimator structure is described in (6). However,
it remains difficult to rapidly distinguish between plant
disturbance transients and parameter changes.

For those control systems based on Z transform

controllers (section 14.4.3 rather than plant models, recent years have seen the rapid development of self-tuning controller algorithms. As with adaptive Kalman filtering/state feedback controllers, the principal problem is to ensure sensible behaviour under all conditions. Thus much effort is being directed towards the so called 'jacketing' arrangements in order to achieve robustness.

A self tuning controller has been run at Didcot experimentally on the secondary air control loop. After initial teething problems the controller, developed by Wilkinson (8) in conjunction with CEGB S.W.Region achieved good control in a noisy environment and in the presence of significant actuator backlash. Significantly more work is required before the technique can be fully implemented on all control loops at Didcot or other sites. The first major application is likely to be in the commissioning/maintenance situation since there are several hundred major loops to be set up. Full implementation will only follow when confidence has been built up over a period of time.

14.8 LOGICAL CONTROL FUNCTIONS

14.8.1 Multiple Mode Control

The emphasis in this paper has so far been on modulating control algorithms of one type or another. However, the high level language facilities now available support powerful logical variable facilities which are of considerable use in complex power station applications. It is now common practice to select from a variety of modulating control algorithms. The selection within a loop is based on such factors as the status of other interrelated loops, changes in plant conditions, changes in loop performance requirements or changes in measurement and actuator availability. A typical example pertaining to loop status is the boiler pressure control loop at Pembroke and Didcot. Assuming the loop is on 'auto', the boiler pressure may be controlled by use of the governor valve, total boiler heat input or a combination of both. The modulating control algorithms which apply in each case are different (4) and the appropriate algorithm must be selected accordingly.

An impressive example of the conditions cited above is the air-fuel ratio control loop at Pembroke (see section 14.5.2). The loop operates in a multiplicity of different modes, as is illustrated in figure 8.1 although, as far as the operator is concerned, it is a single loop running under auto control.

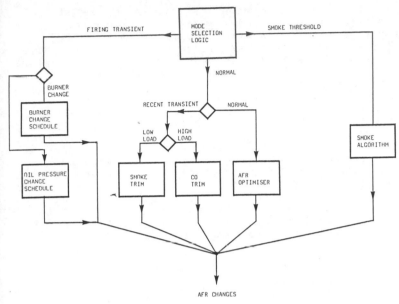

Figure 8.1   Air fuel ratio control

14.8.2 Logical Look-Up Tables

A quite separate role for logical functions is in the
oil burner pattern management at Pembroke. Each unit is
equipped with four rows of eight oil burners, with a modul-
lating range that is restricted to 5-20% depending on load.
The number of firing burners is computed in accordance with
the total boiler firing demand. The burner in order to:
a)   Meet the number of burners (N_b) required

b)   Achieve the preferred burner pattern for N_b burners
     or the best approximation in a geometric sense.

c)   Optimise the preferred patterns in accordance with
     metal temperature and combustion product data.
     In order to achieve b) a preferred pattern is tabulated
for each value of N_b. Furthermore, for each burner in
each pattern there is apreferred order of subisdiary burner
options should the selected burner not be available or
refuse to fire.

14.9 CONSTRAINT HANDLING

However simple or sophisticated the control sytem, the
designer concerned with real plant inevitability faces the
problem of handling constraints in a safe and predictable
fashion. The problems and solutions are often highly
specific and it is difficult to develop a generic overview.
However, a brief review of some typical problems may be of
interest to the reader:

Amplitude Limits. It is common for a control loop output to be constrained by amplitude limits which can cause clipping and hence noise-rectification in the control action. This is not serious transiently, but can cause steady state errors if the bias caused by noise rectification is significant. Amplitude limits also apply to controlled plant variables. Generally speaking, it is better to constrain the loop set point than the control output.

Rate Limits. These are an essential part of any real control system. However, care must be taken with incremental algorithms to execute the correct control action under large transient situations. For example much of the proportional action in a PI algorithm may be lost if the remainder after rate limiting of the control output is not stored. In some loops this is desirable and in other it is not.

A subtle form of integral wind-up can also occur if the rate limits applied in the computer software are less severe than those imposed by the plant actuator. The moral is to be aware of actuator characteristics.

Ordering of Constraints. The order in which control calculations and the application of constraints takes place is of importance, particularly when items such as controller gains are variable. The designer should be clear minded about the objective of applying constraints.

Interprocessor Communications Links and Data Skew. In distributed computer control systems it is necessary to transmit control commands using data links of one form or another. Handshaking applications software is used to verify correct transmission and reception. Where constraints are dynamic in one processor it is important that data skew due to sampling and transmission times does not lead to misinterpretation of the contraints in other processors.

Loop Stability. The application of constraints can degrade control system stability and lead to limit cycling or even instability. The only effective way to check out such conditions is by simulation of the plant and control system, preferably using the actual final control system software.

14.10 INITIALISATION PROCEDURES

As with constraint handling, initialisation procedures must be considered as a part of the complete digital control system. There are essentially two separate cases to consider. These are the introduction of auto controls and the recovery from failure.

With respect to the introduction of controls, it is normal practice in the CEGB to ensure tha all algorithms are initialised such that bumpless transfer occurs when operators transfer a loop from manual to auto. As a part of this process, loop set points usually track the measured value so that when 'auto' is engaged plant conditions remain unaltered. However, there are occasions when this is undesirable. These are where the loop set point is predetermined for operational reasons and in these circumstances special initialisation procedures must be adopted. Typical examples include furnace suction pressure control and final superheater temperature control.

Recovery from failure situations also poses some subtle problems. In particular, there are difficulties in the resetting of a measurement after a failure has been detected and cured. Generally, the reset action is operator-initiated through the VDU/desk interface. It is desirable that the operator should manually control the plant to bring the process variable within defined limits in order to demonstrate that the measurement is no longer defective. This can and is achieved with measurements such as combustion products but with others, such as boiler pressure, the limits have to be so wide for operational reasons that there is little merit in enforcing the procedure.

As a matter of policy, the 'auto' status is never engaged except as a result of operator action. Thus, for example, if after a power fail, loops are tripped off auto by the software interlocks checking hardware watchdog status, the 'auto' status is not reinstated when power re-start occurs. This applies whether or not the loops were in a healthy state prior to power fail.

Reinitialisation of the Kalman filter algorithms after measurement failure-reset also requires some care. At Pembroke, the covariance equations are recursively solved and hence the covariance matrix is reinitialised after a measurement is re-instated. However, where fixed gain algorithms are employed alternative procedures are required if unwanted transients are not to occur.

14.11   CONCLUSIONS

In this paper the author has attempted to summarise the experience that he and his colleagues have accumulated in digital control techniques over the past few years. The principal conclusions are:

1.   There is considerable benefit in developing digital techniques rather than merely mimicking analogue control functions.

2.   There are many different facets to digital algorithm implementation, all of which must be considered if the system is to achieve its full potential safely and reliably.

14.12   ACKNOWLEDGEMENTS

This paper has been produced by permission of the Director General of South Western Regionof the Central Electricity Generating Board.

14.13 REFERENCES

1. Wallace, J.N., 'Design Concepts and Experience in the Application of Distributed Computing to the Control of Large CEGB Power Stations.' IAEA Specialist Meeting on Distributed Systems for Nuclear Power Plants, May 1980.

2. Sorensen, H.W., 'Kalman Filter: Theory and Application.' IEEE Press Selected Reprint 1985.

3. Bierman 'Measurement Updating Using UD Factorisation' Automatica Vol 12 pp 375-382, 1976.

4. Wallace, J.N., and Clark, R., 'Load Control of a 500 MW Oil-Fired Boiler/Turbine', IEE Conference on Control and its Applications, Warwick, May 1981.

5. Marshall, J.E., 'Extension of O.J.Smith's Method to Digital and other Systems', Int. J. Control, 1974, Vol.19.

6. Wallace, J.N., and Clarke, R.C., 'The Application of Kalman Filtering Estimation Techniques in Power Station Control Systems', IEEE Transactions on Automatic Control. Vol.AC.28 pp 416-427, March 1983.

7. Jazwinski, A.K., 'Adaptive Sequential Estimation with Applications', Automatica Vol. 10, 1974.

8. Wilkinson A. 'Application of Self Tuning Control to Power Station Control Systems' PhD.Thesis, University of Kent - to be submitted.

Chapter 15

An expert system for process control

E.H. Higham

15.1 INTRODUCTION.

It is interesting to reflect on the fact that the first successful process controllers incorporating proportional, integral and derivative actions were developed more than five decades ago. They were based on the now unfashionable pneumatic methods of transducing temperature, pressure, flow and level measurements - which still represent more than 90% of the measurements made for process control purposes. The control actions were generated by a combination of pneumatic and mechanical devices which are equivalent to the resistor, capacitor, summing junction and operational amplifier, in today's electronic analog systems.

It is also interesting to note that, at that time, the theory on which analog control systems are based had hardly been established in the adjacent disciplines for controlling rotating machinery, gun laying and the steering of ships. In the subsequent evolution, pneumatic measuring systems have largely been superseded by electronic systems, principally because pneumatic signals cannot be transmitted over long distances without serious delay and distortion but also because only relatively simple signal conditioning and computation can be implemented with pneumatic signals.

The advent of digital computers not only facilitated the signal conditioning and the computations associated with the control functions, but also provided an opportunity to co-ordinate the operation of several process plants and then to optimise their overall performance by adjusting the operating conditions in individual loops. However, it was recognised from the early stages that it was difficult to identify values for the proportional band, together with the integral and derivative actions, which ensured acceptable performance under normal operating conditions for perhaps as many as 10% of all control loops.

As microprocessors became more powerful and less expensive, the possibility of applying the optimisation at individual loop level became a reality, as is evidenced by the plethora of papers which have appeared on the subject during the past two decades. It is also interesting to note that most recently the emphasis has been on proving the robustness of various algorithms, which is equivalent to bridging the gap between theory and practice.

It is not entirely surprising that amongst those who have had many years of experience and involvement in implementing difficult and complex control schemes, there have remained a few who felt that the theoretical approaches lacked a touch of realism, and that the techniques and experience of commissioning engineers, as well as the staff who subsequently operate the plants, should provide the basis of an alternative approach to self-tuning or adaptive control.

15.2   EARLY ADAPTIVE CONTROLLERS.

As long ago as 1967, E.H. Bristol [1] described an adaptive control scheme for the heat exchanger shown in Fig. 15.1. This is a widely used method for generating hot water for process purposes from a site steam supply. The system is characterised by a large dead time which, with conventional feedback control, results in poor response to transient load changes. It is, however, an ideal application for feed forward control, and Bristol describes an adaptive system in which the flow rate of the water is measured and used as the load signal (L) to the feed forward model. The model computes the required steam flow rate and valve setting (V) from the equation $V = aL + b$. The adaptive system then determines, continuously and automatically, the best values of (a) and (b) so that the output temperature is held at the desired value, irrespective of load variations. Fig. 15.2 shows how the convergence is effected and Fig. 15.3 shows the actual system, which used pneumatic equipment.

By present day standards, it is a rather trivial example but, from the viewpoint of process control, it established a pragmatic approach which has continued to evolve during the past two decades. Some of the thinking behind this approach has been described in the cited references, and one can perceive the progressive introduction of what we now refer to as 'expert systems'. These exploit the experience and knowledge of skilled control engineers, who observe the behaviour of the process plant and then apply the criteria first identified by Zeigler and Nichols [2], for establishing the optimal controller settings.

Shinskey [3] has pointed out that a complex multi-capacity process can be considered to have an effective dead time plus an effective time constant, the sum of which equals the total lag in the process. Such a single-capacity-plus-dead-time process can be made to represent any degree of difficulty, from one extreme to the other, simply by varying the ratio of dead time to capacity lag. It is this similarity of complex processes under feedback control that has enabled PID type controllers to be used so widely and to provide stable operation of processes by applying knowledge based rules, but without having to derive mathematical models of the actual processes which, in themselves, are almost always

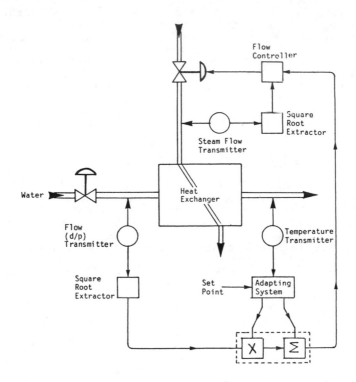

Fig. 15.1. Functional diagram of heat exchanger
with adaptive control scheme.

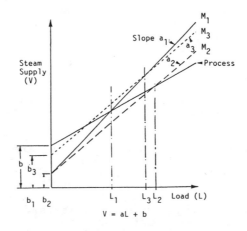

Fig. 15.2. Convergence effected by control scheme.

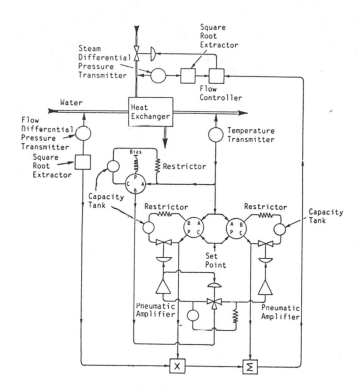

Fig. 15.3. Pneumatic implementation of control scheme.

non-linear and time varying. Furthermore, the physical principles involved are often not fully understood, and the process parameters cannot always be measured.

15.3 EVOLUTION OF AN EXPERT SYSTEM.

The Foxboro EXACT* algorithm is an 'expert system' for self-tuning that relies heavily on the heuristic approach developed by control engineers over several decades. It is important to recognise this lineage and to appreciate that self-tuning controllers are well suited for applications in which a standard fixed parameter PID controller is stable but, due to changing process dynamics, requires constant re-tuning to maintain optimal conditions. It is equally important to recognise that self-tuning controllers are no substitute for good process design and good measurement technology, and although they may help to identify malfunctioning process equipment, they can only reduce the adverse effect it may have.

The customary method for bringing an automatic controller into service is for the commissioning engineers to set the proportional band (P), integral action (I), and

* Trademark of The Foxboro Company.

derivative action (D) at arbitrary values, based on their own knowledge of the type of process, or previous experience. Then, with the control loop closed, a step change is applied to either the input, the load or the set point and the resulting transient response is observed.

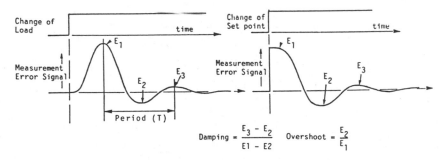

Fig. 15.4.   Transient response of a closed loop.

As shown in Fig. 15.4, the distinctive features of this transient are:

(1) the presence (or absence) of peaks and/or troughs,
(2) the time interval between successive peaks and
(3) the steady state error, from which
(4) the overshoot
(5) the damping and
(6) the period can be identified.

For most processes, the damping and overshoot are not independent and the period of the closed loop must be included to define the response uniquely. The period can be expressed in non-dimensional terms using the values of the controller integral action (I) and the derivative action (D) to produce ratios similar to those proposed by Zeigler and Nichols[2] and Shinskey[3]. Fig. 15.5, shows that the ratio of integral action (I) to loop period (T) defines the lead angle of the controller whilst the ratio of derivative action (D) to loop period (T) defines the lag angle. Based on the period (T), but constrained by the allowable overshoot and damping, unique values of (P),(I) and (D) can be determined.

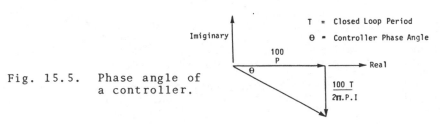

Fig. 15.5.   Phase angle of a controller.

In implementing a continuously adapting control system, it is important to recognise that the noise component of the measurement signal does not contain any useful information from a control viewpoint. Hence the control algorithm must not start to determine new values for (P),(I) or (D), until the deviation between measurement and set point has moved decisively outside the noise band. An appropriate value of the deviation to trigger the algorithm is twice the noise band.

All such deviations must cause a transient to which the basic controller responds. If they remain small, the controller corrects them using the values of (P),(I) and (D), which were previously installed, and under these conditions it is not particularly important for the values to be optimal. When the deviation becomes large and the algorithm is brought into operation, the controller must respond by applying corrective action during the present transient according to the values of (P),(I) and (D) that were established by the algorithm during the previous significant transient.

Whilst it is doing so, the algorithm starts to search for the first peak (E_1 in Fig. 15.4). Once this has been detected and verified, its value is memorised and the algorithm proceeds to search for, and then verify and memorise, the second peak (E_2). It continues by searching for the third peak (E_3) and immediately this has been identified and verified, the algorithm determines the period (T), the damping factor $(E_3-E_2)/(E_1-E_2)$ and the overshoot $-(E_2)/(E_1)$, from which it calculates new values for (P),(I) and (D). Provided that the changes come within the limits identified in the 'safety jacket', the revised values are introduced into the basic controller, in readiness for dealing with the next transient. This concept is implemented in the Foxboro EXACT* self-tuning controllers, where the algorithm includes other features based on 'expert systems'.

To put one of these controllers into service, it is necessary to enter some initial values for (P),(I) and (D), and a general understanding of the process should enable some sensible values to be identified. However, it is not particularly important for these initial values to be accurate, because the first transient brings the algorithm into operation and, as a result, new values are determined. These are further improved following each of the succeeding transients.

There are several other parameters which have to be entered. As mentioned previously, the 'noise band' (NB) should be set at a value which prevents the algorithm from attempting to extract tuning information from the measurement signal when, in fact, none is present. The algorithm also requires an approximate value for the maximum time it must wait (WMAX) for a second peak to appear after it has detected and verified the first peak, so that it can determine whether or not the response is over-damped, as illustrated in Fig. 15.6.

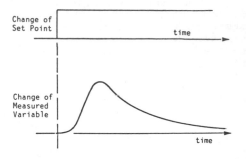

Fig. 15.6. An overdamped
response of a
control loop.

15.4   ADDITIONAL OPERATIONAL FEATURES.

From an application viewpoint, it is desirable to set four further parameters, namely the 'maximum allowable damping' (DMP), the 'permissible overshoot' (OVR), the 'derivative factor' (DFCT) and the 'change limit' (CLM) or the maximum change that can be made to the initial values set for (P) (I) or (D) as a result of operation of the algorithm. However, the self-tuning action can be implemented without setting values for these parameters, in which case the default values, installed during factory tests, are operative.

Because the limits for damping and overshoot are interdependent, the selected values should represent the maximum that can be accepted from an operational viewpoint, in which case the algorithm selects the limit that comes closest to being exceeded although, in practice, the best control is achieved when the maximum allowable damping is applied. This allows the influence of the derivative action to be reduced by multiplying the derivative time (D) by the 'derivative factor'. Setting this parameter to zero eliminates the derivative action in the controller, which is usually desirable for processes where there is a large dead time or a high level of noise in the measurement signal. In some applications, it is desirable to limit the total change that the algorithm can make to the initial or previous values of (P),(I) and (D). Thus, if a value of 4 is selected for the 'change limit' and 100 for the proportional band (P), then the algorithm will not change the value to more than 400 or less than 25 and will optimise the corresponding values of (I) and (D).

A further feature of the algorithm is that it monitors the controller output and identifies the components having a period shorter than the basic period of the loop. If the average peak-to-peak value exceeds the 'output cycling limit' over a period of a few minutes, the algorithm increases the proportional band (P) and reduces the derivative action (D) gradually until the amplitude of these oscillations has been reduced to an acceptable level.

The features described in the previous paragraphs illustrate the 'expert system' aspects of the algorithm, but it has also been assumed that sufficient knowledge or experience is available to select some initial values for

(P)(I) and (D). However, the EXACT* algorithm includes a further feature, namely 'Pre-Tune' (PTUN), which, once the process has reached steady state conditions under manual supervision, enables the user to introduce a step change in the controller output. The algorithm observes the resultant transient from which it determines the optimum values for the various parameters.

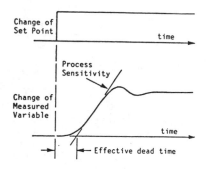

Fig. 15.7. Reaction of a process to a step change.

As shown in Fig. 15.7, such a response identifies the effective dead time and process sensitivity. From the former, the algorithm determines the integral action time (I), the derivative action time (D) and the maximum time to wait (WMAX) for the appearance of the second peak in the transient response. The proportional band (P) is calculated from both the sensitivity of the process reaction curve and dead time, whilst the noise band is determined by estimating the peak-to-peak amplitude of the noise in the measurement signal having a frequency spectrum above the cut-off of the closed loop.

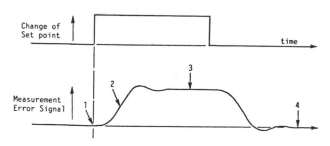

Fig. 15.8. Four main phases of 'Pre-Tune'.

The four main phases of the 'Pre-Tune' (PTUN) are shown in Fig. 15.8. The process is perturbed by a step change to the controller output which takes place at point (1). This is followed by a transient response (2) and the algorithm waits for the system to stabilise. When the steady state is established, the algorithm calculates the (P),(I) and (D) values and returns the controller output to its initial value at (3). Finally, the 'noise band' and 'derivative factor' are calculated during (4).

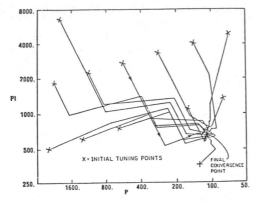

Fig. 15.9. Convergence plot for a process having a pure dead time of 2.0 sec as well as five lags of 1.2 seconds each.

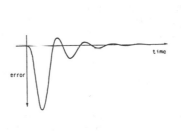

Fig. 15.10. Loop response to a step change of load, after optimisation.

To show how the algorithm optimises the controller parameters, it has been operated in conjunction with a process simulator. Fig. 15.9 shows the performance of the algorithm for a simulated process comprising five lags of 1.2 seconds each and one dead time of 2.0 seconds. The initial controller parameters were deliberately set at inappropriate values and the status of the simulator was disturbed to bring the algorithm into operation. It can be seen that the algorithm established the optimal values after five or six disturbances, and the response with the final settings is shown in Fig. 15.10.

15.5 OPERATIONAL CONSIDERATIONS.

In reviewing the performance of standard two- or three-term controllers in established process plants, one arrives at a concensus opinion that probably 85% of the control loops function satisfactorily with (P+I) or (P+I+D) controllers, and they are seldom troublesome. About 8% require some skill and attention initially, to establish good controller settings, whilst a further 3% are troublesome and require regular attention. Finally, about 4% are so troublesome that they are usually under manual control. It is difficult to make a valid assessment of the improvement that might be achieved by ensuring that the controller parameters are set at, or maintained close to, their optimal values. This is usually because the necessary information is either not available or cannot be obtained, but sometimes there are operational and other difficulties which cannot be overcome.

There are, however, examples where it has been possible to substitute an EXACT* controller for a conventional controller and, as a result to demonstrate very substantial savings as well as improvements in quality

by the continuous adaption of the controller parameters.
One example is the temperature control of a catalytic
naptha reformer where the (P,)(I) and (D) settings were
changed from P = 300%, I = 15 min and D = 1 min to P = 38%,
I = 20 min and D = 4 min. The improved transient response
resulted in fuel saving, the avoidance of the need to bring
additional process plant into operation and an improvement
in yield which, together, have led to an attributable
annual saving of £500,000.

In another example, the small changes in temperature
in a degasser column led to large changes in column product
composition. This, in turn, gave rise to excessive
concentration of the purge gas which caused frequent
tripping of a large compressor. To avoid this, the column
was operated with the concentration of the purge gas
significantly below the optimal value. The application of
an EXACT* controller achieved sufficiently tight
temperature control to avoid tripping of the compressor and
the column itself to be operated with the concentration of
of the purge gas much closer to the desired value. The
consequent improvement in product quality and the reduced
losses which occured each time the compressor was tripped
represented an annual saving of £94,000.

The most successful applications of the EXACT*
controller have been in effluent treatment plants. In
general, these plants have to deal with large and irregular
changes in the rate of throughput as well as in pH, and are
required to ensure that 'acid spikes' do not take the pH
below 5.5 or allow it to exceed 10.0. The plant
configuration is seldom ideal from the process control
viewpoint and the transfer characteristic of the measuring
system is highly non-linear. There are numerous examples
where the introduction of the EXACT* controller has not
only reduced the variability of pH in the discharge, but
also effected substantial economies in the usage of the
neutralising agent.

In an entirely different context, there are many
examples of how an EXACT* controller has been able to
demonstrated that the shortcoming in the operation of a
process was attributable to defective or malfunctioning
process equipment, rather than the control equipment.

15.6 IMPLEMENTATION OF THE ALGORITHM.

The EXACT* algorithm was first made available in a
controller as a component in the SPEC 200* system and there
were two main reasons for doing so. In the first place,
there are probably more than a million SPEC 200* control
loops in service, and it is almost certain that significant
operational improvements could be achieved in 5% of them
simply by substituting the EXACT* controller card for the
existing (P+I) or (P+I+D) controller card. The second
advantage is that the power supplies are already available
within the system and all the input and output signals are
standardised at 0 to 10 V dc, corresponding to 0 to 100% of
the input and output signals. Thus there is a wide range of

process applications in which the algorithm can be evaluated without the need to reconfigure the existing plant or systems.

The components involved in the system are shown in Fig. 15.11. The input component provides the power for the field mounted transmitter, and converts the measurement signal into the 0 to 10 V dc for the system. The controller card compares the input signal with the set point and, if set to operate in the manual mode, the functions in the basic controller on the card generate an output signal according to the previously established values of (P), (I) and (D). This signal is also in the range 0 to 10 V dc and passes to the output component where it is converted into 4 to 20 mA dc, or some other signal, for actuating the final operator. If the controller card is set to self-tuning mode, then the EXACT* algorithm becomes operative and determines the values for (P),(I) and (D) to be applied in the controller, as previously described.

From the outset it was recognised that, once the integrity of the algorithm had been established, a free-standing version of the controller would be required. This is now available as the Model 760 Single Station Micro*, shown in Fig. 15.12. It combines the basic components from the SPEC 200* system into a single unit, which can be mounted either in a panel in a shelf. The front panel incorporates an alpha-numeric display and a key pad, which together provide all the functions available in the hand held configurator. The data, however, is presented in a sequential roll-over manner, controlled by operating the appropriate keys on the key pad. There are also three bar-type displays on the front panel which display continuously the measurement and output signal, as well as the set point. Each of these displays comprises 50 discrete segments and hence provide a discrimination of 2% of span, but the signals can be selected individually and displayed in engineering units on the alpha-numeric display, to four significant figures. Both over-range and under-range indicators are associated with each of the bar type displays. Finally there are an alarm indicator and three status indicators for the Workstation (W) or Panel (P), Remote (R) or Local (L), Auto (A) or Manual, (M) mode of operation.

The functional diagram, Fig. 15.13, shows provision is made for up to four input channels, which may be selected as follows:

1. Analog current   (a)   4 to 20 mA dc   for any or all of
                     (b)   10 to 50 mA dc   the input channels
              or (c)   1 to 5 V dc   but only two of the
                                       4 to 20 mA powered
                                       from the internal
                                       24 V dc supply.

2. Frequency signals   (a)   0 to 1.0 Hz   for one or two
   in the range   and (b)   0 to 7.0 kHz   channels only

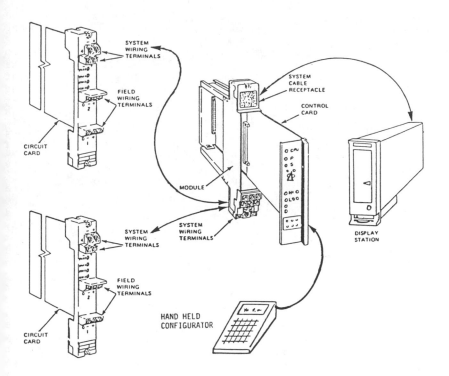

Fig. 15 11. The EXACT* Controller Card as a
Component in the SPEC 200* System.

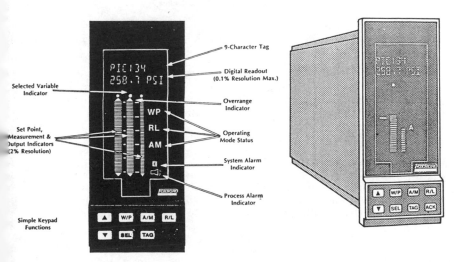

Fig. 15.12. Front panel of the Single Station Controller.

3. Platinum RTD -   one channel only

4. Thermocouples -   Types J,K,or E,
                     (via 4 to 20 mA converters)

5. Contact inputs - two channels.

There are two non-isolated outputs. Output 1, the controller output signal, is a current of 4 to 20 mA into a load of 500 ohms maximum. Output 2 is an auxiliary output of 1.0 to 5.0 V dc, which can be assigned to retransmit either the measurement, set point, output 1, or the remote set point signal. There are also two transistor switch outputs, rated at 50 V dc and 250 mA, which can be assigned to one of the status indications or to the alarms.

Any of the following control functions can be selected:

EXACT*, (P+I+D), (P+I), (P+D), (P), or (I),

and there are the following associated functions;

1 Ratio, adjustable between 0 and 5,
2 Calculations on the inputs, using 'add', multiply' or 'divide'
3 Gain adjustments from -10 to +10,
4 Bias adjustments from -100% to +100% of span,
5 Filtering of the measurement signal via a second order Butterworth filter, adjustable between 0 and 10 minutes,
6 Signal conditioning for square law, square root law, RTD linearisation, thermocouple linearisation, two characterisers, one of which can be used as a (P I D) non-linear extender.

The alarms include two sets of dual absolute alarms assigned to the measurement signal, another set assigned to the measurement/set point deviation and a further set assigned to the output. Each set can be adjusted to respond to the H/H, H/L, or L/L conditions. The unit is also equipped with an RS 485 serial port so that, using available protocols, it can be connected to a computer via an RS 232 port, or via a special converter. Up to 3 controllers can be interconnected, in parallel, to a two wire common bus, and this enables the measurement, set point, output, alarms and status of A/M and R/L to be read. It also enables the A/M and R/L to be changed so that, when operating in the 'Local' mode, the set point can be adjusted, and when operating in the 'Manual' mode the output can be changed. The power requirement is 30 V maximum, at 24, 120, 220 or 240 V rms, at 47 to 63 Hz, or 24 V dc.

In this way, most of the feature and function previously available as separate components in the SPE 200* system have been combined into one self-conntained unit, but with the addition of a communication facility.

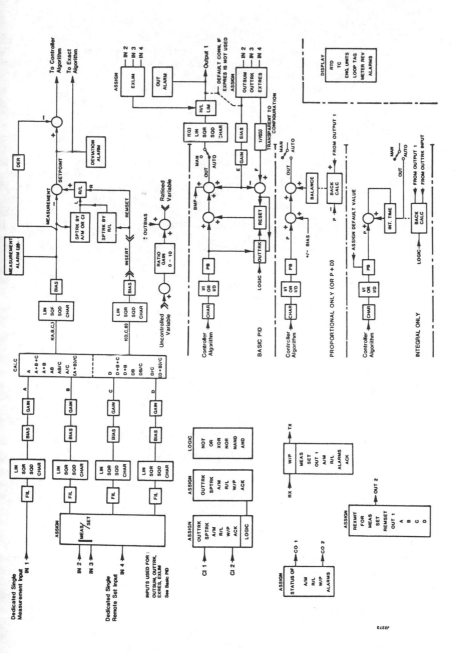

Fig. 15.13   Functional diagram of the
Single Station Controller.

15.7   CONCLUSIONS.

The EXACT* control algorithm demonstrates that an 'expert system' can provide a viable alternative to the mathematical model approach for self-tuning or adaptive control. Quite appart from the fact that it has been used successfully in a wide variety of applications, it incorporates important operational features which reflect its heuristic origins and which are either not recognised by or difficult to incorporate in control schemes based on mathematical models.

It is seldom possible to quantify the benefits which could result from substituting a self-tuning controller in place of a conventional PID controller and consequently there is little incentive to accept the cost and inconvenience of making the change. However, as the concept becomes more mature and confidence is established in both the method and the associated equipment, it will come into more general use. This is particularly relevant in view of the trend towards batch, rather than continuous, processes because accurate control of the transients which occur at start-up, shut-down or during the progression through the various phases of the process can have a significant effect on product quality, as well as the utilisation of both raw materials and energy.

The EXACT* algorithm relies heavily on the heuristic rules which control engineers have developed over the last four decades in tuning and applying conventional PID ccontrollers. It important to recognise this lineage when identifying the application in which the EXACT* algorithm can be used effectively. If a Conventional PID controller does not provide acceptable control, neither will an adaptive controller, and it is no substitute for good process design or good measurement technologies and no remedy for defective plant.

The applications for which the adaptive controller are most suitable and effective are those in which the process dynamics vary irregularly, so that a conventional PID controller would require frequent adjustment of its parameters to avoid instability or an over damped response.

15.8   ACKNOWLEDGEMENTS

The evolution of the self-tuning controller described in this paper has taken place over a long period, during which many members of the Foxboro Company have been involved. Particular mention should be made of the contributions of E.H. Bristol, F.G. Shinskey, T.W. Kraus, T.J. Myron, P. Badavas, G. Diomandes, J.W. Eva, J. Fahey, N. Fondeneau, L. Gordon, P. Hansen, P. Levesque, R. Rys, G. Tucciatrone, and others to the concept and its implementation as a viable design. The author acknowledges all this work and also the permission of The Foxboro Company to publish the paper.

5.9 REFERENCES

1. Bristol, E.H. "A Simple Adaptive System for Industrial Control". Instrumentation Technology, June 1967.

2. Ziegler, J.G. & Nichols, N.B. "Optimum Settings for Controllers". Trans ASME, November 1942.

3. Shinksey, F.G. "Process Control Systems". 1979, McGraw Hill.

4. Goodman, T.P. and Rewick, J.B. "Determination of System Characteristics from Normal Operating Records". ASME IRD Conference, April 1955.

5. Pessin, D.W. "Investigation of a Self-Adaptive Three-Mode Controller" ISA, September 1963.

6. Bakke, R.M. "Adaptive Gain Tuning Applied to Process Control" ISA, October 1964.

7. Daklin, E.B. "On-line Identification of Process Dynamics". IBM Journal, 1967.

8. Inagaki, T. "An Adaptive Direct Digital Control". Case Western Reserve University Report, SRC 68-6.

9. Caldwell, W.I. "Control System with Automatic Response Adjustment", U.S. Patent No. 2,517,081.

0. Bristol, E.H., Inalogu, G.R. and Steadman, J.F., "Adaptive Process Control by Pattern Recognition". JACC, August 1969.

1. Bristol, E.H., et al, "Adaptive Process Control by Pattern Recognition". Instruments and Control Systems, March 1970.

2. Bristol, E.H. "Adaptive Control Odyssey". ISA, October 1970.

3. Shinskey, F.G., and Myron, T.J. "Adaptive Feedback Applied to Feedforward pH Control". ISA, October 1970.

4. Bristol, E.H. "Adaptive Process Control: A Versatile On-line Tool". Control Engineering, April and June 1973.

5. Shinskey, F.G., "Adaptive pH Controller Monitors Non-linear Process". Control Engineering, February 1975.

16.   Bristol, E.H.   "Pattern Recognition; An Alternative To Parameter Identification in Adaptive Control. Automatica, Vol 13, pp 197-202 1977.

17.   Field, M. and Wilhelm, R.G.   "Self-Tuning Regulators - The Software Way".   Control Engineering, October 1981.

18.   Clarke, D.W.   "The Application of Self-Tuning Control - The Software Way".   Trans. Inst. MC., April/June 1983.

19.   Rohrs, C.E., Valavani, L., Athans, M., and Stein, G. "Robustness of Adaptive Control Algorithms in the Presence of Unmodelled Dynamics.   MIT Industrial Liaison Program, Publication 01-016,1983.

20.   Kraus, T.W. & Myron, T.J.   "Self-Tuning Controller Uses Pattern Recognition Approach".   Control Engineering, June 1984.

21.   Shaw, R.   "Expert Systems in Process Control". 'Processing', Control and Instrumentation Supplement, October 1984.

22.   Myron, T.J. and Thurston, C.W.   "Performance and Evaluation of a Self-Tuning PID Controller.   40th Annual Symposium on Instrumentation for Process Industries, Texas A+M University, January 1985.

Chapter 16

Microcomputer control
case study I
P.A. Witting

This chapter presents the approach used by the author to
implement a computerized 3-term control scheme on a small
process. The computer used was a standard Commodore CBM
4032 connected to the plant via a proprietary interface
unit. The program was written in BASIC for convenience and
clarity but this gave severe timing problems which were,
however, overcome. Because of the stringent speed
requirements no modifications to the three term algorithm
were attempted. Had a more efficient programming language
been available, the techniques outlined in chapters four and
five could have been implemented.

16.1 THE PROCESS

The process to be controlled consists of a fan heater
supplying warm air to a process tube. The temperature of
the air is monitored in the tube some 11" downstream of the
heater, the measurement being made with a thermistor-and-
bridge arrangement (16.1)

The air flow velocity in the tube is controlled by a
manually adjustable throttle valve. A separate controller
can be incorporated into the system by switching to
"external" and connecting the new controller between "error"
and "actuation signal" Fig 16.1. Both the heater coil and
the thermistor unit behave as first-order lags while the air
tube is modelled as a pure transport lag. All other time
constants in the system are negligible. Thus the plant
model is

$$\frac{Ke^{-s\tau}}{(1 + sT_1)(1 + sT_2)} \qquad (16.1)$$

The gain and all of the time constants are a function of the
air velocity and hence the throttle setting. In practice
all of the time constants (τ, T_1, T_2) are of the order of a
few hundred milliseconds. The static gain, K, being
approximately 2.0. The error signal is fed to the
controller input and the resulting correcting signal
supplied to the power controller.

For the purpose of this case study it was decided to investigate only the efficiency of a digital 3-term controller and to this end, the Zeigler and Nichols tests were used to obtain initial estimates for the controller settings.

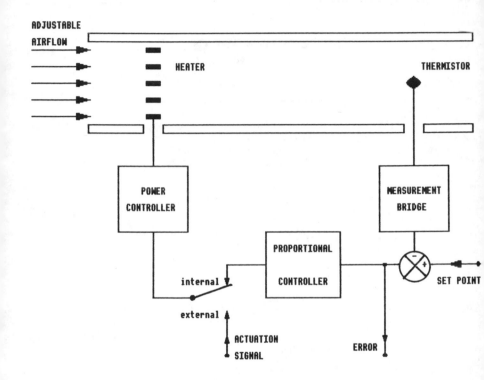

Fig 16.1   Arrangement of Process to be Controlled

16.2   ZEIGLER-NICHOLS TESTS

An adjustable-gain proportional controller was included in the system shown in figure 16.1 and the gain adjusted until sustained oscillations were achieved. This proved to be a very difficult procedure since, once the system became lightly damped, the noise disturbances gave the appearance of sustained oscillations even when there was some positive damping present.

Figure 16.2 shows the oscillations finally achieved. The period of the oscillations, Tu, is 1.4 seconds and the gain required was Ku = 1.7.

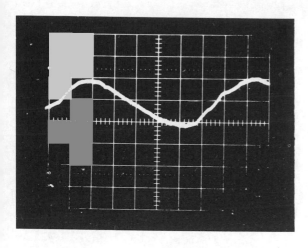

scale: 0.5 volt/cm
         vertical

         0.2 s/cm
         horizontal

Figure 16.2   Zeigler-Nichols Closed-Loop Test

Bearing in mind the difficulty in obtaining these figures, an attempt was made to confirm them via the open loop test. The proportional controller was removed and a 1.5v, 0.1 Hz square wave applied to the power controller. The response of the temperature signal to this stimulus is shown in Figure 16.3.

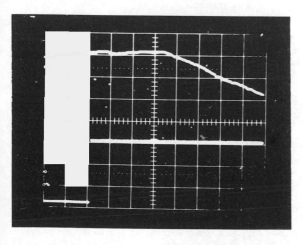

scale:

y_1 (step) = 0.2
         volt/cm

y_2 (response) =
         0.5 volt/cm

x = 0.1 s /cm

Figure 16.3

From this figure the pure delay of the system
may readily be estimated.

$$L = 350 \text{ msec}$$

This figure agrees closely with that obtained from the
closed loop test since Zeigler and Nichols give the
approximate equivalence:-

$$Tu = 4L$$

and hence Tu = 1.4 seconds.

The scales in figure 16.3 were chosen to give a good
estimate of L.   To estimate the "reaction rate", R,
different scales were used and a smaller input voltage was
employed to avoid saturating the system and getting
misleading results.   Figure 16.4 shows the results of using
a 0.5v, 0.1 Hz square wave.

Measurements on this response indicate that the reaction
rate is

$$R = 0.6 \text{ volt/sec}$$

Thus the unit reaction rate is $R' = R/(\text{input voltage})$

$$R' = 1.2 \text{ volt/sec/volt}$$

This gives an estimate for Ku of

$$Ku = \frac{2}{R'L} = \frac{2}{1.2*0.35} = 4.75$$

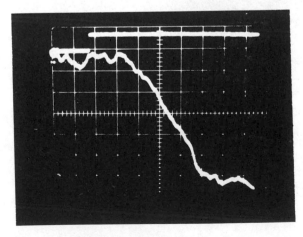

scale:

y₁ (step)
= 0.5 volt/cm

y₂ (response)
= 0.1 volt/cm

x = 0.2 s/cm

Figure 16.4   Zeigler-Nichols Open-Loop Test I

This figure is at considerable variance with the closed loop test. Furthermore the static gain deduced from Figure 16.4 is only 1.2 which is very much less than the figure measured in a direct DC calibration test (not shown). Finally the rise shown in Figure 16.4 is suspiciously straight, indicating that the system might have saturated.

Consequently the test was repeated using a 0.25 volt input and the corresponding response is shown in Figure 16.5. From this the static gain is found to be about 2.3 which agrees more closely with the DC calibration. Also the curve has lost its saturated appearance.

The reaction rate is found to be

$$R = 0.75 \text{ volt/sec}$$

giving a unit reaction rate of

$$R' = 3.0 \text{ volt/sec/volt}$$

The estimate value for Ku is then

$$Ku = \frac{2}{R'L} = \frac{2}{3*0.35} = 1.9$$

This is reasonably close to that of the closed loop test.

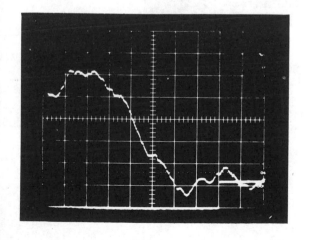

scale:

y_1 (step)
= 0.2 Volt/cm

y_2 (response)
= 0.1 volt/cm

x = 0.2 s/cm

Figure 16.5 Zeigler-Nichols Open Loop Test II

To give an assurance that the effects of saturation had been completely eliminated, the test was repeated with an even smaller input (0.16 volts). The response for this test is shown in Figure 16.6 and is beginning to be severely noise corrupted.

From Figure 16.6 it is possible to measure

$$R = 0.4 \text{ volt/sec}$$

$$R' = 2.5 \text{ volt/sec/sec}$$

$$K_{11} = 2.3$$

The static gain is seen to be 2.0. All of these figures confirm those obtained from Figure 16.5 and from the closed loop test.

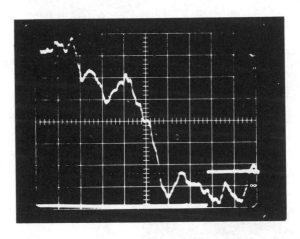

scale:

y_1 (step)
= 0.1 volt/cm

y_2 (response)
= 0.05 volt/cm

x = 0.2 s/cm

Figure 16.6   Zeigler-Nichols Open Loop Test III

16.3   ASSESSMENT OF THE SUITABILITY OF THE CBM COMPUTER

From the foregoing discussions, it will be appreciated that a sampling period of about 140ms (Tu/10) will be required in the final system. The next task in the design was to determine whether the necessary computations and input-output (I/O) operations could be completed in the time available.

From the outset it was decided to use high level language programming since this speeds system development and avoids the necessity for developing special mathematics subroutines, these being an integral part of the language. The CBM computer utilises a Rockwell 6502 microprocessor, and 8-bit device, and writing all of the necessary routines in assembly code would have been tedious. This option was not, anyway, immediately available to the author as he did not have access to a 6502 assembler. It was therefore decided that the BASIC language provided as an integral part of the CBM should be used.

The CBM computer may be connected to peripheral devices in a number of ways. The most popular of these being the "user port" and the "IEEE bus port". In this investigation an IEEE connected interface was used, a PCI6300. This particular device provides for up to 8 input channels (with individually programmable gains), 4 output channels and 4 relays. The IEEE bus consists of 16 signal lines (plus grounds). Eight of these lines are used for bi-directional data transfer on a byte-by-byte basis and the remainder are used for control. A maximum transfer rate of 250,000 bytes/second is specified.

To assess the possible I/O speed of the bus and peripherals used in this exercise a simple program was written. This program consists of a loop containing 2 I/O instructions and runs continuously when started. The program first outputs +10 volts to channel 0, then to channel 1. Following this -10 volts is output to both channels. The process is repeated indefinitely and the skew between the waveforms is a measure of the time taken for one output instruction. This time was measured to be 7.5 ms. Thus the I/O time (consisting of data conversion, data transmission and program execution) is unlikely to be a limiting factor in this study. The program used is shown below.

```
10 OPEN 1,9              REM OPEN CHANNEL TO PERIPHERAL
20LS$=CHR$(255)          REM DEFINE CONSTANTS
30 MS$=CHR$(15)
40 NS$=(7)
50 PRINT#1,"OO";MS$;LS$  REM SET OUTPUT O TO + FULL SCALE
60 PRINT#1,"01";MS$;LS$  REM SET OUTPUT 1 TO +FULL SCALE
70 PRINT#1;"OO",NS$;LS$  REM SET OUTPUT O TO -FULL SCALE
80 PRINT#1;"01";NS$;LS$  REM SET OUTPUT 1 TO -FULL SCALE
90 GOTO 50
```

16.3.1 The Control Program

Following this the complete control program was written so that its execution speed could be assessed. This is shown below in both flow chart form fig 16.7 and in program form, fig 16.8.

Lines 5 to 70 allow the user to input his chosen control parameters from which the coefficients of an incremental control algorithm are computed and printed out. The variable TMIN is the minimum sampling time allowed for this program. The variable T% calculated on line 35 is used in a "delay loop" later, to control the sampling time. Since the program itself takes 144 ms to execute, the required sampling time, T, is reduced by this amount and then divided by the time taken to go once round the delay "padding" loop (1.82 ms) in lines 3035 and 3036. If the user selects a value of T less than the minimum allowed he is told so and asked to re-input his data, (defensive programming).

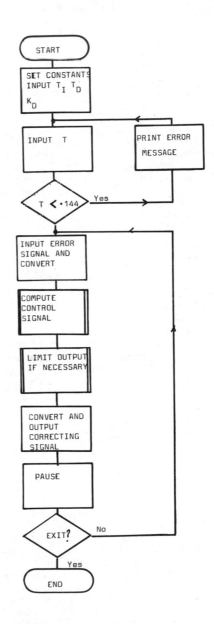

Figure 16.7   Control Programme Flowchart

```
5 TMIN=.144
10 INPUT"INTEGRAL TIME(SEC)=";T1
20 INPUT"DERIVATIVE TIME(SEC)=";TD
30 INPUT"SAMPLE TIME(SEC)=";T
31 IF T=>TMIN GOTO 35
32 PRINT"MINIMUM SAMPLE TIME=",TMIN,"SEC"
33 GOTO 30
35 T%=INT((T-TMIN)*1000/1.82)
40 C1=1+TD/T+T/T1
50 C2=(-1)*(1+2*TD/T)
60 C3=TD/T
65 PRINT C1;C2;C3
70 INPUT"PROPORTIONAL GAIN=";KP
80 DIM E(3)
90 PRINT"HIT SPACEBAR TO STOP"
1000 OPEN 1,9
1010 PRINT#1,"G0,I0"
1020 GET#1,H$,L$
1030 IF H$="" THEN GOTO 1060
1040 H=ASC(H$)
1050 GOTO 1070
1060 H=0
1070 IF L$="" THEN GOTO 1100
1080 L=ASC(L$)
1090 GOTO 1110
1100 L=0
1110 IF H>7 THEN H=H-8:GOTO 1150
1130 H=-H
1140 L=-L
1150 E(1)=256*H+L
2000 GOSUB 4000
2010 GOSUB 5000
3000 H=INT(ABS(OP)/256)
3010 L=ABS(OP)-256*H
3020 IF OP>=0 THEN H=H+8
3030 PRINT#1,"O0";CHR$(H);CHR$(L)
3035 FOR I=1 TO T%
3036 NEXT I
3040 GET SB$
3050 IF SB$="" GOTO 1020
3060 GOTO 6000
4000 DELTA=KP*(C1*E(1)+C2*E(2)+C3*E(3))
4010 OP=OP+DELTA
4020 E(3)=E(2)
4030 E(2)=E(1)
4040 RETURN
5000 IF OP<=2047 GOTO 5030
5010 OP=2047
5020 GOTO 5050
5030 IF OP>=-2048 GOTO 5050
5040 OP=-2048
5050 RETURN
6000 CLOSE 1
6010 END
```

Figure 16.8   Control Programme Listing

Line 1020 reads the 2 bytes containing the 12-bit error signal from the peripheral, placing them in the variables HS (most significant byte) and LS (least significant byte).

Lines 1030 and 1040 convert the data from the character form in which they were read over the IEEE bus into numeric form.

The peripheral used employs a code for the input data in which the most significant bit is one for positive numbers and zero for negative ones. Figure 16.9.

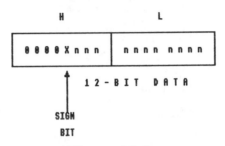

Figure 16.9

Thus if the sign bit is zero H is in the range 0 -> 7 and the data is negative.

i.e.
$$\text{Data} = -\left[256H + L\right] \qquad (16.2)$$

However, if the sign bit is one, H is in the range 8 -> 15 and the sign bit must be stripped off to give the true value of H, H',

i.e.
$$H' = H - 8 \qquad (16.3)$$

Then
$$\text{data} = 256H' + L \qquad (16.4)$$

This conclusion is achieved by lines 1110 to 1150.

At line 2000 the control routine is called. This resides at lines 4000 to 4040. Line 4000 calculates the ΔV of the incremental algorithm while line 4010 adds this to the current output to form the new output. Lines 4020 and 4030 merely update the data table ready for the next calculation, when the current input will be the "immediately previous" input e_{k-1}.

On return to the main program another subroutine is called. This resides at line 5000 and exists to check that the output has not exceeded its limits (+2047 to -2048; equivalent to ± 10 volts). If it has, the output is set to the limiting values. Failure to do this results in a "wrap around" phenomenon. For instance, +2048 would be output as 0 volts and this would lead to instability. This is another example of defensive programming.

Lines 3000 to 3030 convert the output to a form acceptable to the interface and transmit the data to it. This is effectively the inverse of lines 1020-1150. Lines 3035 and 3040 are executed T% to waste time. Thus padding out the control loop delay to the required sampling period. By inserting several different values in this loop and measuring the change in execution time, it was determined experimentally that each execution of the loop would take 1.82ms.

Finally line 3040 interrogates the keyboard to see whether the space bar has been pressed. If so the program jumps to line 6000 and closes down. If not execution continues and a new input sample is fetched.

16.3.2 Testing the Control Program

In order to assess the speed of this program, a special test version was written. In this there was an extra line inserted.

$$4 \text{ OP} = 2000$$

In addition two lines were modified.

$$35 \text{ T\%} = 1$$

$$4010 \text{ OP} = -\text{OP}$$

These have the effect of making the program perform all of the calculations necessary but instead of outputting a correction signal the program toggles its output between (approximately) +10 volts and -10 volts. Also the delay loop is effectively eliminated.

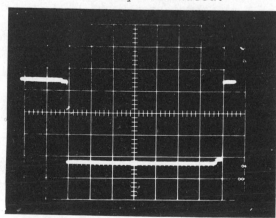

scale:

y = 5 volt/cm

x = 20 ms/cm

Figure 16.10 Output of System During Measurement of Calculation Time

Figure 16.10 shows the resulting output from which it may be seen that the control algorithm and housekeeping functions require 144 ms (hence Tmin = 0.144 on line 5 of the program). This timing is dangerously close to the value predicted as the required sample time and it is pertinent to ask at this stage whether anything can be done to speed matters up.

(i)   Input-Output operations; Each of these takes 7.5 ms approximately. The use of the user port could reduce this to a few hundred microseconds provided that the appropriate analogue-digital and digital-analogue convertor circuits were used.

(ii)   The majority of the time is spent in various calculations using BASIC. As implemented in the CBM computer (and most others) BASIC is an interpreted language. That is, a program is only ever stored in its high level form. When executed each line is translated, it overwrites the preceeding line. Thus each line has to be re-translated every time it is used. This is very wasteful. A considerable saving in time can be achieved by performing a once-and-for-all translation, with the machine code version being stored as a complete entity. This is called COMPILIATION. BASIC compilers are available for the CBM machine and tests indicate that a speed improvement of between four and six times is possible in this application.

(iii)   Another reason for the relatively slow execution speed is the 6502 microprocessor itself. This is considerably slower in its operations than a typical minicomputer. Further it lacks the special hardware for multiplication and division etc. which is quite common in mini-computers. Finally, and by no means least, the 6502 microprocessor uses an 8-bit data word. BASIC integers are stored to a precision of 16 bits thus requiring 2 words of storage. In a typical minicomputer only one word would be required thus speeding up both the transmission of the data and its arithmetic manipulation.

To demonstrate the correct operation of the controller it was subjected to a 1Hz square wave input and configured as a proportional-only controller (Kp = 2 Td = O Ti = 10), as an integral-only controller (Kp = 0.05, Td = 0, Ti = 0.05) and as a derivative-only controller (Kp = 0.05, Td = 20, Ti = 10^6). The results are shown in Figures 16.11, 16.12, and 16.13, respectively, these are as expected, giving square waves, dipole pulses and ramps.

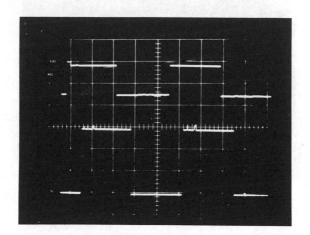

Figure 16.11   Controller with Proportional Action Only

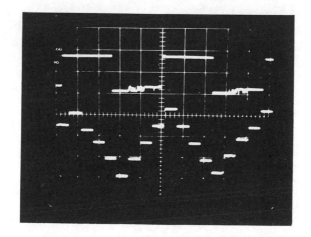

Figure 16.12   Controller with Integral Action Only

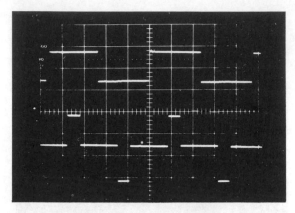

Figure 16.13   Controller with Derivative Action Only

16.4   IMPLEMENTATION OF THE CONTROLLER

In an ideal world the control algorithm would calculate its output once the input was available.   However, as has been shown above, it takes some 144ms to achieve this result. This is not insignificant when compared with the time delay element of the system and allowance must be made for it.

One way of doing this is to model the controller as an ideal instantaneous algorithm plus a pure delay of 144ms.

This additional delay must be added to the process lag of 350ms to give an overall effective time lag of

$$L = 494 \text{ ms}$$

The unit reaction rates inferred from the closed loop test data measured in the valid open loop tests were 3.3, 3.0, 2.5. Taking the largest value as a conservative estimate

$$R' = 3.3$$

and

$$Ku = \frac{2}{R' L} = \frac{2}{3.3*0.494} = 1.22$$

and

$$Tu = 1.98 \text{ sec.}$$

using the revised value for L.

This leads to sampling period = 0.2 sec

$$Ti = 1.0 \text{ sec}$$

$$Td = 0.25 \text{ sec}$$

$$Kp = 0.74$$

The response of the system with these settings is shown in Figure 16.14.

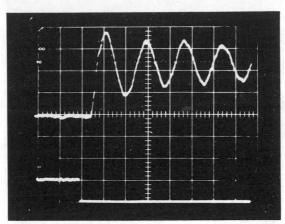

Figure 16.14   Plant Closed Loop Response with Initial Settings of PID Controller

It will be seen from Figure 16.14 that the response is rather more oscillatory than might be considered satisfactory, even for a Zeigler-Nichols design.

This can be accounted for by postulating an additional delay in the system which has not been accounted for. The output of the controller is a simple number which would appear as a pulse at the output unless "stretched" by a "HOLD" circuit. The Digital-Analogue convertor provides just such a holding action, in fact a "Zero Order Hold" with a transfer function.

$$\frac{1 - e^{-sT}}{s}$$ (16.5)

It is readily shown (1) that this transfer function provides a pure time delay of T/2 seconds plus a low pass gain characteristic with a sampling period of 200 ms there is an additional lag of 100 ms to account for.

Thus L = 0.594 seconds

$$Ku = \frac{2}{R\ L} = \frac{2}{3.3*0.594} = 1.02$$

Figure 16.15 shows the step response of the system with K set to 0.6 (i.e. 0.6Ku). The values of Ti ,and Td and T being left unchanged for simplicity.

It will be noted that the damping is now much better and approaches the value assumed as "Ideal" by Zeigler and Nichols.

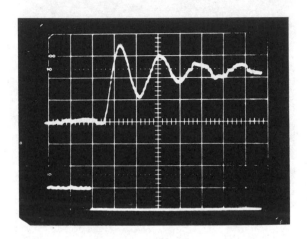

Figure 16.15   Closed-Loop Response of Plant with Revised
Settings of PID Controller

16.5   CONCLUSION

This case study has shown how a successful digital controller may be realised using quite simple equipment. It has been shown that computational delays can, on occassions, be quite significant and need to be considered during the design.

No attempt has been made to investigate the effects of finite word length arithmetic since the computational overheads involved in simulating this would have made the controller too slow. In the circumstances floating point arithmetic, implemented by the BASIC language, has been used throughout.

Where appropriate the need for "defensive programming" has been demonstrated, this is a necessary technique which must be employed to ensure that unusual data, or operator errors, do not cause serious malfunctions.

References

Katz, P. "Digital Control Using Microprocessors", 1981, Prentice Hall International. p31.

Chapter 17

A comparison of DDC algorithms
case study II

Dr. D. Rees

17.1 INTRODUCTION

This case study describes the implementation of a number of discrete algorithms - phase advance, PID, velocity feedback, state feedback using observers and finite-time settling controller - to control an electro-mechanical system (Rees and White (1)).

The system is mathematically well defined and is in open-loop unstable. The system is used within a teaching environment and originally the control function was implemented using an analogue controller. To aid the teaching of discrete-time control systems this was later replaced by a microcomputer. This particular system gives the student the opportunity of designing, implementing and critically assessing the performance of discrete algorithms which are the counter parts of well established continuous algorithms (three term, phase advance etc) and also those which are less well established like state variable feedback using observer estimates. It also allows the investigation of discrete controllers which have no analogue counterparts such as dead-beat and finite time.

17.2 SYSTEM DESCRIPTION

The electro-mechanical system has been described in detail by Wellstead et.al (2). The basic system, shown in Fig.17.1 consists of a rigid beam which is free to rotate in one plane about a centre pivot with a solid steel ball which is free to roll along the top of the beam on two steel support wires which are electrically insulated.

The control task is to position the ball at a desired point on the beam using a torque or force applied to the beam as the control input. The beam angle is adjusted, via a universal joint coupling, by a vertically mounted moving coil actuator. The angle of the beam is measured by a precision servo-potentiometer mounted on the pivot axis. The position of the steel ball along the beam is measured by a potentiometric method in which a small voltage is generated across the wires, proportional to the position of the ball.

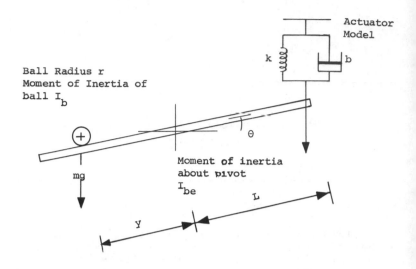

Fig.17.1 Ball and beam system

The digital computer used is a CBM PET microcomputer which is also used for the acquisition of ball position data, this data is stored on a floppy disc and provides the facility of tabulating the position data for subsequent post mortem analysis. The linking of the CBM PET to the beam control box is done through a multipurpose computer interface (Fig.17.2). The two-way exchange of data between the micro and interface takes place over the IEEE data bus at the sampling interval which is set under software control.

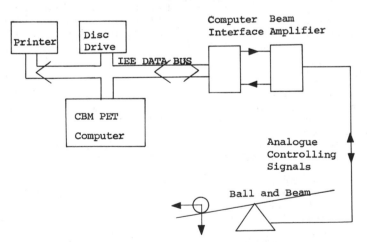

Fig.17.2 System configuration for direct digital control.

17.3 BALL AND BEAM MODEL.

An essential requirement in the development and implementation of the control algorithms is the determination of the system dynamics. With a knowledge of these dynamic characteristics a suitable control strategy can be devised.

Wellstead et.al.(2,3) has developed a complete mathematical model for the ball and beam and has shown, provided certain assumptions are made which are valid for the equipment in question, that the system can be described by the equations.

$$\text{Ball} \qquad \left[m + \frac{I_b}{r^2} \right] \ddot{y} = mg\Theta \qquad (17.1)$$

$$\text{Beam} \qquad I_{be}\ddot{\Theta} + bL^2\dot{\Theta} + kL^2\Theta = -u(t)L \qquad (17.2)$$

The constants are defined in the diagram of Fig.17.1. The approximations made result in a convenient set of equations where the dynamics of the ball and beam are separated. The system model using equations (17.1.) and (17.2.) may be represented in transfer function form as shown in Fig. 17.3. where

$$C = \frac{mg}{m + I_b/r^2} \qquad (17.3)$$

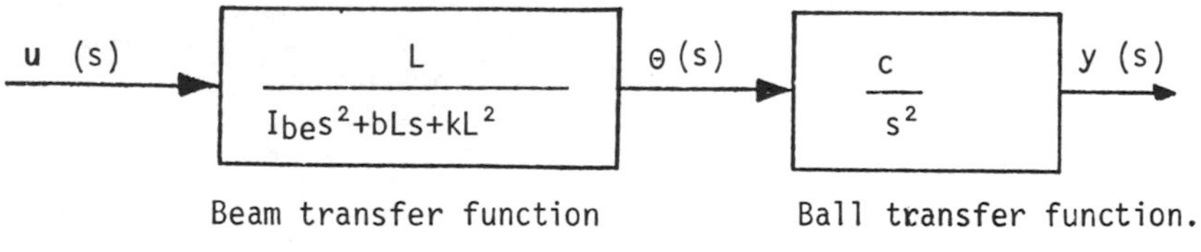

$$u(s) \longrightarrow \boxed{\dfrac{L}{I_{be}s^2 + bLs + kL^2}} \xrightarrow{\Theta(s)} \boxed{\dfrac{c}{s^2}} \longrightarrow y(s)$$

Beam transfer function   Ball transfer function.

Fig.17.3 Ball and beam transfer function

A further approximation can be made with the assumption that the beam dynamics are negligible compared with the ball dynamics. This reduces the system transfer function to a double integrator which is of the form

$$G(s) = \frac{Y}{u}(s) = \frac{b^1}{s^2} \qquad (17.4)$$

where b^1 is a gain constant and equals c/kL and can be
determined experimentally as follows:

The system is put into closed loop control using a
negative feedback loop from the ball position y output to
the beam actuator input, the ball is then set into simple
harmonic motion (in practice there is a slow build up in the
amplitude of oscillations due to the dynamics of the beam
(Fig.17.4). The period of oscillation is given by $T = 2\pi/\sqrt{b^1}$
and from the experimental results is equal to 6.5. seconds.

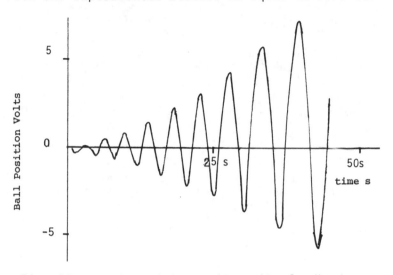

Fig. 17.4 Ball position under unity feedback

The beam dynamics were investigated by carrying out a
frequency response test between the actuator input and
beam angle potentiometer output. The frequency response
plot yielded the following parameter values.

    Peak Magnitude - Mp = 11.5 dB
    Damping Ratio ξ = 0.15
    Natural Frequency ω_n = 14.59 rad s^{-1}

The beam transfer function is therefore given by

$$\frac{\theta}{u}(s) = \frac{212.8\ k_1}{s^2 + 4.38s + 212.8} \tag{17.5}$$

where k is the gain constant relating the output beam angle
to the input force. It can be seen from the root locus plot
of the uncompensated system that the beam transient rapidly
decays compared with that of the ball (Fig.17.5.)

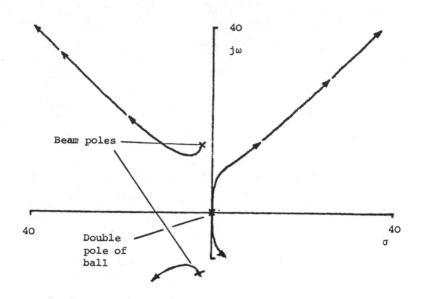

Fig.17.5 Root locus plot of the ball and beam

17.4 ALGORITHMS FOR DIGITAL CONTROL

17.4.1 Phase Advance Compensation

The transfer function of a phase lead network is given by

$$\frac{u}{e}(s) = K \left[\frac{s-z_1}{s-p_1} \right] \tag{17.6}$$

where z_1 and p_1 are the zero and pole of the compensator and $|z_1| < |p_1|$

One approach to obtain a discrete version of eqn.17.6. is to use a first order backward difference transformation (many other transformations can be used like the Bilinear, modified z transform etc (Ben-Zw, A and Preizler, M (4)), so that

$$\frac{de}{dt} = \frac{e(n)-e(n-1)}{T} \qquad ; \qquad \frac{du}{dt} = \frac{u(n)-u(n-1)}{T} \tag{17.7}$$

where T is the sampling interval.

The z transform of the backward difference operation is

$$Z(s) = \frac{1-z^{-1}}{T} \tag{17.8}$$

which results in the pulse transfer function

$$\frac{u}{e}(z) = K\left[\frac{A - z^{-1}}{B - z^{-1}}\right]$$ (17.9)

where $A = 1 - z_1 T$ and $B = 1 - p_1 T$

    The difference equation, at the nth sampling interval is

$$u(n) = \left[\left[Ae(n)-e(n-1)\right]K+u(n-1)\right]/B$$ (17.10)

and can be represented by the programming block diagram shown in Fig.17.6.

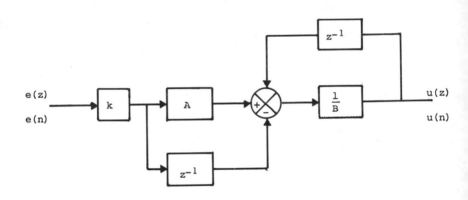

Fig.17.6 Discrete phase advance compensator

17.4.2   Three Term Controller

    The three term controller, proportional-integral-derivative (PID), has been covered in some detail in Chapter 3. It has been shown that the digital PID control algorithm can be written in the velocity form as follows

$$\Delta u(n) = K_p\left[y(n-1)-y(n)\right] +K_I\left[r(n)-y(n)\right] +K_d\left[2y(n-1)-y(n-2)-y(n)\right]$$ (17.11)

or in the position form

$$u(n) = K_p y(n) +K_I \sum_{i=0}^{n}\left[r(i)-y(i)\right] + K_d\left[y(n)-y(n-1)\right]$$ (17.12)

where $y(k)$ and $r(k)$ are sampled data of the system output and setpoint, $\Delta u(n)=u(n)-u(n-1)$ is the incremental change of the manipulated variable, $u(n)$, and

$$Kp, \quad K_I = \frac{TK}{T_I} \text{ and } K_d = \frac{T_d K}{T}$$

are respectively the proportional, integral and derivative control gains. To avoid the so called "set point and derivative kick" the set point signal $r(k)$ is included only in the integral action term of eqn.17.11 and 17.12. This representation of the PID controller, results in the proportional and derivative actions forming a minor loop in the control scheme (fig.17.7) where only the integral action is left in the main loop.

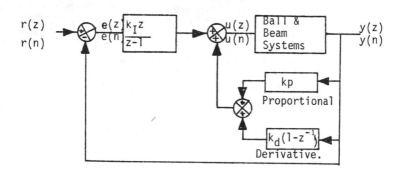

Fig.17.7 Discrete PID controller

17.4.3 Velocity Feedback

With velocity feedback, stability is introduced to the electro-mechanical system by taking the derivative of the ball position and providing both velocity and position negative feedback. The controller output in its continuous form is

$$u(t) = ke - k_v \dot{y} \tag{17.13}$$

Using the backward difference transformation of eqn.17.8 and letting $r(n) = r(n-1)$ then in its discrete form.

$$u(n) = (k+k_v/T)\,e(n) - (k_v/T)\,e(n-1) \tag{17.14}$$

The block diagram realisation of this algorithms is shown in Fig.17.8.

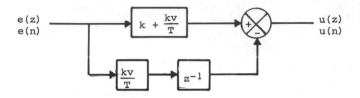

Fig 17.8 Discrete velocity feedback controller

17.4.4   State Feedback using observers

The use of state feedback in control systems (Kuo (5)) often encounters the practical problem that not all the state variables are accessible, and even if they were it would often be economically impractical to install all the transducers that would be necessary to measure the states.

This problem however, can be overcome if a reliable mathematical model is available to generate an estimate $\hat{x}$ of the state vector of the controlled plant. The reconstructed state vector is then used to implement a state feedback control law $u = \underline{k}\ \hat{x}$.

The linear state vector differential equations are written as

$$x = \underline{A}\ \underline{x} + \underline{B}\ \underline{u} \qquad (17.15)$$

$$\underline{y} = \underline{C}\ \underline{x}$$

where for the ball and beam system

$$A = \begin{bmatrix} 0 & 1 & 0 & 0 \\ 0 & 0 & 1 & 0 \\ 0 & 0 & 0 & 1 \\ 0 & 0 & -\omega n^2 & -2\xi\omega n \end{bmatrix}$$

$$ \qquad (17.16)$$

$$B = \begin{bmatrix} 0 \\ 0 \\ 0 \\ \omega n^2 b^1 \end{bmatrix} \qquad C = \begin{bmatrix} 1 \\ 0 \\ 0 \\ 0 \end{bmatrix}$$

The discrete state and output equations can be written as   (for greater detail refer to chapter 6)

$$\underline{x}(n+1) = (T\underline{A}+\underline{I})\ \underline{x}(n) + T\underline{B}\ u(n) \qquad (17.17)$$

$$y(n) = \underline{C}\underline{x}(n)$$

where, for the ball and beam system

$$(TA+1) = \begin{bmatrix} 1 & T & 0 & 0 \\ 0 & 1 & bT & 0 \\ 0 & 0 & 1 & T \\ 0 & 0 & -\omega n^2 T & (-2\xi\omega nT+1) \end{bmatrix}$$

(17.18)

$$TB = \begin{bmatrix} 0 \\ 0 \\ 0 \\ T\omega n^2 b^1 \end{bmatrix} \text{ and } C = \begin{bmatrix} 1 \\ 0 \\ 0 \\ 0 \end{bmatrix}$$

Here, the assumption is made that the time increment T is small compared with the dynamics of the system.

These equations are simulated in the computer with the input u being the same as that applied to the actual physical system. Since the physical system may be subjected to unmeasurable disturbances and the mathematical model is at best a good linear approximation of a non-linear system, the difference between the actual plant output y and the simulated output $\hat{y}$ is used as another input to the simulated equation. The observer model must also be given the initial condition values of the state variables. So that

$$\hat{\underline{x}}(n+1) = (TA+1)\ \hat{\underline{x}}(n) + T\underline{B}u(n) + L(y(n)-\hat{y}(n))$$

(17.19)

where L is the observer matrix, which for the ball and beam system was a one dimensional vector.
Since $\hat{y}(n) = C\hat{x}(n)$, then

$$\hat{\underline{x}}(n+1) = (T\underline{A}+I-\underline{LC})\ \hat{\underline{x}}(n)+T\underline{B}u(n)+\underline{L}\underline{y}(n)$$

(17.20)

This equation generates the states x which are estimates of the plant states x. The re-construction error between the plant states and the observer states x satisfies the equation

$$e(n+1) = (T\underline{A}+\underline{I}-\underline{LC})\ e(n)$$

(17.21)

The control law is chosen so that

$$u(n) = r(n)-k\hat{x} = r-k(x(n)-e(n))$$

(17.22)

Therefore,

$$x(n+1) = (T\underline{A}+\underline{I}-T\underline{k}\underline{B})\ x(n) + T\underline{B}ke(n) + T\underline{B}r(n)$$

(17.23)

where r(n) is the reference input at the nth sampling interval.

An important feature of eqn.17.21 and eqn.17.23 is that the eigenvalues of the closed-loop and error equations can be assigned independently by the selection of appropriate matrices for k and L. The eigenvalues of (TA+I-LC) are usually selected so that they are to the left of the eigenvalues of (TA+I-TkB) thus ensuring that the observer states rapidly approach the plant states. The representation of this algorithm is shown in Fig. 17.9.

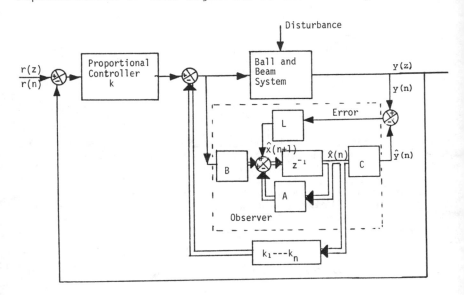

Fig.17.9 State feedback controller using observers

17.4.5 Finite-Time Settling Controller

The design approach for a finite-time settling controller falls into the category of direct synthesis methods, covered in chapter 4, where the controller is designed for a specific closed-loop pulse transform. The primary purpose of the digital controller is to cancel any undesirable poles and zeros of the uncompensated system and replace them with poles and zeros which give the desired response. In the ideal case, this can be accomplished by designing the controller so that its zeros correspond to the poles of G(z) which lie on or outside the unit circle of the z plane. This method of compensation however is not practical, since any slight change in the plant or controller parameters may result in imperfect cancellation of the poles and zeros of G(z) that lie on or outside the unit circle. Shinners (6) describes an approach that overcomes these problems, that produces a controller that is realizable and that cancels any non-minimum phase zeros in the plant transform G(s). Using the simplified model of the ball and beam as a double integrator, it can be shown (6) that a finite time settling controller for such a system is of the form:

$$\frac{u(z)}{e(z)} = \frac{(1-k_1z^{-1})}{(1+k_2z^{-1})} \frac{k_3}{T^2}$$

(17.24)

where $k_1 = 0.6$, $k_2 = 0.75$ and $k_3 = 2.34$.

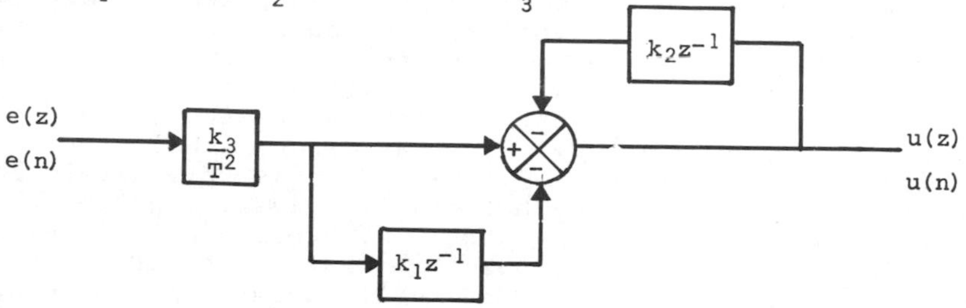

Fig.17.10 Finite-time settling controller

17.5 SUITABILITY OF THE CBM COMPUTER

The two factors that dictate whether the CBM PET computer is suitable for the control application are the sampling interval and the computational time requirements of the algorithms.

In the choice of sampling period T there are two primary considerations, namely that of loop stability and noise rejection. Provided the sampling rate is sufficiently high, the principal effect of a sample-and-hold in the control loop is to introduce a time delay, of $T/2$ seconds (this assumes a zero order hold representation for the sampler). This means that at a frequency of $2/T$, the zero-order hold device introduces a phase-lag of $360°$. In order that the closed-loop performance should not be degraded by the introduction of the sample-and hold element then as a general rule, not more than $10°$ to $20°$ of phase lag should be introduced at the loop natural frequency.

The other consideration is that the sampling period should be as long as possible in order to minimise the effect of noise in the loop. An analysis on the proper choice of sampling interval T has been carried out by Lee et.al (7) who concludes that T should be chosen to lie in the range:

$$1/8f > T > 1/16 f$$

where f is the loop natural frequency in Hertz.

Using this relationship, T for the ball and beam should be in the range $0.8s < T < 0.4s$. In practice however it proved necessary to reduce T to $0.1s$ to achieve satisfactory control.

The next task in the design was to determine whether the necessary I/O operations and computations could be completed in the time available. From the outset it was decided to use high level language programming since this reduces considerably the time required for system development and removed the necessity for developing special mathematic subroutines since these are an integral part of the language. It was therefore decided to use the BASIC language. To assess the maximum possible I/O speed of the computer interface a program was written consisting of a loop containing two instructions in BASIC which ran continually when started. It was found that the I/O operation took approximately 74 ms. This is well within the upper limit suggested by Lee et.al. Available to the authors was a software package 'PET SPEED' which is an optimising Basic compiler for the CBM. It provides a maximum possible execution speed for programs written in PET BASIC, without producing a prohibitively large object program. A compiled version of the I/O timing routine completed a program loop in approximately 20 ms. The control strategies were each implemented separately to produce a series of independent programs which the user could then call up from disc. The programs are structured to allow the user to enter via the keyboard the parameters of the controller, before the start of each run. It was found that provided PET SPEED was used, the control action could be evaluated for each of the algorithms within the 0.1s interval between samples.

17.6 SELECTING CONTROLLER PARAMETERS

In estimating initial values for the parameters of each controller the assumption was made, with the one exception of the direct synthesis method of the finite-time settling controller, that the sampling rate was sufficiently fast compared to the system bandwidth for it to be considered as a continuous system thus enabling the use of classical methods.

For the three term controller two separate approaches were adopted. Firstly the method of Ziegler and Nichols, (8) was used to obtain values for the controller settings. These were based on the estimates of Ku (gain at which the system goes into continuous oscillation) and Tu (period of oscillation) obtained from the system open loop Bode plot. Clearly, this approach was an oversimplification as the plant model used by Ziegler-Nichols to obtain their optimum settings was open-loop stable. Nevertheless it was felt that this approach did provide some basis for initial estimates. The second approach was to use settings based on a specific performance index. The performance index chosen was "integral time absolute error" (ITAE) and using the simplified double integrator model of the ball and beam the optimum coefficients for the system characteristic equation based on ITAE criteria for a ramp input (D'Azzo and Houpis

(9).) were determined. Thus enabling the evaluation of initial estimates for the controller parameter.

The design of the phase-lead compensator was done in the frequency domain with the help of a control systems design package which was available on a VAX 11/780 computer system. The pole and zero positions of the compensator were designed to meet the system specification that the settling time should be less than 6 seconds and that the maximum overshoot was less than 15%.

The parameters for the state feedback controller were obtained by selecting values that minimised the integral-square-error ISE performance index for specific initial condition values of the state vector. The procedure is detailed by Dorf (10) and was readily applied when using the second order state model, but the mathematics within the time scale of the project proved too intractable when using the fourth order state model.

The parameters of the finite settling time controller are determined directly from the direct synthesis design procedure (6).

17.7 SYSTEM RESPONSES

To assess the relative merits of each of the control algorithms, step response tests were performed on the system. This was obtained by placing the ball at the right end of the beam. The demanded position, entered at the keyboard was set at the beam centre. This procedure effectively applied a step input of magnitude 10 volts to the system. The ball position data was sampled and stored for subsequent analysis, with about 400 measurements taken. The responses obtained for the range of controller configurations and parameters are shown in Fig. 17.11. The response using the Ziegler and Nichols setting for the PID controller are completely unsatisfactory, with the first overshoot tending to exceed the beam length, and so it proved necessary to do some hand tuning. The response using the parameter estimates from the ITAE criterion was a considerable improvement from those using the Ziegler and Nichols settings, but still fine tuning was necessary to reduce overshoot and settling time. The phase-advance controller using the designed parameters while exercisig good control failed to meet the desired specification with the settling time rather prohibitive. Fine tuning of the zero and pole positions of the compensator improved the response.

With the use of velocity feedback the system was readily stabilised, and provided the velocity constant was carefully tuned an acceptable response profile is obtained.

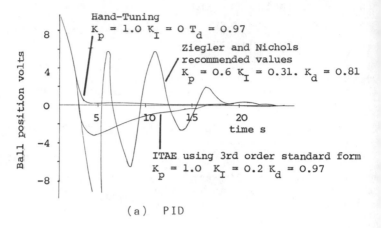

(a)   PID

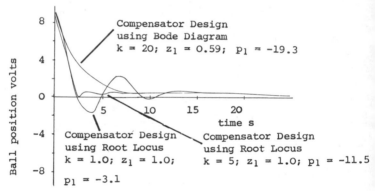

(b)   Phase advance

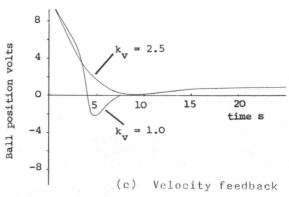

(c)   Velocity feedback

Fig. 17.11   System responses using discrete controller

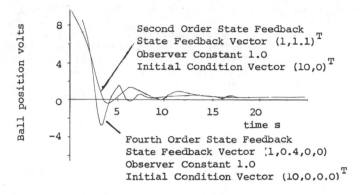

Second Order State Feedback
State Feedback Vector $(1,1.1)^T$
Observer Constant 1.0
Initial Condition Vector $(10,0)^T$

Fourth Order State Feedback
State Feedback Vector $(1,0.4,0,0)$
Observer Constant 1.0
Initial Condition Vector $(10,0,0.0)^T$

(d)   State feedback using observers

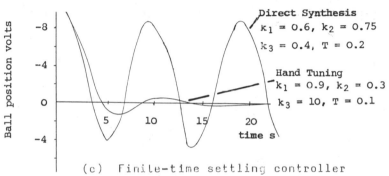

Direct Synthesis
$k_1 = 0.6$, $k_2 = 0.75$
$k_3 = 0.4$, $T = 0.2$

Hand Tuning
$k_1 = 0.9$, $k_2 = 0.3$
$k_3 = 10$, $T = 0.1$

(c)   Finite-time settling controller

Fig.17.11   System responses using discrete controllers

The use of state feedback using the observer model was successful once the problem of computing the observer constants had been overcome. An approach found useful was to display on the PET screen during program execution the generated observer and actual ball position when the ball was fixed at the centre of the beam by trapping it between the guide wires. The observer matrix (L) was chosen to minimise the convergence time between actual and generated position, with a minimum of oscillation.

The finite-settling time response using the coefficients based on a second order model proved unsatisfactory and it was found necessary to change these. In general the performance of this algorithm was found to give a jerky controlling action. It is apparent from the responses obtained that the presence of non-linear friction in the beam meant that if zero offset is to be achieved then the error must be integrated by the controller. This means that of the control algorithms used only the three term controller satisfied this condition. The system using state-variable feedback could be readily modified to incorporate an additional integrator in the forward path, which then would satisfy the zero offset specification.

In general the performance of the controller schemes were limited by the mechanical constraints of the beam, and to a lesser extent by the choice of controller settings. The actuator only rotated the beam by +1.8 degrees and -4.2 degrees. If a controlling action was produced which demanded a greater beam tilt then saturation occurred.

18.8 CONCLUSIONS

This case study has shown how discrete algorithms, to control an electro-mechanical system, may be successfully realised using quite simple equipment. A number of controller implementations have been investigated, and it has been shown that they all were able to stablise an unstable system, provided the controller parameters were carefully chosen.

References

1.   Rees D and White M.J.,1984.
     "The control of an unstable open loop system by means of discrete algorithms using a microcomputer" IERE Conference Proceedings No. 58.

2.   Wellstead, P.E.,et.al.,1978.
     "The Ball and Beam Control Experiment".
     Int.J. Elect.Enging.Educ.Vol 15 pp. 21.-39

3.   Wellstead, P.E.,1979.
     "Introduction to Physical System Modelling". Academic 1979.

4.   Ben-Zw,A.and Preizler,M.,1979.
     "Comparison of Discretization Methods" Rafael, Israel MOD,1979.

5.   Kuo, B.C.,1980
     "Digital Control Systems" HRW.

6.   Shinners, S.M.,1964.
     "Control System Design John Wiley & Sons"

7.   Lee, W.T. et.al,1985.
     "Direct Digital Control".Society of Instrument technology Symposium

8.   Ziegler, J.G. and Nichols, N.B.,1942.
     "Trans.A.S.M.E." Vol.64 pp 759-68.

9. D'Azzo, JJ and Houpis C.H. 1981
    Linear Control Systems-Analysis & Design," McGraw Hill

10. Dorf, R.C.,1980.
    "Modern Control Systems," Addison-Wesley.

Chapter 18

Controller implementations using novel processors
case study III

Dr. D. Rees and P. A. Witting

18.1 INTRODUCTION

The application of conventional 8 and 16 bit microcomputers to control systems is now well established. Such processors have general purpose (usually Von Neumann) architectures which make them applicable to a wide range of tasks, though not remarkably efficient in any. In control applications such devices may pose problems such as inadequate speed, difficulties with numerical manipulation and relatively high cost for the completed system. This latter being due to both the programming effort and the cost of the peripheral hardware. These problems may be overcome by the design of specially tailored architectures, provided that there is a sufficient volume of production to carry the overheads inherent in this approach. Special purpose I/O processors and signal processors are examples of applications where dedicated design has been successfully applied.

It has recently been shown (1,2,10) that such processors can be readily used to implement discrete algorithms for control system applications. The use of digital signal processors (DSP) for controller realisation overcomes some limitations associated with their design using microprocessors, such limitations include having to work in fixed point two's complement arithmetic where the consequent scaling of all the variables and coefficients represents a substantial part of the overall design effort. Programmable DSP use architectures and dedicated arithmetic circuits which provide high resolution and high speed arithmetic, making them ideally suited for use as controllers. Two devices the Intel 2920 and the British Telecom FAD are considered in this case study and their relative merits and demerits for control application outlined. It will be shown that a PID controller with three adjustable parameters, a Smith predictor and a Dahlin controller can be readily mechanised using both devices. Conclusions concerning the likely application areas and levels of complexity possible with each device are considered.

18.2 SYSTEM DESCRIPTION USING THE 2920 IMPLEMENTATION

18.2.1 The 2920 Architecture

In the last five years several programmable monolithic DSP's have become available, and the 2920 is the earliest of these and not the most powerful in terms of speed and size of instruction set. It was decided however to use the 2920 as a controller as this is the only processor currently available which has A/D and D/A convertors on chip. For this particular application the limited speed and size of instruction set proved no real restriction, whilst having a single chip controller with no external circuitry required, apart from a clock and power supplies, was a decided advantage.

The 2920 analog signal processor (3) is a single chip micro-computer designed specifically to process real-time signals. The functional block diagram of the 2920 is shown in Fig. 18.1. The chip constitutes a complete digital sampled-data system. It has on-board program memory (EPROM), scratch pad memory, D/A and A/D circuitry, and a digital processor which includes an arithmetic unit. The architecture and instruction set has been developed to perform precise high speed signal processing using digital sampled - data techniques to implement continuous analogue functions.

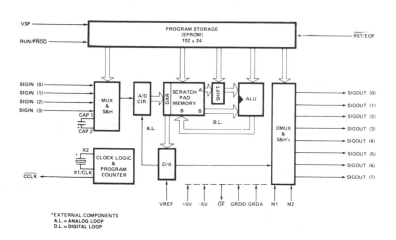

Fig.18.1 Block diagram of Intel 2920.

Under program control, one of four possible inputs is selected and then sampled and held. The signal is then converted to a digital word with 9-bits resolution. An internal digital to analogue register (DAR) accumulates the digital word giving the required sampled value. The word may then be transferred into scratch pad memory to be used in subsequent calculations. The DAR is also used to drive the D/A converter to any of eight analogue outputs via the output demultiplexer. The EPROM is arranged as 192 words of 24 bits each, each word corresponding to an instruction. During the RUN mode the EPROM acts as the system controller and the memory locations are accessed sequentially with no program jumps allowed. The EPROM program counter returns to location '0' upon executing the instruction in word '191' or when an end of program (EOP) instruction is encountered.

Data within the arithmetic logic unit (ALU) is processed using 25-bit two's complement arithmetic and the normal range of any variable X is within the range ± 1.0, with a resolution δ where

$$\delta = 2^{-24} = 5.96 \times 10^{-8}$$

The storage array consists of RAM with two ports (port A and B) organised as 40 words of 25 bits each. The A port is read only with the data read from it being passed through a scaling unit before it is input to the arithmetic logic unit (ALU) as the source operand. The B port passes data to the second ALU input, and receives the ALU results. The constants array consists of 16 'pseudo-locations' in the RAM address field. These constants are accessed only from the A port. Each constant is a multiple of one-eigth and has a magnitude of $\pm$ (n/8) where n is an integer in the range 0 to 8. In practice a wide range of constants is required and this is achieved by passing the selected constant through the scaling unit. The shifts can be a maximum of two positions to the left (multiplier of 2^2) or a maximum of thirteen positions to the right (multiplier of 2^{-13})

18.2.2 Thermal Process

The 2920 was applied to the control of a thermal process. (Fig.18.2) The process consists of a heater container in an aluminim block which is surrounded by a water jacket with the rate of flow of water through the jacket being adjustable. The temperature of the block, which is the controlled variable, is measured by means of a platinum resistance thermometer, and the power input to the system is controlled by means of adjusting the timing of the trigger input to a thyristor.

From open loop step response tests, for a specific water flow rate, the transfer function for the thermal plant is

$$G(s) = \frac{0.49e^{-15s}}{(414s+1)(16s+1)} \tag{18.1}$$

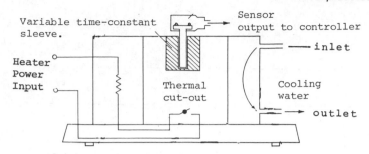

Fig.18.2 Thermal Process.

18.3 CONTROL ALGORITHMS IMPLEMENTED

18.3.1 Three Term Controller

The digital PID control algorithm (Chapter 3) was
implemented in the position form, as given by

$$U(n) = Kp\ e(n) + Ki\sum_{i=o}^{n}e(i) + Kd\left[e(n)-e(n-1)\right]$$

(18.2)

where

e(n)	temperature error at the nth sampling instant
u(n)	actuation signal (d.c. voltage to thyristor triggering circuit) at the nth sampling instant
Kp	proportional gain
$Ki = \dfrac{KpT}{Ti}$	integral gain
$Kd = \dfrac{KpTd}{T}$	derivative gain
T	sampling period
Ti	integral time
Td	derivative time

In implementing eqn.18.2 on the 2920 one of the primary
considerations is the choice of T and its real-isation. This
will be considered in the section dealing with
implementation.

A number of guidelines (4,5) have been established for
the empirical settings of the controller constants Ti, Td
and K; the best known being those due to Ziegler and
Nichols. It has been pointed out by Roberts & Dallard (6)
that in a sampled data system the sampling time, T and the
parameters Ti and Td are related, so that a generic
controller with only a single tuning parameter, Kp, may be
implemented and it has also been shown (6) that there is
only a small penalty for so doing.

The resulting pulse transfer functions are given in table 18.1.

	ABSOLUTE	INCREMENTAL
PID	$K_p' \left[\dfrac{1- 1.429z^{-1}+0.51z^{-2}}{1 - z^{-1}} \right]$	$K_p' \left[1-1.429z^{-1}+0.51z^{-2} \right]$
PI	$K_p'' \left[\dfrac{1-0.893z^{-1}}{1 - z^{-1}} \right]$	$K_p'' \left[1-0.893z^{-1} \right]$

$$Kp' = 2.45Kp \qquad\qquad Kp'' = 1.12Kp$$

Table 18.1 Pulse transforms controller with one tuning

18.3.2.   Smith Predictor

The Smith predictor (chapter 5) is a method which utilises a mathematical model of the process in a minor feedback loop around the conventional controller, to overcome some of the difficulties associated with controlling processes with pure time delays. The process has a transfer function $G(s)e^{-s\tau}$ and a controller with transfer function $D(s)$. By modelling the time delay and the system dynamics separately Smith (9) developed the control scheme shown in Fig.18.3.   It can be seen that the closed loop characteristic equation resulting from using a minor feedback loop does not contain the pure time delay and so, $D(s)$ can be designed simply on the basis of $G(s)$. The design problem for the process with delay has thus been converted to one without delay.

An alternative arrangement for the Smith predictor is shown in Fig.18.3 (b) which is a simple re-arrangement of Fig.18.3 (a) for the purpose of implementations with the FAD.

The process to be controlled using the 2920 comprises of two exponential lags and a pure time delay.   Since a discrete controller is being designed then it is necessary to evaluate the pulse transform of the process and this must take into account the sample and hold operation of the D/A converter.   For this specific case, the pulse transform (Chapter 4 eqn.4.6.) is

$$\frac{C}{U}(z) = \frac{Kz^{-k}(b_1+b_2z^{-1})z^{-1}}{(1-e^{-T/\tau_1}z^{-1})(1-e^{-T/\tau_2}z^{-1})}$$

(18.3

where

$$b_1 = 1 + \frac{\tau_1 e^{-T/\tau_1} - \tau_2 e^{-T/\tau_2}}{\tau_2 - \tau_1}$$

$$b_2 = e^{-T(1/\tau_1+1/\tau_2)} + \frac{\tau_1 e^{-T/\tau_2} - \tau_2 e^{-T/\tau_1}}{\tau_2 - \tau_1}$$

For the thermal process under consideration K = 0.49, τ_1 = 414s, τ_2 = 16s and the delay θ = 15s. This must be an integral number of sample periods, say θ = kT for a sample period of T.

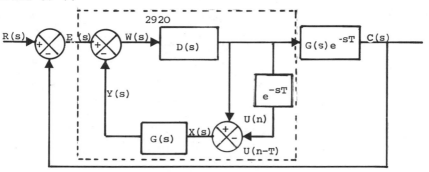

(a) arrangement for 2920 implementation

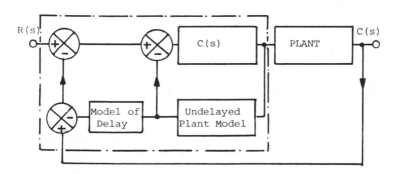

(b) arrangement for FAD implementation

Fig. 18.3 Block Diagram of Smith Predictor.

The block diagram of the Smith Predictor shown in Fig 18.3 (a) indicates the section to be implemented on the 2920. As can be readily seen the control algorithm incorporates three separate equations:

(i)   three term controller

(ii)   process transfer function without pure time delay

(iii) process pure time delay

From Fig.18.3 (a) we have using pulse transform notation,

$$\frac{Y}{X}(z) = K \frac{(b_1 + b_2 z^{-1})}{(1 - e^{-T/\tau_1} z^{-1})(1 - e^{-T/\tau_2} z^{-1})} \tag{18.4}$$

which expressed in difference equation form is:

$$Y(n) = a_1 Y(n-1) + a_2 Y(n-2) + K b_1 X(n) + K b_2 X(n-1) \tag{18.5}$$

where

$$a_1 = e^{-T/\tau_1} + e^{-T/\tau_2} \text{ and } a_2 = e^{-T(1/\tau_1 + 1/\tau_2)}$$

The required pure time delay is 16 seconds (15 seconds due to the process and 1 second due to structure of eqn.18.3). Selecting a sampling period of 2 seconds, means that 8 previous controller outputs $U(1), U(2)$ -- $U(8)$ have to be stored in RAM.

The Smith predictor equations for the thermal process would then be

$$X(n) = U(n) - U(n-8) \tag{18.6}$$

$$Y(n) = a_1 Y(n-1) + a_2 Y(n-2) + b_1 X(n) + b_2 X(n-1) \tag{18.7}$$

$$W(n) = e(n) - Y(n) \tag{18.8}$$

$$U(n) = KpW(n) + Ki\sum_{i=0}^{n} W(i) + Kd\left[W(n) - W(n-1)\right] \tag{18.9}$$

18.3.3 Dahlin Controller

It has been seen from Chapter 4, that Dahlin's method enables the designer to design directly in the discrete domain, and is based on defining a first order closed loop response, as given by

$$K(s) = \frac{\lambda e^{-\theta s}}{s + \lambda} \tag{18.10}$$

or in the z-domain, with the zero-order hold included,

$$k(z) = \frac{\left[1-e^{-\lambda T}\right]z^{-k-1}}{1-e^{-\lambda T}z^{-1}}$$

(18.11)

The reciprocal of the time constant (λ) is used as a tuning parameter with larger values giving increasingly tight control.

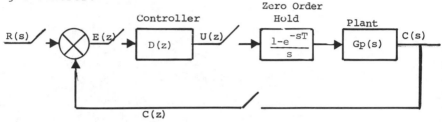

Fig.18.4 Block diagram of sampled-data system for Dahlin design.

From Fig.18.4 the direct synthesis relationship is given by

$$D(z) = \frac{1}{G(z)} \frac{K(z)}{1-K(z)}$$

(18.12)

which gives for the Dahlin controller, the pulse transform

$$D(z) = \frac{1}{G(z)\left[1-e^{-\lambda T}z^{-1}-(1-e^{-\lambda T})z^{-k-1}\right]} \left[1-e^{-\lambda T}\right]z^{-k-1}$$

(18.13)

Using a first-order lag plus time delay model for the thermal process (neglecting the shorter time constant), the pulse transform of the process is

$$G(z) = K \frac{(1-e^{-T/\tau})z^{-k-1}}{(1-e^{-T/\tau}z^{-1})}$$

(18.14)

and

$$D(z) = \frac{(1-e^{-T/\tau}z^{-1})(1-e^{-\lambda T})}{K(1-e^{-T/\tau})(1-e^{-\lambda T}z^{-1}-(1-e^{-\lambda T})z^{-k-1}}$$

(18.15)

Selecting a sampling period of 8 seconds and setting $\lambda = 2.5/\tau$ the controller difference equation becomes

$$U(n) = 0.952U(n-1)+0.0477U(n-3)+2.49(E(n)-0.98E(n-1)) \quad (18.16)$$

The constants of the difference equation can be readily represented to a high degree of precision using the constant array of the 2920, for example 0.952 can be represented as $2^0 - 2^{-5} - 2^{-6} - 2^{-10}$ which represents using the arithmetic barrel shifter codes the operation R00 - R05 - R06 - R10 on the U(n-1) data.

18.4 IMPLEMENTATION ON THE 2920

18.4.1 Sampling Interval

An essential consideration in any discrete controller implementation is the choice of sampling interval. This is necessary to ensure realistic ranges for the controller parameters Ki and Kd, and also in the case of the Smith predictor and Dahlin algorithm to provide time delays which match that of the process; as well as ensure a stable numerical solution to the difference equations.

A program consists of a series of 2920 instructions which are executed sequentially at a rate fixed by the clock frequency. The processor allows no internal jumps and the program is of fixed length and execution time, and therefore it is not possible to implement a delay using a subroutine. The sample interval is determined by the time taken for one pass through the program which is given by

$$T = 4 \times \frac{\text{Number of instructions}}{\text{Clock frequency}} = \frac{4N}{f} \qquad (18.17)$$

The factor '4' is due to the program counter which increments its count value every four clock cycles and the maximum number of instructions is 192. The system design kit used had a clock frequency of 5MHz, which gives, for one program pass, a maximum period of 0.1536 milliseconds. This can be extended using sub-multiple sampling, which is a software technique which makes use of the conditional load operation. By this method the calculation of the controller equations (for example eqn.18.16 in the case of the Dahlin controller) is made conditional on the sign of a constant, say SUB in the DAR. SUB is decremented by a small constant, say DEL, during each pass of the program and, when it becomes negative, the controller equations are evaluated. Using this procedure the effective sampling period is given by

$$T = 4(n+1) \frac{N}{f} \qquad (18.18)$$

where n = SUB/DEL. For a sampling interval of 8 seconds, as required by the Dahlin controller, n is 52082.

One consequence of sub-multiple sampling is that the program length has to be increased. This mainly arises from the fact that instructions which are conditionally executed cannot have I/O operations simultaneously occurring because both operations use the "analogue" field of the instructions.

18.4.2. Controller Parameters

With the three term controller and Smith predictor implementation all the analog inputs on the 2920 are utilized, with provision for the user to set the proportional gain (Kp) integral constant (Ki) and the derivative constant (Kd) by means of potentiometers whose wipers are connected to SIGIN(1), SIGIN(2) and SIGIN(3) inputs of the 2920. To ensure adequate ranges for the parameters each input was multiplied by a constant as follows:

$$Kp = SIGIN(1) \times K_1$$

$$Ki = \frac{TxKp}{Ti} = SIGIN(2)xK_2 \qquad (18.19)$$

$$Kd = \frac{TdxKp}{T} : SIGIN(3)xK_3$$

Giving, for the integral and derivative action times, the expressions:

$$Ti = \frac{T \times SIGIN(1) \times K_1}{SIGIN(2) \times K_2} \qquad (18.20)$$

$$Td = \frac{T \times SIGIN(3) \times K_3}{SIGIN(1) \times K_1}$$

By making a suitable choice of values for K_1, K_2 and K_3 it is possible to have an adequate range of values for the controller parameters. For example with T = 2s, K_1 = 16. K_2 = 0.25 and K_3 = 16, then it is possible to set Ti in the range 1 to 200s and Td in the range of 2 to 50s. The Ziegler and Nichols settings for the thermal process are Kp = 10, Ti = 63s and Td = 16s, which can be readily achieved using the parameter ranges provided.

18.4.3 Programming

The philosophy adopted in implementing the controller was to consider the 2920 in terms of the functions it can implement, and then to break the controller equation into individual building blocks which embody these functions. For example, for the Smith predictor these functions are shown in Table 18.2, which also gives the number of instructions, and RAM locations required to implement them. The functions to be implemented can be readily identified by considering eqn.18.6 to eqn.18.9. In terms of the total instructions used for this implementation 56% was required for I/O activities, 21% for evaluating the multiplications and sensitivites of the controller parameters, and the remaining 23% for implementing the controller equations, including the process quadratic function, pure time delay and submultiple sampling.

Function	Number of Instructions	No.of RAM Loc.	Comment
Submultiple Sampling	4	1	Sub multiple sampling = SUB/DEL Ratio
Pure Time Delay	8	8	Store 8 previous values u(n)-u(n-8)
A/D Conversion for E,Kp,Ki,Kd	100	4	Read error e(n) and controller constants SIGIN(1),SIGIN(2) & SIGIN (3)
Minor Loop Feedback Input	2	1	Evaluate eqn.18.6
Process Quadratic Function	24	2	Evaluate eqn 18.7
Controller Input	2	2	Evaluate eqn.18.8
Controller Sensitivity Coefficients	3	-	Implement K_1 K_2 and K_3 by shift operations on W(n) ΣW(i) and W(n)-W(n-1) respectively
4 Quadrant Multiplications for P + I + D	36	4	Evaluate eqn.18.9
Controller Output	3	1	
D/A Conversion for controller output	7		Output u(n)

Table 18.2 Smith predictor function implementations
on the 2920.

Of the three control algorithms programmed the Dahlin controller was the most efficient, requiring only 63 instructions for its implementation. The PID and Smith predictor implementation used far more coding, and required considerable programming ingenuity to fit into the 192 EPROM memory.

Since the normal range of any variable Q in the 2920 is $-1.0 < Q < 1.0$, it proved necessary to minimise the possibility of airthmetic overflow when evaluating terms in the difference and controller equations. For example when calculating the contribution of the proportional term to the controller output, it proved necessary to first evaluate the produce KW(n) before multiplying by the proportional controller constant SIGIN(1)

18.5 APPLICATION OF THE 2920 TO THERMAL PROCESS

To demonstrate the performance of the 2920 as a controller the results for the three controller imlementations are shown in Fig.18.5 to Fig.18.7 respectively. The process setpoint was at 40°C and the parameters for the PID and Smith Predictor Controller were set at the Ziegler and Nichols settings which were obtained from the open loop step response test. Figure 18.5 gives the step response of the closed loop system under PI control - it proved unnecessary to use the derivative term. Fig. 18.6 gives the corresponding response when using the Smith Predictor and it can be readily seen that for the same controller settings the response, as one would expect, has less overshoot and an improved settling time. Figure 18.7 gives the step response of the system using the Dahlin controller, which was designed for $\lambda = \tau/2.5$response. Of the three controllers considered the Dahlin controller was the most satisfactory in terms of response profile and ease of coding. It should be observed that since $(1-z^{-1})$is a factor of the denominator polynomial for the Dahlin algorithm then the controller has integral action which ensures that there is zero steady-state error between the plant output and the desired setpoint.

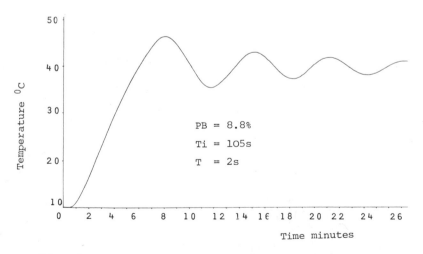

Fig. 18.5 System response using PID controller. (Ziegler & Nichols settings)

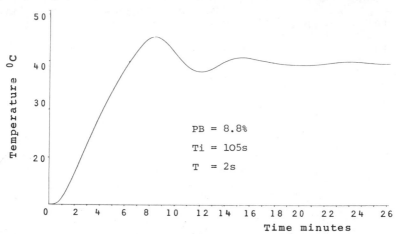

Fig. 18.6   System response using Smith predictor.
(Ziegler and Nicholas settings)

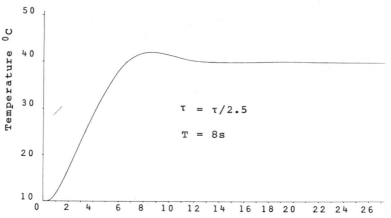

Fig. 18.7   System response using Dahlin controller.

18.6   THE   ARCHITECTURE   OF   THE   FAD

The FAD (7,8) is designed to implement the standard second order digital algorithm:-

$$K \frac{1 + AZ^{-1} + BZ^{-2}}{1 - CZ^{-1} - DZ^{-2}} Z^{-1} \qquad (18.21)$$

The processing occupies a whole computation cycle thus giving rise to the z^{-1} term.

A Canonic structure is used to implement the processor as   shown   in   Fig.18.8   which   represents   the   logical

architecture rather than the detail of the electronics. The
coefficients are held with a maximum error of 1.22×10^{-6} and
are arranged so that poles and zeros may be placed anywhere
within the unit circle.

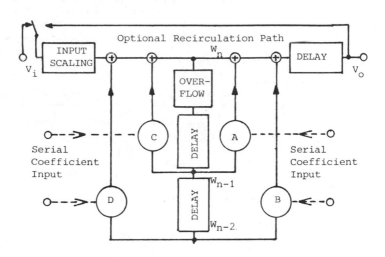

Fig. 18.8 Architecture of FAD circuit.

The internal structure of the FAD permits its
output to be recycled to the input so that cascades of
second order sections may be implemented. Because the
output may be sampled at the same time that it is being fed
back it is possible to implement a 'Smith Predictor' scheme
in a very simple manner.(Fig. 18.10)

The coefficients for the algorithm are supplied in
serial format, as is the data, each time a section is
implemented. Although it is usual for the coefficients to
remain constant between each instance of a section it is
easy to arrange for these to vary. Because of the serial
nature of the coefficient data these changes may be
implemented by low cost serial arithmetic hardware to
produce a very simple adaptive controller.

The clock range for the FAD is in the range 500 KHz to
2 MHz which means that each section may be computed in 64 µs
to 16 µs. Where multiple biquadratic sections are used the
intermediate data may be held on chip (for up to 8 sections)
or off-chip for more than 8 sections. Because of the serial
nature of the data the storage required is merely a shift
register of length 32 (N-1) bits where N is the number of
stages.

For control engineering applications the FAD arithmetic capability and overflow protection are particularly attractive. Multiplication and addition are carried out to full precision before the results are truncated to 16 bits.

Thus the multiply-add errors do not build up in the same way as for a typical fixed-point arithmetic implementation in a general purpose computer. Overload is detected by the use of additional range for the internal representation of the feedback variable, W. The details of this feature and its implications for controller design are discussed later.

18.7. APPLICATION OF THE FAD TO CONTROLLER IMPLEMENTATION

It will be observed that the PID and PI controllers can be readily implemented by a single section of the FAD. Many other simple controllers may also be implemented in this fashion merely by adjusting the coefficients. Because the bilinear sections may be cascaded, essentially without limit, it is also possible for arbitrarily complex controllers to be implemented. No additional hardware is required over and above that needed for a single section except for coefficient storage (32 x 2 bits per section). The maximum achievable sampling rate is reduced in proportion to the number of bilinear sections implemented.

Adaptive and self-tuning controllers offer significant benefits over fixed-parameter controllers especially in applications such as batch reactors where the process dynamics are often subject to considerable variation during operation. Because the coefficients have to be entered serially at each computation time the FAD offers a paticularly simple way of implementing such controllers as illustrated in Fig.18.9.

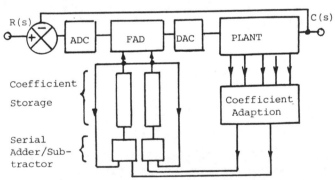

Fig.18.9 Use of FAD in adaptive control

It has been pointed out above that the FAD implements a cascade of bilinear sections by recyling the current output back to the input, the internal timing being designed with this in mind. Although the data stream part way along a cascade is normally of no interest, it is made available at the output of the device and may be used if required. This opens the way to a rather simple implementation of the 'Smith Predictor' with one group of bilinear sections realising the controller while a second group realises the plant model. The output of the last controller section is 'captured' and applied to the plant as well as being recylced to the first section of the plant model. The transport delay required in the controller can be realised by proving a 32M-bit shift register externally (when the delay realised is M I. When T is the calculation period for one section i.e. 32 clock periods (Fig. 18.10)

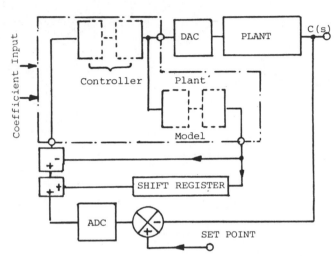

Fig. 18.10 Use of FAD as a Smith Predictor

18.8 OVERLOAD BEHAVIOUR OF FAD-BASED CONTROLLERS

There are three mechanisms whereby the FAD may go into overload and produce anomalous results.

18.8.1 Input overload

This mechanism is more closely associated with the input ADC than the FAD itself. It is common to virtually all processors and occurs when the analogue input exceeds the maximum capability of the device, in such circumstances the limiting digital code is produced thus giving a hard saturation characteristic.

18.8.2 Output overload

This occurs when the computer output exceeds the range which can be represented. On normal processors this is handled by software which again implements hard limiting. The FAD incorporates no such overload correction and the output generated corresponds to the least significant 16 bits of the data. That is the output "rolls-round" from (say) full positive to full negative. In control applications this non linearity is potentially disastrous and must be avoided (except transitorily). It is possible to place a bound on the coefficient values so that output overload does not occur.

$$V_o = W_n + A * W_{n-1} + B * W_{n-2} \qquad (18.22)$$

In a steady state $W_n = W_{n-1} = W_{n-2}$ and a worst case condition will occur when W is maximised i.e. W=1 (see below.) Hence:

$$|V_o| = |1 + A + B| \qquad (18.23)$$

and for no output overload

$$|1 + A + B| \leq 1 \qquad (18.24)$$

This condition is satisfied for both the PI and PID controllers given above.

However, there are conditions in which controllers satisfying eqn.18.24 may exhibit output overload. This is because for any controller with integral action a 'steady state' occurs with constant input giving a constant rate-of-change-of-W.

Assume W changes by Δ at each sample we have

$$W_n = W$$

$$W_{n-1} = W-\Delta \qquad (18.25)$$

$$W_{n-2} = W-2\Delta$$

and

$$V_o = W(1+B+B) - \Delta (A+2B) \qquad (18.26)$$

clearly $|W|$ is constrained to have a maximum value of unity and $|Vo| < 1$ to avoid output overload. Thus a bound on Δ is given by

$$\Delta \leq \frac{1 - 1 + A + B}{A + 2B} \qquad (18.27)$$

This bound breaks down if internal overload (see below) occurs since in such circumstances there is a discontinuous change in W. Because internal overloads are dealt with automatically by the FAD the anomaly is only possible for one sample period.

In order to estimate whether output overload is likely to occur one needs to estimate Δ. Clearly

$$W_n = K * V_i + C * W_{n-1} + D * W_{n-2} \qquad (18.28)$$

using 18.28 with 18.25 gives

$$\Delta = V_i \ \frac{K}{C+2D} \ + \ W_n \ \frac{C+D-1}{C+2D}$$
(18.29)

where K is the input scaler.

Clearly some degree of control over is achievable via the input scaler and, for the controllers in table 18.1 (C = 1 D = 0), total control is possible.

18.8.3 Internal Overload

The FAD contains special circuitry to deal with an overload on the internal variable W. This can be in the range $\pm$ 4 and the result is held temporarily to an exact value in this range. If the value computed exceeds the $\pm$1 range which is used for all other internal storage a reset action is initiated which takes effect on the next set of calculations. This reset involves a zeroing of all of the internal storages, W, thus avoiding the potential for instability which exists if grossly wrong values are passed into the feedback loops via coefficients C and D. So far as the current calculation is concerned the value of W is passed to the output calculation with the most significant two bits missing, thus giving the gross nonlinearity in the output calculation referred to above.(i.e a value of W = 1.5 is passed on as an approximate value of -0.5). This error only persists for a sample period.

18.8 4 Performance of typical controllers under overload

Fig.18.11 shows the response of two FAD-implemented absolute controllers (PID and PI - Table 18.1) with constant error signal inputs.

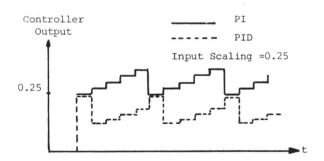

Fig. 18.11 Response of FAD-based absolute controllers

It will be seen that the output is cyclic each cycle including a proportional element, a derivative "spike" (for the PID only) and an integral ramp which continues until internal overload occurs when the whole cycle is repeated. With the usual choice of sampling rate this cycling will occur at a relatively high frequency and will be smoothed by the plant. Thus the plant sees an effectively constant output.

Fig.18.12 shows how the PID controller response changes with the size of the input error. Two characteristics are evident. Firstly cycle frequency increases with increasing error magnitude. Secondly the mean value of the output, to which most plants will respond, rises approximately in proportion to the input error. Thus the FAD internal overload mechanism automaticaly causes the controller to behave like a simple proportional controller under circumstances where there is an error present for a long period (Fig.18.13). The output signal drives the plant towards its designed target and once the set point is approached and the error begins to fall away PI or PID behaviour will re-assert itself, as appropriate, to eliminate the error. Thus 'integral wind-up' is avoided.

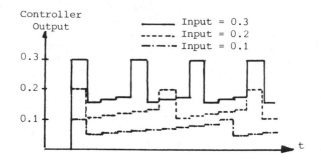

Fig.18.12 Response of FAD-based PID controllers for various input error sizes.

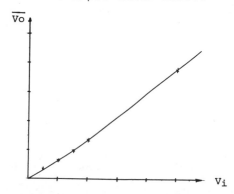

Fig.18.13 Relationship between input and average controller output in overload.

Fig.18.14 shows how the controller behaves when the error changes sign. Although the controller output is likely to be averaged by a typical plant it is important to ensure that no control elements are capable of responding to the sampling frequencies as this would cause unacceptable "chatter" and wear.

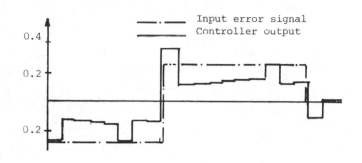

Fig. 18.14 Response of FAD-based PID controller to change of sign of error.

The synchronisation of the FAD requires particularly careful attention when implementing 'Smith Predictor' controllers and similar model-reference schemes. Under normal circumstances the FAD does not commence overload reset until the next synchronisation pulse, this is normally in the first clock period of the first bilinear section of the controller. Resetting continues until the next synchronisation pulse. This is perfectly satisfactory for simple controllers but, in the case of 'Smith Predictor' systems it results in overload resetting of both the controller and the plant model. Thus the model is reset and its state no longer represents that of the real plant. This can be avoided by injecting an additional synchronisation pulse into the first clock period of the model simulation if, and only if, the immediately preceding controller sections were reset. The FAD IC has no output on which 'Overload Reset' is signalled and so this must be arranged externally. This could be triggered by detecting the value of the internally stored variables W_{n-1} and W_{n-2} as they appear in serial form on one of the pins of the FAD. (Fig. 18.15). If all the bits of all the variables in the controller are zero it is highly likely that overload reset has occurred and so the additional synchronisation pulse may be passed to the FAD. On the rare occasions when the variables are zero due to 'natural' causes this additional synchronisation pulse will cause no harm. The only circumstances where problems could arise are

(a)   an overload which leaves W_n in overload but with its 16 least significant bits set to zero.

These will emerge and cause a synchronisation pulse at the start of the start of the model thus resetting the mode <u>not</u> the controller.

(b)   A noise 'glitch' upsetting the zero detector thus causing it to wrongly suppress the additional reset pulse and so permitting both the controller and the model to go into 'overload-reset'.

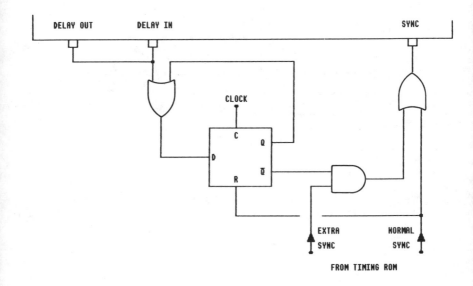

Fig.18.15   Suggested Reset Circuitry for Smith Predictor System.

18.9   IMPLEMENTATION OF FAD-BASED CONTROLLERS

In order to operate as a controller the FAD requires certain support circuitry to provide signal conversion, timing etc. A typical system is shown in Fig.18.16.

The analogue input signal is captured by an analogue-digital convertor (ADC) and transferred in parallel to a Parallel-in-serial-out (PISO) shift register for entry to the FAD in serial format as required.

Data is output from the FAD in serial form and is captured by the serial-in parallel-out (SIPO) register before being transferred to the digital-analogue convertor (DAC).

The controller coefficients have also to be entered in serial form and this is achieved by storing A & B (Fig. 18.8) in one bit position of thirty two successive addresses of the EPROM. The C & D coefficients (Fig.18.8) are also stored in a similar manner. The EPROM is also used to generate all of the timing waveforms required by the circuitry. To reduce the work involved in programming the EPROM, and to eliminate errors, an intractive computer programme has been written. This requests the coefficient values and assembles the timing and coefficient bits required at each memory location.

The complete FAD-based controller is shown in Fig.18.17.

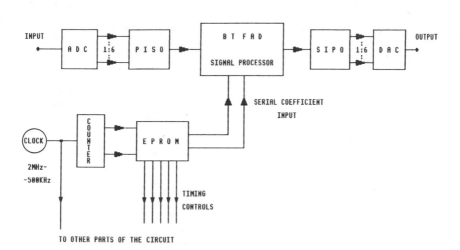

Fig. 18.16 Block diagram of FAD based controller hardware.

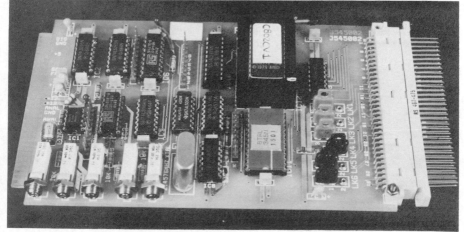

Fig. 18.17 Prototype of FAD-based controller hardware.

The FAD makes use of dynamic storage cells internally. These impose a minimum clock frequency constraint of 500 KHz which gives a maximum sampling period of 64Nµs where N is the number of sections required to implement the controller. This sampling rate of 15.625/N KHz is much too high for virtually any control scheme and the work to date has centred on finding a mechanism whereby the sampling rate may be made arbitrarily low. A number of possibilities exist.

Firstly, the value of N may be increased. the FAD itself includes the provision for two cases N=1 and N=8 (15.625 KHz and 1.953 KHz sampling), by the use of a pair of on-chip shift registers. In addition facilities are provided for the user to provide external memory. For low sampling rates however the length of the shift register becomes somewhat cumbersome (500 Kbit for 1Hz) and the method is inflexible.

Instead, use is made of a 32 bit static shift register along with some additional timing circuity. This captures the internal data of the FAD and permits the 'state' of the circuit to be preserved virtually indefinitely. This is satisfactory for the realisation of single PI and PID controller stages. For more complex arrangements such as the Smith Predictor more stages are required in the algorithm and a larger external shift register must be provided. For an algorithm requiring M 2nd order stages a 32M-bit state shift register will be required.

The closed loop response of a plant having the transfer function given below

$$\frac{e^{-sT}}{(1+sT_1)(1+sT_2)}$$

and controlled by a FAD-realised PID controller is shown in Fig. 18.18.

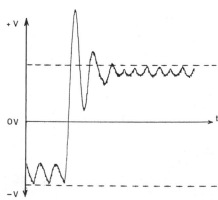

Fig.18.18 Process response under FAD PID controller

18.10 CONCLUSIONS

In this chapter the application of two quite different processor architectures to the design of digital controllers has been discussed. The Intel 2920 can be used to implement PID, Smith Predictor and Dahlin type algorithms. However, the amount of memory available does place a limit on the complexity of the algorithms that can be programmed.

The FAD-based controller requires rather more external logic but is capable of dealing with plants and controllers of essentially unlimited complexity. The external logic is needed both to provide data conversion facilities and to overcome the problems of slowing the circuit down to sampling speeds appropriate for control systems.

The strengths and weaknesses of the two processors, in the context of controller design are summarised in table 18.3.

CRITERION	Intel 2920	BT FAD
Hardware	Single Chip & Clock	FAD & 11 chips (MSI)
Cost	£105 (one off)	£60 (one off)
Sample Rate (PID 2nd Order Plant Model)	130 Hz - 13 KHZ (basic) No lower limit with submultiple sampling	4 KHz - 16 Khz (basic) 1 Hz with external shift No lower limit when START/STOP mode used.
Complexity Limit	PID/Smith Predictor with 2nd order model	Essentially unlimited.
Coefficient Resolution	6×10^{-8}	10^{-4}
Overload Protection	NONE	Reset stored values to zero.

Table 18.3 A comparison of the Intel 2920 and British Telecom FAD Processors for controller realisation.

One particular advantage of FAD-based controllers is the ease with which adaptive control can be implemented. The inbuilt overload protection mechanism also gives controllers built with this device interesting properties when large errors are present.

ACKNOWLEDGEMENTS

The authors wishes to thank the staff of the British Telecom Research Laboratories at Martelsham Heath, Suffolk, U.K. for their help during this project. Also Mr. D. Hajiaghaee, Mr. S.R. Smith and Mr.H. Jenkins on the FAD-based controllers who carried out much of the practical work.

Thanks are also due to Mr. R. Dixon who carried out much of the work on the Intel 2920 controllers.

REFERENCES

1.   Rees, D and Witting P.A."Use of novel processor achitectures for controller realisation". 1984. Proceedings of EUROCON 84 conference on computers in Communication and Control.

2.   Rees D. "Controller implementation using a monolilthic signal processor", 1985, International Journal of Microcomputer Applications, Volume 4, No.3.

3.   Intel"2020 Analog Signal Processor Design Handbook" 1980, Intel Corporation.

4.   Ziegler J.G. & Nichols, N.B. "Optimum Settings for Automatic Controllers"Taylor Instrument Company Bulletin TDS-10A100.

5.   Cohen & Coon "Theoretical Consideration of Retarded Control" Taylor Instrument Company Bulletin TDS-10A102.

6.   Roberts & Dallard, "Discrete PID Controller with a single tuning parameter",1984, Measurement Control Vo. 7,Dec, 1974, T97-T101.

7.   Adams, P.F.,Harbridge, J.R., Macmillan,R.H. "An MOS Integrated Circuit for Digital Filtering and level Detection",1981,IEE Journal of solid state circuits, Vol. SC16 No.3.

8.   Adams, P.F., Macmillan, R.H. "An LSI Circuit for Digial Filtering and Level Detection (367) Functional description and user guide",1979, British Telecom Research Labaoratories.

9.   Smith O.J.M.,'A controller to overcome dead time',1959 ISA Journal Vol.6,No.2,28-33.

10. Witting P.A. "Controller Realisation via VLSI Signal
     Processors", 1986, ISMM Journal of Microcomputer
     Applications, Vol.5,No.1.

Chapter 19

Optimal control of a thermal plant
case study IV

R.A. Wilson

R.A. Wilson

19.1   Introduction

This study concerns optimal control of a thermal
plant; the plant is slow and thus digital control using a
high level language and sophisticated control techniques was
possible because of the low sampling rate. It is also shown
that "classical" control knowledge can be applied to the
design of optimal controls.

19.2   The Plant

The plant is a tank of oil, gently stirred, with an
immersed electrical heating coil. Cold water cooling flow
is also provided but uncontrolled; Figure 19.1 is a
schematic of the plant.

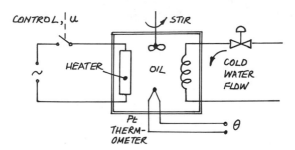

Figure 19.1 The plant

Experiments were conducted to establish the major
features of the plant open-loop response - the dominant
effects were an exponential lag time of about 12 minutes,
and a delay of between 10 and 30 seconds over the 20-50°C
working range. The thermal cpaacity doubled over the same
range whilst the thermal resistance was less variable but
difficult to quantify exactly.

19.3   Specification of Performance

In closed loop we might expect to reduce the lag time to 3 or 4 minutes and on this basis a 40 second sampling time was chosen. This is long enough to create a truly discrete control problem rather than a discretised continuous one, and also to permit the use of a high level language in performing the control calculations.

These and other considerations led to the selection of following controlled system specification:

. Settling time to $\pm$ 1°C of 720 seconds, for a step demand of 20°C to 50°C

. Steady state error less than $\pm$ 1°C at 50°C.

. Disturbance rejection: less than 1°C change for a cooling flow change of .2 1/min.

. Overshoot to be severely limited, since cooling flow is uncontrolled and excess temperature can only cool naturally.

With a proportional-only controller, it was readily shown that at 50°C, stability and speed requirements led to a controller gain than results in a steady state error of 1.25°C. Thus an integral term is necessary and all the controllers investigated were assumed to include a pole at z = 1.

19.4   Modelling the Plant

The plant is $G(s) = \dfrac{K\,e^{-sT_d}}{1 + sT}$ which, with a zero-order hold and sampling period of 40 seconds, transforms to:

$$G(z) = \frac{kz^{-1}\,(z + b)}{(z - 0.97)}$$

where b accounts for the mismatch between the pure delay T and the sampling period T. Including the integration and estimating k and b from output measurements this becomes:

$$G(z) = \frac{0.00028(z + 2.5)}{(z - 1)(z - 0.97)} \,°C/W$$

and in state space form:

$$\underline{x}_{n+1} = F\,\underline{x}_n + G\,u_n, \quad \theta_n = H\,\underline{x}_n$$

where $\underline{x}_n = \begin{bmatrix} x1_n \\ x2_n \end{bmatrix}$, $F = \begin{bmatrix} 0 & 1 \\ -.97 & 1.97 \end{bmatrix}$, $G = \begin{bmatrix} 0 \\ 1 \end{bmatrix}$

and $H = \begin{bmatrix} 0.70 & 0.28 \end{bmatrix} \times 10^{-3}$.

Here u_n is the heater power and θ_n the oil temperature at time $t_n = nT$.

19.5   Control Practicalities

Whatever controller is chosen, the 5kW heater can only
be turned on or off. With the system losses being 700W at
50°C, then strictly u_n can only be 0 or 4300W. We are using
linear analysis and thus a device is required to allow u_n to
be any value in $0 \leqslant u_n \leqslant 4300W$.

We chose a time division mode of control, where the
sampling period was split into 4 sub-periods each of 10
seconds. During these, the heater was turned on for
$10u_n/4300$ seconds and off for the remainder. Overall, the
plant is so slow that the accumulated energy input was then
equivalent to a single pulse of u_n watts over the whole
sampling periods, as the theory assumes.

Further, step demands on such a plant are unrealistic,
so set point changes were ramped, at 2.5°C per minute. This
also avoided 'integral wind-up'. Lastly, any demands for a
u_n outside $0 \leqslant u_n \leqslant 4300W$ were prevented by replacing them

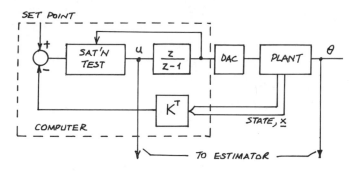

Figure 19.2 The controlled system

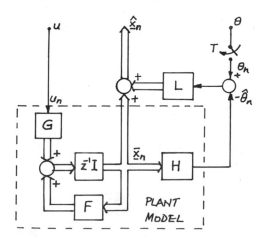

Figure 19.3 The Estimator is a plant model,
driven by the same control and
corrected by the output error.

with the extreme values, using a 'saturation test'.

19.6 Optimal Control

Optimal control theory demands state feedback control, rather than just output feedback as classically used. In fact, any state feedback, if it results in a stable closed loop, is optimal, though quite what performance index is so optimised is not readily determined! Thus the controller is:

$$u_n = -\underline{K}^T \underline{x}_n = -k_1 x1_n - k_2 x2_n$$

and the choice of $\underline{K}$ is central to the design. This control system is shown in Figure 19.2; note the integration, the 'saturation test' and most importantly, that the states $x1$, $x2$ are not readily available.

19.6.1 State estimation.
The plant states $x1_n$ and $x2_n$ are somewhat artificial variables and not measurable by real instruments. We used an estimator to estimate them as $\hat{x1}_n$ and $\hat{x2}_n$ and then relied on the theories that guarantee that using

$$u_n = -K^T \hat{\underline{x}}_n \text{ rather than } u_n = -K^T x_n$$

will not matter, (the Separation Theorem) (1).

Such an estimator is a model of the system, driven by the control u_n and by the difference between θ_n and $\hat{\theta}_n$, the true estimated outputs.

The 'current estimator' is shown in Figure 19.3 and has predictor and estimator eqns:

prediction: $\bar{\underline{x}}_{n+1} = F \hat{\underline{x}}_n + G u_n$, a system model

and: $\hat{\underline{x}}_n = \bar{\underline{x}}_n + L(\theta_n - H \bar{\underline{x}}_n)$, the estimation.

The estimator system matrix L can be chosen to place its poles anywhere within the unit circle, $|z| = 1$, and then Ackermann's formula (2) used to calculate L. Here, placing the estimator poles at the origin will give a 'deadbeat' response, and the formula yields:

$$L = \begin{bmatrix} 745 & 1789 \end{bmatrix}^T$$

19.6.2 The Kalman filter.
Such estimators are in fact designed to operate in a noise free situation; if the system noise is significant, then the optimal estimator is the Kalman Filter, which minimises the variance of the error, $\theta_n - \hat{\theta}_n$. Such a steady state (Weiner) filter has the same structure as the above estimator, but requires that L be reduced somewhat, to:

$$L = \begin{bmatrix} 673 & 1473 \end{bmatrix}^T.$$

19.6.3 Determination of Optimal Feedback Vector.

The feedback vector $\underline{K}$ is now to be determined; however the optimal control literature does not admit of a $\underline{K}$ that will directly satisfy the specification of section 19.3. We posed the LQP problem with

$$J - \sum_{n=0}^{N} (\underline{x}_n^T Q \underline{x}_n + u_n^T R u_n) \quad , \quad N \rightarrow \infty$$

where Q and R are appropriately sized matrices.

This cost function reflects our concern that the state x_n should be driven to the origin (equivalent to 50°C) and that any large x_n are therefore penalised and secondly, that the control magnitude should remain within bounds. R = 0 is therefore not permitted, whilst Q = 0 would be if we are not concerned about large x_n.

19.6.4 Weighting matrices.

The choice of these weighting matrices Q and R now faces us; since u_n is scalar, R can be set to 1, which just scales J. Then, noting that $x1_{n+1} = x2_n$ (the system is in 'phase variable' form) we need only weight against $x2_n$ and so

$$Q = \begin{bmatrix} 0 & 0 \\ 0 & q \end{bmatrix} \text{ is all that is necessary}$$

19.7 Problem Solution

The optimal control problem has thus been reduced to:

What q gives the $\underline{K}^T$ that satisfies the original (classical) specification of section 19.3?

We saw in Chapter 8 that the feedback vector required is given by:

$$\underline{K}^T = [R + G^T PG]^{-1} G^T PF$$

Here $P = \begin{bmatrix} p_{11} & p_{12} \\ p_{21} & p_{22} \end{bmatrix}$ and is the solution of the MRE:

$$P = Q + F^T PF - F^T PG [R + G^T PG]^{-1} G^T PF.$$

Performing these calculations algebraically, we found that:

$$\underline{K}^T = \frac{1}{1 + p_{22}} \left[p_{22} f_{22} \mid \frac{p_{22} f_{21} f_{22}}{1 + p_{22} - f_{21}} + p_{22} f_{22} \right]$$

We avoided the choice of q now, and worked directly with p_{22}. Approximating the system matrix F as $\begin{bmatrix} 0 & 1 \\ -1 & 2 \end{bmatrix}$ (the numerical accuracy of the model is unrealistic) then:

$$K^T = \begin{bmatrix} \frac{-p}{1+p} & \frac{2p}{2+p} \end{bmatrix} \quad (p = p_{22}),$$

the closed loop characteristic equation is:

$$\det \left[zI - \left[F - G K^T \right] \right] = 0$$

and this has roots:

$$z = \frac{-2}{2+p} \pm j\frac{p}{(2+p)(1+p)}^{\frac{1}{2}} \quad (p \geqslant 0).$$

These roots are plotted in the root locus of Figure 19.4 and allow the selection, via p_{22}, of suitably damped closed loop pole positions so that the specification is satisfied.

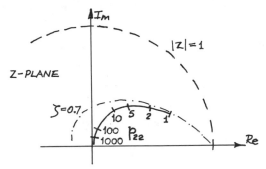

Figure 19.4 The closed loop root locus: the parameter is effectively the weight against the state x_2.

19.8   Simulation and Results

Table 19.1 gives $\underline{K}$ and the closed loop poles for some p and Figure 19.5 the simulated system response for these p. The choice $p = p_{22} = 5.0$ is seen to be satisfactory; little is gained by increasing p yet the response offers a significant speed improvement over the lower values of p.

Table 19.1 Closed loop poles

p_{22}	k_1	k_2	Poles
1	-.490	.664	.657 ± j.240
2	-.653	.995	.492 ± j.290
5	-.816	1.42	.281 ± j.291
10	-.890	1.65	.163 ± j.250

Figure 19.6 shows the actual system response and steady-state disturbance rejection behaviour; there is excellent agreement with the predicted response. The jitter in steady state exceeds the theoretical minimum of one A-D quantisation step, 0.2°C, but is well within specification.

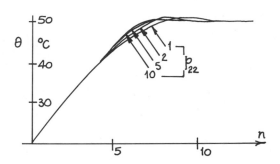

Figure 19.5 Simulated responses

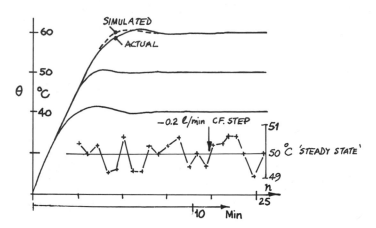

Inset is the steady state response around 50°C and showing the effect of a step in cold water flow of -0.2 1/min.

Figure 19.6 Actual responses

19.9 Conclusion

This optimal controller satisfied the complete specification, which was difficult to achieve using standard forward path compensators or the classic 3-term controller. Even better results were obtained using the so-called Extended Kalman Filter which models the adaptive plant; however the computing load becomes rather high. (4)

Optimal control, then, teaches us that state feedback is highly advantageous, rather than just output feedback as used in classical design.

However, classical performance criteria and design concepts need not be abandoned in using modern control theory; the modern control techniques do give improved control and are well worth attempting.

REFERENCES

1. Tagahashi, Rabins, Auslander: 'Control', Wiley 1972

2. Franklin, Powell: 'Digital Control' Addison 1981

3. Bryson: 'Applied Optimal Control' Halstead 1976

4. Morfett, Wilson: 'Optimal Control of a Thermal Plant'; Proc. IASTED Conf:MIC 85, Feb.1985

Acknowledgements are due to A J Morfett of BAe Communications Division, Stevenage, who performed the work reported above as part of his studies for the MSc in Control Engineering at the Hatfield Polytechnic during 1983.

Chapter 20

Self-tuning control applied to an environmental system
case study V

Dr. A.L. Dexter

20.1 Introduction

The control of the environment inside a large building is a complex problem which has received considerable attention since the cost of energy became a significant factor in the design and operation of heating, ventilating and air-conditioning (HVAC) plant. The control problem is concerned with two conflicting requirements. The internal environment (air temperature and, in some cases, relative humidity) of the various zones in the building must be maintained at a desired 'comfort' condition; and the cost of operating the HVAC plant must be minimised.

Two basic types of HVAC systems are in common use:

(a) Wet systems which use hot water as a medium to transfer heat from a central boiler plant to the conditioned air spaces of the building. A typical multi-zone, stored-energy wet heating system is shown in Fig. 1.

The electrode boilers are operated overnight using off-peak electricity to generate hot water which is held in the storage vessels for use during the following day. The heat output to each zone of the building is adjusted by varying the position of the three-port mixing valve that controls the amount of return water which is recirculated to the heat emitters.

(b) Dry systems which use an air stream to heat, cool or simply ventilate the various zones of the building. In some cases, the air must also be humidified or dehumidified to control its relative humidity. The main features of a typical central plant air-conditioning system are shown in Fig. 2.

Outside air flows through the cooling coil where it is cooled and dehumidified before it passes through the heating coil to be reheated to the desired supply temperature.

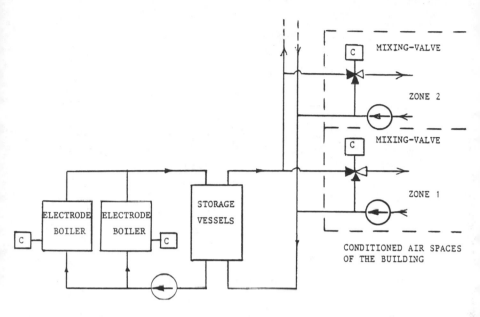

Figure 1   A typical wet heating system

A three port, diverting valve varies the flow of cold water through the cooling coil to control the temperature of the air leaving the coil. The cold water is supplied from the chiller (the evaporator of the refrigeration plant) at a nominally constant temperature. The absolute humidity of the air leaving the coil is determined by the air temperature if the temperature is below the dew-point.

A similar three-port, mixing valve varies the temperature of the hot water supplied to the heating coil to control the temperature of the air leaving the coil. The hot water is supplied from the boiler at a nominally constant temperature.

Under certain circumstances (for example when the outside air is very hot and humid), the cost of operating plant can be reduced by mixing some of the air returning from the zones of the building with the outside air before passing it through the heating and cooling coils; or by allowing some air to bypass the cooling coil if dehumidification is taking place.

Conventionally, HVAC plant has been controlled using a number of separate, but interacting control loops; each

dealing with a particular aspect of the overall control
problem.

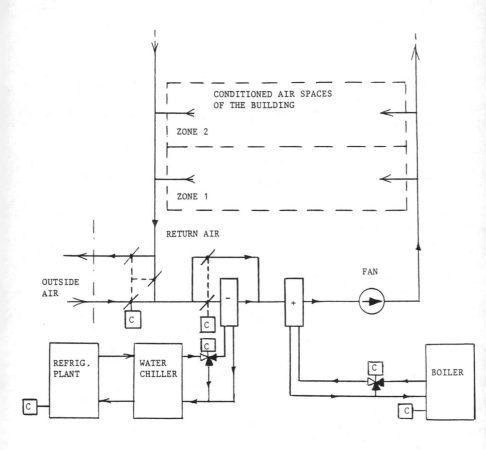

Figure 2   A typical air-conditioning system

Typical control loops are concerned with:
(1)   the operation of the boiler
(2)   the control of the refrigeration plant
(3)   the sequence control of multi-unit central plant
(4)   the management of the heat store
(5)   the regulation of the air temperatures in the
      zones of the building
(6)   weather compensation
(7)   intermittent heating control
(8)   optimum start and stop control
(9)   humidity control
(10) the operation of solar panels or heat pumps

The multi-loop approach has much to commend it in practice; particularly when it is structured hierarchically as shown in Fig. 3.

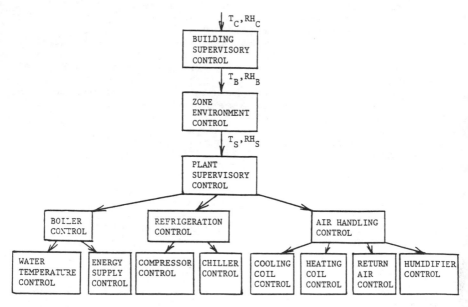

Figure 3   The hierarchical control structure for a dry system

It is seldom economic to custom design heating controllers to suit the characteristics of a specific building or heating plant since the the thermal behaviour of each building (even each zone in the building) and each heating plant will be different. Heating engineers must manually tune conventional control systems, after they have been installed, to obtain satisfactory control performance. On-site tuning of controllers is frequently time consuming and often difficult. Commissioning costs are usually high since it requires the attention of an experienced HVAC control engineer. Even after the initial tuning has been completed successfully, periodic retuning of the controllers may be necessary to take account of changes in the behaviour of the heating system caused by modifications to the building or plant, or through seasonal changes in its mode of operation. The application of self-tuning control techniques offers a way of reducing commissioning costs and improving long term control performance whilst reducing operating costs.

Although some HVAC controllers have been proposed which are based on integrated, adaptive optimal control schemes (1,2), much attention has been given to the use of single-input, single-output self-tuners within the conventional, multi-loop context (3,4,5,6,7,8). The first commercial application of self-tuning controllers in the HVAC field was in the area of intermittent heating control. Adaptive optimum start controllers made their appearance in the late seventies. Initially, manufacturers claimed significant increases in the energy savings though recent work (9) has shown that improvements are not always as large as was first suggested. Self-tuning water and air temperature controllers, and even weather compensators, which use only feedforward control action to adjust the output of the heating system according to the external temperature, are now marketed. A variety of explicit and implicit self-tuning control schemes have been applied although, to date, very little comparative data has been presented to justify the claims of improved temperature control or reduced energy consumption. In practice, the most successful controllers would seem to be those based on relatively simple algorithms which incorporate a control law with a structure similar to that of a conventional PI or PID controller.

All of the self-tuning control schemes described in this case study are based on the Clarke-Gawthrop Generalised Minimum Variance algorithm (see chapter 10). The reasons for choosing this algorithm and the necessity for simplification are first explained.

20.2 A simple self-tuning control algorithm

Since the controllers must be implemented in the out-stations of a building management system or as stand-alone controllers which will be retro-fitted to existing plant to replace relatively inexpensive analogue electronic control equipment, attention has focussed on implicit algorithms which require modest computing power and a minimum of commissioning effort. A simplified version of the – incremental form of the generalised-minimum-variance algorithm has been developed (10) for applications in environmental control. The resulting controller has a low-order control law and uses a simple recursive, scalar gain estimator to find its parameters so that it is feasible to implement the algorithm on a dedicated single-board microcomputer or even a single-chip microcontroller, in integer arithmetic, should this prove necessary.

The parameter estimator is of the form:

$$\hat{\underline{\theta}}(t) = \hat{\underline{\theta}}(t-1) + K(t)\underline{X}(t-k)\epsilon(t)$$

where: $K(t) = \dfrac{\bar{a}}{r(t)}$ is a scalar, time varying gain;

$\bar{a}$ is a predefined constant ($0 < \bar{a} < 2$);

$r(t) = \beta(t)r(t-1) + \underline{X}^T(t-k)\underline{X}(t-k)$;

$\beta(t)$ is the time varying forgetting factor ($0 < \beta \leqslant 1$);

typically,

$$\underline{\hat{\theta}}^T(t) = \left[\ \hat{f}_0,\ \hat{f}_1,\ldots,\hat{g}_0,\ \hat{g}_1,\ldots\hat{s}_0,\ \hat{s}_1,\ldots\right] \text{ is the parameter}$$

vector;

$$\underline{x}^T(t-k) = \left[\Delta_K y\ (t-k),\ \Delta_K y\ (t-k-1),\ \ldots\ \Delta_K u(t-k),\ \Delta_K u(t-k-1),\right.$$

$$\left.\Delta_K v(t-k),\ \Delta_K v(t-k-1),\ \ldots\right] \text{ is the data vector;}$$

$$\Delta_K = (1-z^{-k});$$

and $\epsilon(t) = \Delta_K y(t) - \underline{x}^T(t-k)\underline{\theta}(t-1)$ is the prediction error.—

The incremental formulation results in better parameter estimation by ensuring zero mean data in the data vector of the estimator.

The control law minimizes a cost function J, where:

$$J = E\{[Py(t-k) - W(t)]^2 + [Qu(t)]^2|t\}$$

where the weighting polynomials P and Q vary from application to application.

Three different, environmental control applications of self-tuning control are now described.

20.3 Example applications

20.3.1 Self-tuning control of off-peak energy supply

The self-tuning controller was designed to control the supply of electrical energy to the electrode boilers of a stored-energy wet heating system (11). The cost of operating the heating system is reduced if the boilers are used during the off-peak period when electrical energy is cheap and, since all energy stores have some uncontrolled heat losses, if the amount of stored energy is kept to the minimum needed to meet only those heating demands which will occur before the start of the next off-peak period. A "backward charging" algorithm (6) is used to delay the start of the off-peak charging so that the boilers produce just enough hot water to meet the heating demand predicted for the following day. Assuming some correlation between outside air temperatures during the current and following days and using an average outside temperature as a feedforward variable, the controller adjusts the charging delay τ_D to minimise the variance of the deviations in the end-of-day store temperature from some specified value.

If P = 1 and $Q = \dfrac{\lambda(1 - z^{-1})}{(1 - \alpha z^{-1})}$

the generalised minimum variance self-tuner may be interpreted as a self-tuning predictor cascaded with a conventional PI controller.

The parameters α and λ are related to the proportional band (P_b) and the integral time (T_I) of the conventional PI controller by the equations,

$$P_b = \lambda/\alpha \text{ and } T_I = \alpha T/(1-\alpha)$$

where T is the controller sampling interval.

The predictor cancels the plant time delay, simplifies the tuning of the P + I controller and improves the closed-loop performance of the control system. The block diagram of the charge control scheme is shown in Fig. 4.

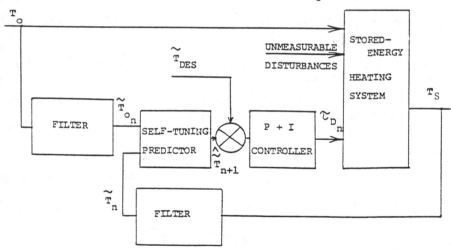

Figure 4   Self-tuning charge controller

Since there is no inherent plant time delay and the predictor has only three parameters f_0, g_0 and s_0, the parameter vector is given by,

$$\theta^T = [\, f_0, \, g_0, s_0 \,]$$

and the data vector by,

$$\underline{x}^T = [\Delta\tilde{T}, \, \Delta\tau_D, \, \Delta\tilde{T}_0] \text{ where } \Delta = (1-z^{-1})$$

The predictor is,

$$\tilde{T}_{n+1} = \tilde{T}_n - f_0 \Delta \tilde{T}_n - g_0 \Delta \tau_{D_n} - S_0 \Delta \tilde{T}_{0_n}$$

where $\tilde{T}_0$ is the mean daily external temperature given by,

$$\tilde{T}_0 = \frac{1}{\kappa_0} \sum_{j=1}^{\kappa_0} T_{0_j}$$

The controlled temperature,

$$\tilde{T} = \frac{1}{\kappa_1} \sum_{j=1}^{\kappa_1} T_{s_j}$$

is calculated from a number of measurements of the water
temperature which are taken at the start of the off-peak
period at different positions inside the storage vessels.
Its value gives some indication of the level of the
uncontrolled heat loss from the store.

$\tilde{T}_{s_{DES}}$ is set at a value which is a compromise between
minimising the uncontrolled heat loss and providing a
reserve of stored energy to compensate for errors in
predicting the next day energy consumption.

In this application, the self-tuner's output directly
controls the operation of the boilers and there is no
supervisor level of control. Those controller parameters
which must be pre-specified are held constant at predefined
values (e.g. sample time = 24 hours, time delay = 0) and
arbitrary values are selected for the parameters of the
cascaded PI controller. The values of the tuned parameters
are chosen so that the initial control action is equivalent
to that of a low-gain integrator.

Figure 5 shows the time variation of the store water
temperature, and the outside and building air temperatures
over a four day period when the self-tuning controller was
used to operate a laboratory pilot-scale stored-energy
heating plant which was linked to a digital simulation of
the building and its external environment. Once properly
tuned, the controller maintained the end of day store
temperature reasonably close to the set-point level.

20.3.2 Self-tuning control of the air temperature in a large
building

A stand-alone, air temperature controller for the
primary wet heating system of a single-zone office building
has been designed and tested on site (12). The controller
operates the main, motor-driven, three-port mixing-valve
which controls the temperature of the hot water supplying
the radiator heating circuits of the building.

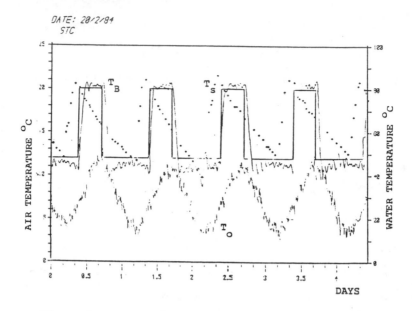

DATE: 20/2/84
STC

DAYS

Figure 5   Operation of the self-tuning charge controller

The controller (Fig. 6) is divided functionally into three basic sections:

(i)   A low-level <u>inner control loop</u> to compensate for actuator non-linearities. The controller directly switches power to the valve motor so as to open or close the valve and regulate the mixed water temperature (T_m). The control algorithm includes backlash compensation to eliminate unnecessary cycling of the mixing valve and reduce the amplitude of oscillations in the flow water temperature. The inner loop controller is equivalent to a simple discrete-time PI controller.

(ii) An intermediate level <u>self-tuning temperature control loop</u> to maintain the air temperature (T_B), at one location in the building, at a desired level. The controller monitors inside and outside air temperatures and generates the desired mixed water temperature for the cascaded inner loop. Once again, the choice of the weighting polynomials of the auxiliary function is made so that the self-tuner is equivalent to a self-tuning predictor cascaded with a P + I controller though, in this application, the parameters of the cascaded PI controller are also tuned; by the next level of control.

Even with the addition of feedforward control action based on measurements of outside temperature, the self-tuning temperature controller has itself only three parameters that are tuned on-line.

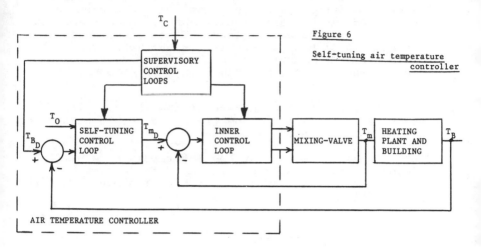

Figure 6

Self-tuning air temperature
controller

(iii) A high-level supervisory control loop which
optimises the controller's overall operation. The
supervisor has a number of separate functions:

(a) Each day it estimates the dead-time of the main
control loop and calculates an appropriate value for
the sampling interval of the intermediate level self-
tuner.

   The dead-time measurement is made at the start of
the first occupancy period. The desired mixed-water
temperature, which will have been at a relatively low
level outside of occupancy, rises quickly to a high
value as soon as the first setpoint change occurs.
Sampling every half minute, the controller calculates
the difference between the current building temperature
and that measured outside of the occupancy period, and
compares it with three, evenly spaced threshold values

of approximately 1.0, 2.0 and 3.0^0C. The time taken to
reach each of the thresholds is measured. Assuming the
building temperature response to be that of a first-
order system with a time delay, each of the three
possible pairs of co-ordinates can be used to find a
point of intersection of the curve and the base line.
An estimate of the system time delay is calculated from
the arithmetic mean of the three intersection values.

   To simplify implementation, the intermediate
level self-tuning control algorithm assumes a time
delay which is a predefined number of sample intervals
(in this case, k = 2) and adjusts the sampling interval
of the self-tuning controller to suit the estimated
time delay of the heating system and avoids the
problems associated with fractional time delays.

The sampling interval, which is re-calculated every day after the system time delay has been re-estimated, is the arithmetic mean of its current and previously calculated values. Time delay estimation is terminated if six consecutive measurements differ by less than one minute. It is then assumed that a sufficiently accurate estimate has been found.

(b)   It tunes the parameters of the intermediate level PI controller to ensure an acceptable closed-loop transient response.

The parameters of the intermediate level PI controller are tuned once a day using a simple deterministic univariate search to minimise the cost function,

$$I = \frac{1}{N_2 - N_1} \sum_{t=N_1}^{N_2} (T_B(t) - T_n(t))^2$$

where $T_n(t)$ is given by:

$$T_n(t) = T_{B_{DES}}(t) \quad \text{for} \quad T_{B_{DES}}(t) > T_o(t)$$

$$T_n(t) = T_B(t) \quad \text{for} \quad T_o(t) \geqslant T_B(t) \geqslant T_{B_{DES}}(t)$$

$$T_n(t) = T_o(t) \quad \text{for} \quad T_o(t) \geqslant T_{B_{DES}}(t)$$

T_o is the outside air temperature, T_B is the building air temperature and $T_{B_{DES}}$ is the desired building air temperature. The sample values of the building temperature are taken every four minutes.

The interval N_1 to N_2 , over which the cost is calculated is a 24 hour period, less the cool-down period from the moment end of occupancy is signalled to the time when the building temperature first reaches the desired night time temperature, or when the new occupancy pre-heat period begins. The cool-down period is not costed because the controller cannot influence the rate at which the building cools.

The value of the cost function at the end of the current day is compared with that of the previous day and a decision is made whether to increase or decrease one of the controller parameters α and λ. The search is started with a high value for λ (i.e. a low value of proportional gain) and an intermediate value for α (α must lie in the range 0.0 to 1.0) so that the controller provides a stable overdamped response at start-up. The parameter λ is first decreased in gradually reducing steps until a local minimum is found. The parameter λ is then held constant and the parameter α is varied in a similar manner until once more a local minimum is found. The parameter α is then held constant and the search is repeated with parameter λ and so on. The step size at each stage is taken as one half the previous step size for the parameter that is being altered, until the minimum step size is reached. The initial values of the parameters are chosen to provide a good compromise between the accurate location of the optimum and the speed of convergence of the search.

(c)   It predicts the preheat time and decides when to start-up the heating plant at the beginning of the day.

The pre-heat time estimation is based on the assumption that the pre-heat time will be reasonably well correlated from day to day. The estimate of pre-heat time, $\hat{t}_p(n)$ is calculated from the previous day's estimate and the measured value for the previous day, $t_p(n-1)$ so that:

$$\hat{t}_p(n) = \hat{t}_p(n-1) + \sigma \left[\hat{t}_p(n-1) - t_p(n-1) \right]$$

where σ is arbitrarily given the value 0.5.

(d)   It conducts a preliminary test to measure the backlash in the actuator linkage.

The supervisor opens the valve in short steps and measures the number of steps required before there is a change in water temperature. The test is repeated for opening and closing until consistent readings from a number of cycles have been made. The measurement is used by the inner loop controller to compensate for the backlash.

(e)   It turns off the parameter estimator of the intermediate level self-tuner whenever abnormal plant behaviour is observed e.g. during a partial boiler failure;   if the inner loop has lost control;   when the mixing-valve is on one of its limits.

The entire control algorithm was programmed using integer arithmetic in Tiny BASIC and implemented on a single-chip microcomputer.

Figures 7 and 8 show typical results obtained during field trials on a real building.

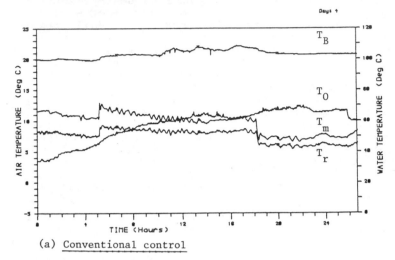

(a) Conventional control

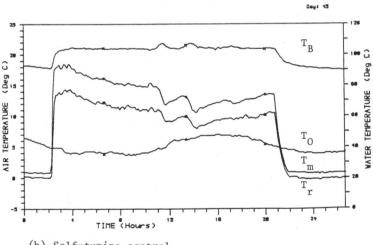

(b) Self-tuning control

Figure 7   Comparison of conventional and self-tuning
                              control of air temperature

Figure 7 compares the operation of the heating plant
before and after the installation of the self-tuning
controller. The variation in the air temperatures observed
in the various zones of the building over a four day period
are shown in Fig. 8.

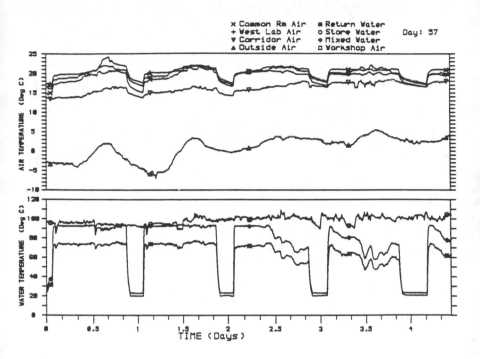

x Common Rm Air	■ Return Water	Day: 57
+ West Lab Air	o Store Water	
▼ Corridor Air	◆ Mixed Water	
▲ Outside Air	□ Workshop Air	

Figure 8   Operation of the self-tuning air temperature control

20.3.3 Multi-loop self-tuning control of air temperature and relative humidity

The third application concerns the design of a multi-loop self-tuning control scheme for the air-conditioning plant of a large building.

The control of the temperature and relative humidity of the air supplied by an air-conditioning plant is a complex problem. The air-conditioning process is highly non-linear; the interaction between the temperature and relative humidity control loops is significant; and the constraints imposed by non-ideal actuator behaviour are considerable. Conventionally, a cascaded, multi-loop control structure has been used where low-level control loops compensate for plant and actuator non-linearities; intermediate level control loops regulate the condition of the supplied air and control the operation of the individual plant items: high-level supervisory control loops monitor the operating conditions and determine the best, feasible control strategy which will minimise plant operating costs.

There is interest in the application of self-tuning control techniques within this multi-loop control structure (13).

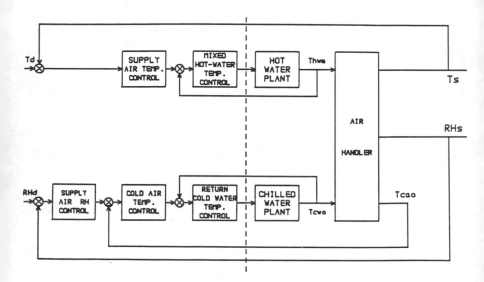

Figure 9   The multi-loop control structure of the air-conditioning plant

As shown in Fig. 9, there are two control loops at the intermediate level:

(i)   The supply air temperature (T_s) control loop.

(ii)   The supply air relative humidity (RH_s) control loop.

The supply temperature controller is cascaded with a hot water temperature control loop. The inner loop regulates the temperature of the hot water (T_{hwm}) supplied to the heating coil to follow the set-point temperature requested by the outer loop.

The supply relative humidity controller is cascaded with a cold air temperature controller which, in turn, is cascaded with a chilled water temperature control loop. The inner loop regulates the temperature of the cold water leaving the cooling coil (T_{cwo}) to follow the set-point water temperature requested by the intermediate loop. The intermediate loop adjusts the chilled water set-point to regulate the temperature of the air leaving the cooling coil (T_{cao}) to follow the set-point air temperature (Td_p) requested by the outer loop.

The single-input/single-output self-tuning controllers in the lower levels of control use feedforward to compensate for disturbances introduced through plant interactions. Both inner loop controllers are conventional fixed parameter PI controllers which have gain Kp and integral time Ti. The other controllers use the same control algorithm which is based on the full incremental form of the 'generalised minimum variance' self-tuning control algorithm. In each case, the P and Q polynomials are given by,

$$P(s) = \frac{P_n(s)}{P_d(s)} = \frac{(1+sT_m)^2}{(1+sT_f)^3}$$

where T_m specifies the desired closed-loop response and T_f is a filter time constant; and

$$Q = \frac{\lambda(1 - z^{-1})}{(1 - \alpha z^{-1})} ;$$

as in the other applications.

The control law is of the form:

$$u(t) = \frac{1}{\hat{G}\Delta_k + Q'} \left[W(t) - \Psi(t) - \hat{F}\Delta_K y'(t) - \hat{S}\Delta_k v(t) \right]$$

where: u(t) is the control output;

W(t) is the set-point;

v(t) is the feedforward variable;

$\Psi(t) = P(s)y(t)$;

y(t) is output of the control loop;

$y'(t) = \frac{y(t)}{P_d(s)}$;

$Q' = \frac{\lambda}{g_0} Q$;

and $\hat{F}$, $\hat{G}$ and $\hat{S}$ are polynomials in z^{-1}, such that

$$\Psi^*(t+k|t) = \Psi(t) + \hat{F}(z^{-1})\Delta_k y'(t) + \hat{G}\Delta_k u(t) + \hat{S}\,\Delta_k v(t)$$

where $\Psi^*(t+k|t)$ is the prediction of $\Psi(t+k)$ at time t.

The control scheme is implemented on a distributed microcomputer system programmed, using floating-point arithmetic, in a real-time, high-level language (Distributed-Pascal-Plus) which has been developed at Oxford for low-cost, dedicated applications of this type (14).

Fig. 10 shows a typical set of results that were obtained during experiments on a pilot-scale air-conditioning plant which is being used as a test-bed for the control scheme. The plots illustrate the behaviour of the two main, intermediate level, self-tuning control loops and demonstrate some of the practical problems which can arise. Note, for example, the interaction between the relative humidity and changes in the air temperature; the difficulty of maintaining good control when control signals are close to their limits; the disturbances caused by poor lower level control.

20.4 General comments on the practical application of the self-tuning controllers

Self-tuning controllers are making a valuable contribution in the area of environmental control, though there are a number of points which must be considered before they are adopted as the solution to all HVAC control problems.

The practical performance of a self-tuning controller is seldom as impressive as that suggested by the results of simulation. Since most of the practical applications have been based on controller designs using a first or second-order linear model of the process, the resulting control law has a similar if not identical structure to that of a PI or PID controller. With relatively few exceptions, the practical performance of a self-tuning controller is unlikely to be better than that of a properly-tuned conventional controller. Predictive self-tuning control schemes are a notable exception since these controllers can eliminate the effects of time delays (the dynamic element of the process which conventional controllers find most difficult to handle) and provide tighter control. In practice, however, tighter control is only feasible if it is both possible and desirable to allow rapid and large perturbations in the input to the process.

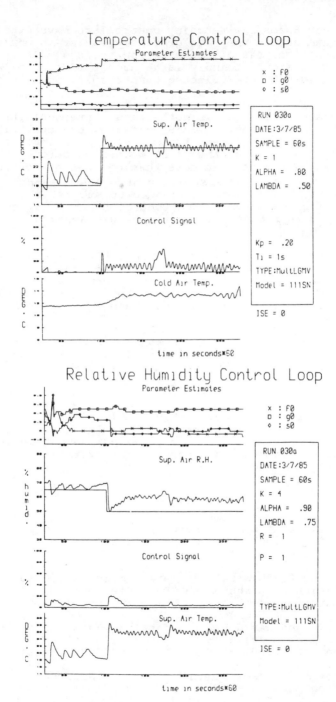

Figure 10   Self-tuning control of the air-conditioning plant

In most cases, self-tuning controllers have little to commend them unless they are genuinely easier to commission than conventional fixed parameter controllers. It is often the case that a surprisingly large number of controller parameters must be pre-selected before self-tuning can begin.

Accurate estimation is not always a prerequisite to good control performance. In some cases, the control action is so cautious that correct tuning is unimportant; in others, the dynamics of the plant and the nature of the disturbances are such that more than one set of controller parameters will provide a satisfactory closed-loop performance. There is some evidence to suggest that, in practice, implicit self-tuning control schemes may be inherently more robust than many explicit schemes.

Care must be taken when both self-tuning feedback and feedforward action is used in the same controller. Difficulties can arise during periods when the feedback control action substantially eliminates disturbance effects or when the measured disturbances are not persistently exciting.

Currently, few self-tuning control algorithms are able to cope directly with variable time delays, or strong plant non-linearities, or the dynamic interactions which can occur in a multi-variable plant (Interactions arising from actuators transiently operating on their limits can cause particular difficult problems). In practice, the self-tuner will need to be sandwiched within a multi-level, hierarchical control structure in which cascaded low-level control loops are used to hide the non-linear behaviour, and high level control loops are used to supervise its overall operation("jacketing").

References

(1)   Farris D.R. and Melsa J.L., "A study of the use of adaptive control for energy conservation in large solar heated and cooled buildings", IEEE Conf. on Decision and Control (1980).

(2)   Schumann R, "Digital parameter adaptive control of air-conditioning plant", 6th IFAC/IFIP Conf. on Digital Computer Applications to Process Control (1980).

(3)   Park C. and David A.J., "Adaptive algorithm for the control of a building air handling unit", National Bureau of Standards report NBS1R82-2591 (1982).

(4)   Dexter A.L., "Self-tuning optimum start control of heating plant", Automatica V17 N3 (1981).

(5)   Barney G.C. and Florez J., "Temperature prediction
      models and their application to the control of heating
      systems", Proc. IFAC Symp. on Identification and System
      Parameter Estimation, York (1985).

(6)   Dexter A.L. and Hayes R.G., "Self-tuning charge control
      scheme for domestic stored-energy heating systems", IEE
      PROC. Vol 128, Pt. D, No. 6 (1981).

(7)   Dexter A.L., Danninger W. and Graham W.J., "Design of
      self-tuning controllers for wet heating systems", Proc.
      ASME Symp. on Dynamic Systems: Modelling and Control,
      DSC-Vol 1, Miami Beach, FL (1985).

(8)   De Keyser R.M.C., "Adaptive microcomputer control of
      residence heating", Int. Symp. on Recent Advances in
      control and Operations of Building HVAC Systems,
      Trondheim, Norway (1985).

(9)   Birtles A.B., "A new optimum start control algorithm",
      Int. Symp. on Recent Advances in Control and Operation
      of Building HVAC Systems, Trondheim, Norway(1985).

(10) Dexter A.L., "Self-tuning control algorithm for single-
      chip microcomputer implementation", IEE PROC. Vol. 130,
      Pt. D, No. 5 (1983).

(11) Dexter A.L. and Graham W.J., "A simple self-tuning
      controller for heating plant applications", Int. Symp.
      on the Performance of HVAC Systems and Controls,
      Buildings Research Establishment, Garston (1984).

(12) Graham W.J. and Dexter A.L., "Self-tuning control of
      the heating plant in a large building", Proc. of IEE
      Int. Conf. 'Control 85', Vol. 2, Cambridge (1985).

(13) Dexter A.L. and Jota F.G. "Multi-loop self-tuning
      control of an air-conditioning plant", Int. Symp. on
      Recent Advances in Control and Operation of Building
      HVAC Systems, Trondheim, Norway (1985).

(14) Jota F.G., Dexter A.L., Fertig L.P. and Wooster W.G.,
      "Distributed-Pascal-Plus: a high-level language for
      low-cost distributed, monitoring and control systems",
      Oxford University Engineering Laboratory Report No.
      OUEL 1618/86 (1986).

Chapter 21

Computer aided control system design
case study VI
Dr. D.J. Sandoz

21.1 INTRODUCTION

This chapter presents two case studies to illustrate the VUMAN approach to control system design and to demonstrate the practicality of the mathematical procedures that are reviewed in chapter 11. All of the design results presented have been established using the VUMAN 'Plant Analysis System' package that is described in some detail in chapter 11. It is necessary for the reader to have studied chapter 11 in some detail in order to properly appreciate this chapter, since frequent reference is made throughout to aspects of chapter 11.

The case studies relate to data collected from a multiple effect evaporator. The first study (section 21.2) is very simple and considers the relationship between a steam flow rate and a valve in the steam pipeline. This study restricts to simulation only and serves to illustrate various aspects of the modelling and control system design exercises. The second study (section 21.3) is more complex, involving five process signals altogether with three as process inputs and two as process outputs. The study gives rise to the development of a multivariable controller that incorporates feedforward compensation and compensation for significant time delays associated with all three inputs. The online implementation of the resulting controller is illustrated, this implementation being carried out by the VUMAN 'Real Time System' package that is briefly mentioned in chapter 11.

21.2 CASE STUDY 1, A SIMPLE STEAM FLOW SYSTEM

Fig. 21.1 presents data collected from plant for this simple system. Two signals are illustrated, M7 a steam flow rate and A4 a valve position. Note that the presented scales relate, for each signal, to the stubs on the y axis that define the boundaries of the window within which the signal is displayed. Data is presented for 576 seconds. The variations in the valve signal A4 follow the pattern of a PRBS, with an amplitude of 5% around 49.67% and with a minimum interval between changes of 4 seconds. The steam flow M7 is seen to clearly respond to the disturbances of A4. The variations in steam flow are between 420 and 442 kilograms/hour, and there is a distinct downdrift in the latter stages of the illustrated span.

The recursive least squares procedure of section 11.3 has been applied to the illustrated data. The most simple model structure possible was assumed, with first order dynamics and no time delay.

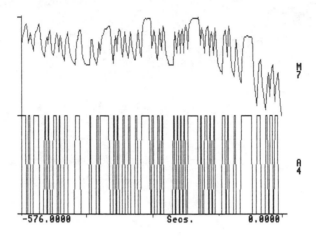

Fig.21.1 Steam system data

M7: Steam flow rate, 420.14 to 441.95 kilograms/hour

The prediction interval c was taken to be 4 seconds and the sampling interval s, 2 seconds.

The four figures of merit that arise from the analysis are as follows (see section 11.3).

Absolute Mean (F1)	-0.4952
RMS Mean (F2)	0.8609
Incremental Mean (F3)	0.0214
Incremental RMS (F4)	0.2240

F1 and F2 are large, which is an indication of significant drift. F4 is relatively significantly smaller and suggests that the plant and model follow quite well.

The normalised model that results has the structure:

$$
\begin{bmatrix} M7_{k+4} \\ M7_{k+2} \end{bmatrix} = \begin{bmatrix} 1.809 & -0.809 \\ 1.485 & -0.485 \end{bmatrix} \begin{bmatrix} M7_k \\ M7_{k-2} \end{bmatrix}
$$

$$
+ \begin{bmatrix} 0.727 \\ 0.358 \end{bmatrix} [A4_k - A4_{k-4}] \qquad 21.1
$$

The accuracy of this model is best appreciated by graphical illustration. Fig. 21.2a presents a direct comparison between plant and model data, with the above model being subjected to the same input sequence A4 that is evident in fig. 21.1.

The dotted line in fig. 21.2a represents the model data and the continuous line that of the plant. In places the model tracks the plant quite well but the drift to the rhs is very clear (hence the high values shown above for F1 and F2). Fig. 21.2b presents the

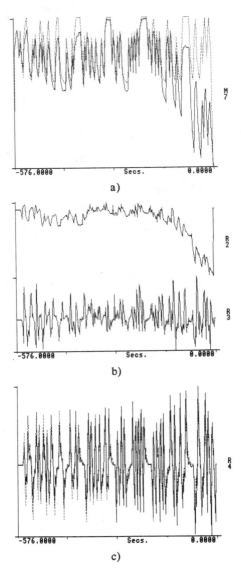

a)

b)

c)

Fig. 21.2 Steam System, Graphical Evaluation of Model
a)   Comparison of absolute plant and model data
     (dotted = model)
     Scale: 420.14 to 441.95 kilograms/hour
b)   Residual Absolute (R2) and Incremental (R3) Errors
     R2 Scale: -13.06 to 1.31 kilograms/hour
     R3 Scale: -0.55 to 0.71 kilograms/hour
c)   Incremental Comparison (dotted=model)
     Scale: -1.85 to 2.03 kilograms/hour

errors associated with fig. 21.2a. Signal R2 is the difference between the plant and model data. The downwards drift in steamflow rate is clearly highlighted. Signal R3 is the incremental error between the plant and model (the difference between the change in the plant data and the change in the model data at each sampling instant). This signal is more difficult to interpret and often has a characteristic that is consistent with noise. Fig. 21.2c presents an alternative and often more effective means for assessing the model. This form compares the actual change in plant data with that predicted by the model at each sampling instant. A close comparison implies that effective control may arise from a control system designed from the developed model. This is because the control systems base their operations upon this form of incremental prediction. In this particular case, fig. 21.2c indicates very good incremental tracking, which augers well for accurate control.

The first order model presented in equation 21.1 now forms the basis for a control system design and simulation exercise. The purpose of the exercise is to determine an acceptable control system that is to manipulate the valve (signal A4) to regulate the steam flow rate (signal M7).

Three control system designs are carried out altogether. Equation 21.1a presents the control system arising from a 'normalised' design, ie with unity weights in the cost function. Equation 21.2b presents that arising with the actuation weight reduced to 0.5. Equation 21.2c presents the Minimum Time control system. The Minimum Time analysis first halts after 3 iterations, with a cost JMT of 0.0109 (see section 7.3.1). A further iteration has to be actioned to reduce the cost to an extremely small value.

$$[A4]_k - [A4]_{k-4} = -0.7176 \ (\ [M7]_k - [S1] \ )$$
$$+ \ 0.3527 \ (\ [M7]_{k-2} - [S1] \ ) \qquad 21.2a$$

$$[A4]_k - [A4]_{k-4} = -1.012 \ (\ [M7]_k - [S1] \ )$$
$$+ \ 0.4776 \ (\ [M7]_{k-2} - [S1] \ ) \qquad 21.2b$$

$$[A4]_k - [A4]_{k-4} = -1.336 \ (\ [M7]_k - [S1] \ )$$
$$+ \ 0.6195 \ (\ [M7]_{k-2} - [S1] \ ) \qquad 21.2c$$

In the above, $[A4]_k$ and $[M7]_k$ represent the values of A4 and M7 at sampling instant k. $[S1]$ is the value of the setpoint of the control loop and defines the value to which M7 is to be controlled.

Note that by reducing the weighting associated with A4, the implication is less constraint on the actuator. The resulting controller therefore has higher gains. The Minimum Time controller is the most urgent possible and is seen to have higher gains still.

Fig 21.3a presents the simulation results that arise for the normalised controller of Equation 21.2a. The figure presents the measured value and the setpoint superimposed (ie M7 and S1) and also the actuation value A4.

An initial setpoint of 420 Kg./hour prevails in each case. The simulation range covers 80 seconds. During the course of the

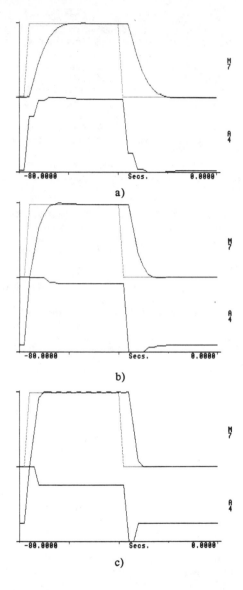

a)

b)

c)

Fig. 21.3   Steam System Control Responses
a)   Normalised
     M7:   steam flow rate, 419.93 to 440.07 kilograms/hour
     A4:   steam valve position, 42.89 to 52.98%
b)   Less constrained
     M7:   419.74 to 440.26, A4: 42.06 to 53.80
c)   Minimum Time
     M7:   420 to 440, A4: 38.44 to 57.42

simulation, the setpoint is adjusted to 440 Kg./hour at the first instant and back to 420 Kg./hour at the second (ie at the half way stage). Figs. 21.3b and 21.3c similarly present results arising for the controllers of Equations 21.2b and 21.2c respectively.

Comparison of figs. 21.3a and 21.3b clearly illustrates the impact of reducing the constraint on the actuator. The magnitude of the initial actuator correction, following a setpoint change, is significantly greater for the case of fig. 21.3b and the measured value response is correspondingly faster. The Minimum Time response of fig. 21.3c is seen to be a very fast transition, with just two control corrections to achieve the demanded new steady state but requiring large amplitude adjustments.

Fig. 21.4a illustrates the response of the controller of equation 21.2b when it is faced with an unreferenced perturbation to the actuator at the half way stage of the simulation. This perturbation is defined to be of magnitude equal to 20% of the actuator deviation arising in the first half of the simulation. It is also defined to be applied across 16 seconds. Since the model update interval is 4 seconds, a 20% deviation is consequently applied to the model at 4 successive update instants, as is apparent from fig. 21.4a which shows A4 adjusting to overcome the disturbances. The setpoint is adjusted from 420 to 440 at the first change and remains constant at 440 at the second change. Fig. 21.4b illustrates the same conditions as for fig. 21.4a but with 1% measurement noise added to the measured value M7. The effect upon actuator activity is significant. Fig. 21.4c presents a comparison between the two measured value responses of figs. 21.4a and 21.4b.

The above simple exercises illustrate the manner in which models may be established and assessed and also the manner in which control systems may be developed and investigated under simulation conditions. Discussions relating to actual implementations are considered for the more complex situation pertaining below with the second case study.

21.3 CASE STUDY 2, A MULTIPLE EFFECT EVAPORATOR

Case Study 2 relates to a much more complex situation. In this case data has been collected from a multiple effect evaporator and five signals in all are of particular interest. These signals divide into three categories

i)   Measured values:
     M4,   the flow rate of the concentrated product from the evaporator; and
     M20, the density of the concentrated product

ii) Actuation values:
     S2,   the setpoint to a flow controller that regulates the supply of raw material to the evaporator (ie the material that is to undergo evaporation); and
     S5,   the setpoint to a flow controller that regulates the supply of steam to the evaporator (ie the energy supply).

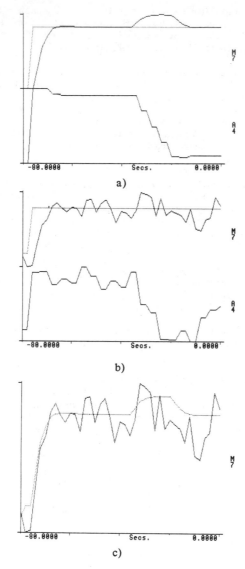

a)

b)

c)

Fig. 21.4   Steam System Control Responses with Noise
                 and Disturbance
a)   Disturbance only
      M7: steam flow rate, 420 to 444.05 kilograms/hour
      A4: steam valve position, 43.09 to 53.80%
b)   1% Noise with disturbance
      M7: 414.26 to 446.71, A4: 38.39 to 55.37
c)   Superposition of M7 for a and b

iii) Disturbance value:

M18, the density of the raw material being supplied to the evaporator

The system is therefore considered to have three inputs and two outputs, with two of of the inputs available for control actuations. In relation to equation 11.1, p=2, m=2 and q=1.

Fig. 21.5 presents data collected from plant. The illustrated data spans 6928 seconds. Signal S2, the material feed rate, varies as a PRBS which has an amplitude of 80 kilograms/hour around 1750. The minimum interval between changes is 64 seconds. Signal S5, the steam feed rate, also varies as a PRBS, in this case with an amplitude of 40 kilograms/hour around 460. Note that these two PRBS's are orthogonal (ie uncorrelated). This is essential if the modelling statistics are to be able to separate the effects of S2 and S5. The PRBS's are also synchronised to the same 64 second intervals.

Signal M18, the feed material density, varies between 1012 and 1022 kilograms/cubic metre. This variation is deliberately induced by injecting water into the feed in order to establish good data for modelling purposes. Signal M18 is a disturbance to the process and the ultimate intention is to establish a feedforward controller to correct for impacts induced by variations in M18..

Signals M4 and M20, the product flow and the product density respectively, illustrate the manner in which the product is influenced by the variations in the three input signals.

To establish a model, initially second order dynamics are selected with a control/prediction interval of 64 seconds (ie the PRBS interval) and a sampling interval of 32 seconds (which corresponds to two samples per prediction interval). The minimum delay is set to zero for all three inputs and the delay spread is set to 3 prediction intervals for each. Signal M18 is considered to vary as a ramp, with S2 and S5 varying as steps. These selections were arrived at following two or three iterations of the identification procedures. Details of these iterations are not presented here. The discussions concentrates upon the best final result achieved. The most important feature to emerge from the studies is that there is a significant time delay (up to 2 minutes) that is associated with all inputs and that there is a strong interaction between signals.

Application of the Least Squares analysis gives rise to the following figures of merit:

	Signal M4	Signal M20
Absolute Mean (F1)	0.71	-0.26
RMS Mean (F2)	1.02	0.42
Incremental Mean (F3)	0.013	0.001
RMS Mean (F4)	0.48	0.17

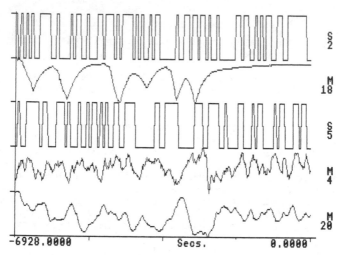

Fig.21.5 Evaporator system data
    S2:   Feed flow, 1710 to 1790 kilograms/hour
    M18: Feed density,1012.19 to 1021.64 kilograms/cubic
          metre
    S5:   Steam flow, 440 to 480 kilograms/hour
    M4:   Product flow, 201.11 to 503.73 kilograms/hour
    M20: Product density,1107.76 to 1194.17 kilograms/
          cubic metre

Interpretation of these figures indicates that M20 has been modelled more effectively that M4. For signal M20, F1 and F2 are relatively much smaller, indicating less drift. Similarly, F4 is much smaller, indicating closer dynamic tracking.

Figs. 21.6 (a,b and c) present the residual error graphs for this model (which correspond to figs. 21.2a, b and c for the steam flow case study). Fig. 21.6a indicates that both signals have been modelled quite effectively, with reasonable tracking throughout. Signal M4 is more noisy than M20, which explains the higher incremental figure of merit (F4). There is an offset in the model to the rhs of fig. 21.6a, which explain the higher mean figures of merit (F1 and F2).

Signals R3 and R4 of fig. 21.6b are the absolute residual errors relating to signals M4 and M20 respectively. They represent the differences between the superimposed graphs of fig. 21.6a. These are interesting in that there is apparent a marked deviation just past the half way point of the display. At this stage of the data collection exercise, there was a problem with the steam supply to the evaporator. This is clearly identified in the residual error traces. Signals R5 and R5 of fig. 21.6b, the incremental residual errors, also highlight this incidence. Fig. 21.6c illustrates the close incremental tracking between plant and model (R7 relates to M4 and R8 to M20).

The full normalised equation for the model may be written down as

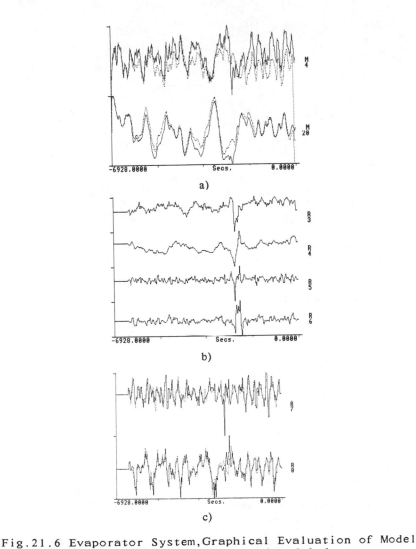

Fig.21.6 Evaporator System,Graphical Evaluation of Model
a) Comparison of absolute plant and model data
    (dotted=model)
    M4 Scale: 201.11 to 503.73 kilograms/hour
    M20 Scale:1107.76 to 1194.17 kilograms/cubic metre
b) Residual (R3 & R4) and Incremental (R5 & R6) Errors
    R3 Scale (for M4): -166.2 to 115.02 kilograms/hour
    R4 Scale (for M20):-30.54 to 14.37 kg./cubic metre
    R5 Scale (for M4): -3.77 to 2.35 kilograms/hour
    R6 Scale (for M20):-0.81 to 0.99 kg./cubic metre
c) Incremental Comparisons (dotted=model)
    R7 Scale (for M4): -3.93 to 1.92 kilograms/hour
    R8 Scale (for M20):-1.25 to 1.34 kg./cubic metre

$$
\begin{bmatrix} M4_{k+64} \\ M20_{k+64} \\ M4_{k+32} \\ M20_{k+32} \end{bmatrix} = \alpha \begin{bmatrix} M4_k \\ M20_k \\ M4_{k-32} \\ M20_{k-32} \end{bmatrix} +
$$

$$
\beta1 \begin{bmatrix} S2_k & - & S2_{k-c} \\ S2_{k-c} & - & S2_{k-2c} \\ S2_{k-2c} & - & S2_{k-3c} \\ S2_{k-3c} & - & S2_{k-4c} \end{bmatrix} +
$$

$$
\beta2 \begin{bmatrix} S5_k & -S5_{k-c} \\ & \vdots \\ & \\ S5_{k-3c} & -S5_{k-4c} \end{bmatrix} +
$$

$$
\beta3 \begin{bmatrix} M18_{k+c} & - & M18_k \\ M18_k & - & M18_{k-c} \\ & \vdots \\ & \vdots \\ M18_{k-3c} & - & M18_{k-4c} \end{bmatrix}
\qquad\qquad 21.3
$$

with

$$
\alpha = \begin{bmatrix} 1.11 & 0.07 & -0.11 & -0.07 \\ -0.05 & 2.32 & 0.05 & -1.32 \\ 1.11 & -0.17 & -0.11 & 0.17 \\ 0.03 & 1.76 & 0.03 & -0.76 \end{bmatrix} \text{ and}
$$

$$
\beta1 = \begin{bmatrix} 0.11 & 0.24 & 0.29 & 0.02 \\ -0.02 & -0.10 & -0.16 & -0.10 \\ 0.00 & 0.01 & 0.16 & -0.03 \\ 0.00 & -0.04 & -0.05 & -0.05 \end{bmatrix}
$$

etc.

Inspection of the model of equation 21.3 indicates that the structure could well be simplified. For example, examination of $\beta1$ suggests that the first and third columns are not significant with respect to the second and third. Thus the terms $S2_k$-$S2_{k-c}$ qnd $S2_{k-3c}$-$S2_{k-4c}$ could be abandoned. The implication is that, for the signal S2 at least, a minimum delay of one prediction interval with a delay spread of one prediction interval would be an acceptable structure specification. Such a structure has been established, although for economy of presentation it is not listed here. There is a slight deterioration in model accuracy although not particularly significant.

Fig. 21.7 illustrates the manner in which model simulations can be used to isolate the contributions from particular input signals. In this case, the residual errors are generated with signals S2 and S5 inactive so that only the density variations of signal M18 drive the model. It is clear from fig. 21.7 that these variations impact very significantly upon the plant.

Consideration now progresses to the design of a control system to

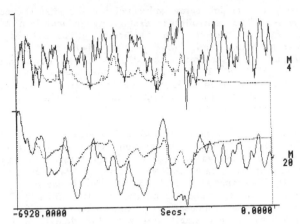

Fig. 21.7 Response of Model to variations in M18 only
M4 Scale: 201.11 to 503.73 kilograms/hour
M20 Scale: 1107.76 to 1194.17 kilograms/cubic metre

regulate product flow and product density in the face of variations in the density of the material feed. The normalised controller that arises from the design procedure has the structure

$$\begin{bmatrix} S2_k - S2_{k-64} \\ S5_k - S5_{k-64} \end{bmatrix} = \begin{bmatrix} -0.55 & 2.32 & 0.08 & -1.58 \\ -0.021 & -2.34 & 0.00 & 1.74 \end{bmatrix} \begin{bmatrix} M4_k - S7 \\ M20_k - S8 \\ M4_{k-32} - S7 \\ M20_{k-32} - S8 \end{bmatrix}$$

$$+ \begin{bmatrix} 1.37 & -0.76 & 1.81 & 11.6 \\ 1.76 & -0.89 & -0.19 & 1.48 \end{bmatrix} \begin{bmatrix} M18_k - M18_{k-64} \\ M18_{k-64} - M18_{k-128} \\ M18_{k-128} - M18_{k-192} \\ M18_{k-192} - M18_{k-256} \end{bmatrix}$$

$$+ \begin{bmatrix} -0.56 & -0.39 & -0.08 \\ 0.11 & 0.10 & 0.06 \end{bmatrix} \begin{bmatrix} S2_{k-64} - S2_{k-128} \\ S2_{k-128} - S2_{k-192} \\ S2_{k-192} - S2_{k-256} \end{bmatrix}$$

$$+ \begin{bmatrix} 0.93 & 1.09 & 0.33 \\ -0.27 & -0.21 & -0.27 \end{bmatrix} \begin{bmatrix} S5_{k-64} - S5_{k-128} \\ S5_{k-128} - S5_{k-192} \\ S5_{k-192} - S5_{k-256} \end{bmatrix} \qquad 21.4$$

S7 is the setpoint for M4 and S8 is the setpoint for M20.

The gains relating to M18 are the feedforward terms and those relating to M4 are the feedback terms. The gains relating to S2 and S5 compensate for the very significant time delays. Inspection of the magnitudes of the various gains indicates that the term $S2_{k-192} - S2_{k-256}$ and perhaps $S5_{k-192} - S5_{k-256}$ could be dropped from the controller with little penalty, although all other elements appear significant.

Fig. 21.8a presents simulation of the operation of the control system of equation 21.4. The simulation covers a span of 6500 seconds, with the model being subjected to the disturbance variations of signal M18. The controller is attempting to maintain M4 and M20 at setpoint in the face of these variations. The setpoint for M4 is 230 kilograms/hour and M4 varies between 253 and 300 kilograms/hour. The setpoint for M20 is 1160 kilograms/hour. The setpoint for M20 is 1160 kilograms/cubic metre and M20 varies between 1153 and 1166 kilograms/cubic metre. To achieve this accuracy, S2 varies by 130 kilograms/hour from 1672 and S5 varies by 52 kilograms/hour from 465.

To provide a measure of the effectiveness of the feedforward control, fig. 21.8b presents a simulation with the feedforward control inactive(in effect, the gains associated with M18 in equation 21.4 are set to zero). In this case, the variations in M4 increase to between 239 and 341 kilograms/hour and in M20 to between 1143 and 1172 kilograms/cubic metre, a significant deterioration . Fig. 21.8c presents a comparison of the responses with and without feedforward control. The improvement is very significant with the density M20 but less so with the flow M4.

Fig. 21.9a presents results for the implementation of the controller of equation 21.4 on plant. This exercise gives rise to interesting considerations which lead to a modification in the approach to controller design. Fig. 21.9a presents 7800 seconds of information with the signals M20, M4, S2 and S5 displayed. Two extra pressure signals are also displayed (M11 the main steam pressure and M13 a pressure in the third effect stage of the evaporator). The density setpoint is varied from time to time through the span and the density M20 is seen to respond quickly and to be maintained accurately at target. The time delay in the response is clearly evident. The control of product flow is less satisfactory. It is clear that both S5 and S2 vary considerably in order to adjust density whilst maintaining a constant flow. It is also clear from the centre region of the span that a small change in the flow setpoint can give rise to large changes in both S5 and S2 in order to maintain constant density. These factors are a problem for normal process operation and, although fig. 21.9a illustrates the successful operation of multivariable control in this difficult process situation, it also clearly indicates that multivariable control is not practicable for permanent installation in this case. The difficulty is that the evaporator feed flow rate S2 is necessarily very limited in its range of operation. If this flow drops below 1600 kilograms/hour, it is possible for deposits to quickly build up on the evaporator walls. If it exceeds 1800 kilograms/hour, there is a likelihood that excess material will foul up the steam heating system. This tight constraint on S2 means it is not really viable as a control input to the process. Note that in figure 21.9a, S2 varies between 1540 and 1900 kilograms/hour for the fairly limited control ranges considered.

Fig. 21.9b presents a further example of online process operation. Because of the experiences gained on the basis of the operations illustrated in fig. 21.9a, the Plant Analysis System was utilised to design a control system of a different structure. This control system has the signal S2 as a disturbance input rather than as a control

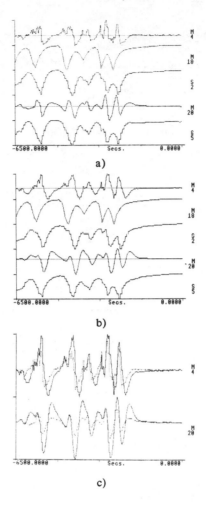

a)

b)

c)

Fig. 21.8   Multivariable control
a)   with feedforward compensation
M4 Scale: 253.13 to 329.32 kilograms/hour
M18 Scale: 1012.19 to 1020.68 kilograms/cubic metre
S2 Scale: 1672.34 to 1801.22 kilograms/hour
M20 Scale: 1153.07 to 1166.26 kilograms/cubic metre
S5 Scale: 465.71 to 506.79 kilograms/hour
b)   without feedforward compensation
M4 Scale: 239.40 to 340.45 kilograms/hour
S2 Scale: 1688.77 to 1798.6 kilograms/hour
M20 Scale: 143.31 to 1171.61 kilograms/cubic metre
S5 Scale: 472.84 to 504.24 kilograms/hour
c)   comparison of responses (dotted=feedforward)
Scales as for b.

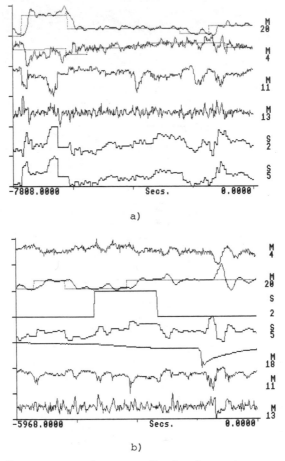

a)

b)

Fig. 21.9 Evaporator System, Control system
          implementation
a)   Multivariable Control
M4 product flow: 263.7 to 499.9 kilograms/hour
M20 product density: 1126 to 1181 kgs./cubic metre
M13 third effect pressure: 73.4 to 76.8 millibars
M11 steam pressure: 5.65 to 6.73 bars
S2 feed flow: 1541 to 1900 kilograms/hour
S5 steam flow: 386.1 to 513.8 kilograms/hour
b) Feedforward Control
M4 289.3 to 462.8 kilograms/hour
M20 1137 to 1169 kilograms/cubic metre
S2 1650 to 1700 kilograms/hour
S5 414.3 to 451.9 kilograms/hour
M18 feed density, 1018 to 1022 kilograms/cubic metre
M11 6.08 to 6.68 bars
M13 74.1 to 75.9 millibars

actuation input. The control system is therefore now required to maintain the density M20 at target in the face of disturbance variations from both M18, the feed density, and S2, the feed flow rate.

Fig. 21.9b presents 6970 seconds of information and illustrates the good control of the density M20 in response to setpoint changes and also in response to alterations in S2 and M18. Two adjustments are made to S2, the first from 1650 to 1700 kilograms/hour and the second back again. The feedforward action of S5 is clearly evident and the impact upon density from the flow change, though evident, is minimal. To the rhs of the span, the feed density M18 is adjusted sharply. This gives rise to a sharp disturbance upon plant, similar to those evident in the simulations. However, the controller quickly restores satisfactory operation. Such a disturbance would normally have required a process shutdown to avoid operating regimes that might give rise to damage to the system.

21.4 DISCUSSION

The case studies presented in this chapter, highlight the practical relevance of the technology of chapter 11 to control system design.

Control systems of the form illustrated are now in continuous operation on a number of industrial processes in Britain. In certain particular cases, the indication is that they can give rise to very significant cost benefits by dramatically reducing standard deviations in certain critical product variables. The benefits arising from steadier process operations, from improved production rates by operation closer to threshold margins and from energy savings, can amount to improvements of as much as 10% in plant productivity.

A difficulty with the new technology is its complexity in comparison with the conventional approach that involves the applcation of three term controllers. Experiences suggest that for control of critical product parameters it is often the case that three term controls cannot be utilised, usually because of large time delays and significant disturbances. The approach for the VUMAN 'Plant Analysis' and 'Real Time' packages has therefore been to tailor the new technology so that it can be used by the process engineer of average experience and does not require a control engineering expert to commission and maintain the required control systems. It is thereby not necessary for the plant engineer to appreciate the intricacies of Least Squares Analysis or Ricatti Equation solutions. On this basis it is anticipated that the technology will progress to establish an accepted role in the process industries and will contribute significantly to the economics of plant operations.

Index